Wave Interactions with Coastal Structures

Wave Interactions with Coastal Structures

Editors

Tomohiro Suzuki
Corrado Altomare

MDPI • Basel • Beijing • Wuhan • Barcelona • Belgrade • Manchester • Tokyo • Cluj • Tianjin

Editors
Tomohiro Suzuki
Flanders Hydraulics Research
Belgium

Corrado Altomare
Universitat Politecnica de
Catalunya—BarcelonaTech (UPC)
Spain

Editorial Office
MDPI
St. Alban-Anlage 66
4052 Basel, Switzerland

This is a reprint of articles from the Special Issue published online in the open access journal *Journal of Marine Science and Engineering* (ISSN 2077-1312) (available at: https://www.mdpi.com/journal/jmse/special_issues/wave_structures).

For citation purposes, cite each article independently as indicated on the article page online and as indicated below:

LastName, A.A.; LastName, B.B.; LastName, C.C. Article Title. *Journal Name* **Year**, *Volume Number*, Page Range.

ISBN 978-3-0365-3040-6 (Hbk)
ISBN 978-3-0365-3041-3 (PDF)

Cover image courtesy of Corrado Altomare

Contents

About the Editors

Tomohiro Suzuki is a researcher at Flanders Hydraulics Research, Belgium. He obtained his Ph.D. from Delft University of Technology. He is a numerical and physical modeler of coastal engineering and has 20 years of experience. He specializes in wave–structure interactions, wave transformation, and nature-based solutions (wave interactions with vegetation). He is the author of more than 80 papers published in international journals, conference proceedings, and book chapters. He has served as a reviewer for more than 10 journals, and he is a member of the editorial board of two international journals.

Corrado Altomare is a postdoctoral researcher and former Marie Curie researcher at Universitat Politècnica de Catalunya in Barcelona (Spain). He is a coastal engineer with expertise in experimental and numerical modelling of wave–structure interaction problems. He is a core developer of the mesh-less DualSPHysics code, based on the Smoothed Particle Hydrodynamics method. He is an expert in wave overtopping and is a member of the Steering Committee of the second edition of EurOtop (2007). He is the author of more than 40 journal papers and 60 works presented in international conferences and participates as an editor or guest editor for four different peer-reviewed scientific journals.

Journal of
Marine Science and Engineering

MDPI

Editorial

Wave Interactions with Coastal Structures

Tomohiro Suzuki [1,2,*] and Corrado Altomare [3]

1 Flanders Hydraulics Research, 2140 Antwerp, Belgium
2 Faculty of Civil Engineering and Geosciences, Delft University of Technology, 2628 CN Delft, The Netherlands
3 Maritime Engineering Laboratory, Department of Civil and Environmental Engineering, Universitat Politecnica de Catalunya—BarcelonaTech (UPC), 08034 Barcelona, Spain; corrado.altomare@upc.edu
* Correspondence: tomohiro.suzuki@mow.vlaanderen.be

Due to the ongoing rise in sea level and increase in extreme wave climates, consequences of the changing wave climate, coastal structures such as sea dikes and seawalls will be exposed to severe and frequent sea storms. Even though much research related to wave–structure interactions has been carried out, it remains one of the most important and challenging topics in the field of coastal engineering. The recent publications in the Special Issue "wave interactions with coastal structures" in the Journal *Marine Science and Engineering* contain a wide range of research, including theoretical/mathematical, experimental, and numerical works related to the interaction between sea waves and coastal structures. The research is related to conventional coastal hard structures in deep water zones and ones located in shallow water zones, such as wave overtopping over shallow foreshores with apartment buildings on dikes. The presented research findings increase knowledge of hydrodynamic processes, and new approaches and developments presented here will be a good benchmark for future works. The research papers published in the Special Issue are highlighted below with discussions of the approaches and developments.

Maiolo et al. [1] investigated wave propagation and transformation over a type of submerged breakwater. In particular, the intervention consists of protected nourishment, an environmentally friendly submerged structure aiming to protect the shoreline. The authors developed a simplified methodology employing a mathematical model to calculate the transmission of the incoming waves over such submerged structures. The study provides a reasonable estimation of the wave transmission, while the computational cost is significantly smaller than conventional numerical models. Such a methodology is helpful for engineering practices when many cases need to be simulated for different storms and sea-level rise scenarios, in combination with different coastal interventions. Furthermore, it can be used for further interventions such as nature-based solutions.

Altomare et al. [2] investigated the overtopping risk for pedestrians on sea dikes using experimental modelling carried out in a small-scale wave flume facility at Universitat Politecnica de Catalunya–BarcelonaTech, Spain. The structural setup comprised a 1:1 sloping smooth dike with foreshore slopes of 1:15 and 1:30, representing typical layouts of the urbanized coastal area north of Barcelona. Along this area, bike paths and railways run very close to the shoreline and have been exposed frequently to extreme storm conditions and overtopping events in the last decade. The flow depth and velocity characterize the stability of pedestrians; thus, these values were measured in the physical model. The overtopping flow velocity was measured by redundant systems composed of two high-speed cameras and two ultrasonic sensors. It is often a problem to measure the overtopping bore velocity on the dry dike since the conventional measurement system using an electric velocimeter does not work correctly at a location that changes wet and dry conditions by each overtopping event. The results indicate that average wave overtopping discharge and maximum individual overtopping volume are necessary but not sufficient variables to assess whether a scenario is safe or unsafe. Instead, the pedestrian hazard is proved to be linked to the combination of overtopping flow velocity and flow depth.

Citation: Suzuki, T.; Altomare, C. Wave Interactions with Coastal Structures. *J. Mar. Sci. Eng.* **2021**, *9*, 1331. https://doi.org/10.3390/jmse9121331

Received: 15 November 2021
Accepted: 23 November 2021
Published: 26 November 2021

Dan et al. [3] investigated the reduction of wave overtopping and force impact on storm walls on dikes and quays due to very oblique waves, where the wavefront forms an angle of $\geq 45°$ with the structure. Such oblique waves need to be assessed when the quay and dike zones inside a harbour are located along the primary wave direction. The authors carried out the test in a wave basin equipped with a uni-directional wave generator (i.e., only long-crested waves were generated). The target structure is vertical storm walls on a dike with 1 in 2.5 slope and a vertical quay wall. Very oblique waves (angle between 45° and 80°) have been tested, measuring overtopping and force reduction with respect to perpendicular wave attacks. The study proposed coefficients of wave overtopping reduction and formulations of wave force reduction, both for very oblique wave cases. Physical models are still a major tool to investigate overtopping and force in engineering applications, not only due to the accuracy but also to the high demand of time and space (e.g., 1000 waves and 2D/3D domain), as can be seen in this work, while the application range of numerical models on wave–structure interaction is growing. Below, a review of six works are related to numerical modelling.

Suzuki et al. [4] investigated the overtopping risk on a dike in shallow foreshores using a non-hydrostatic wave model, SWASH; the topic is in line with Altomare et al. [2]. The model was first validated with a physical model to understand wave overtopping characteristics using different wave conditions and bathymetries. The results show that the overtopping risk is characterized by the time-dependent flow depth and velocity rather than maximum flow depth and velocity. If maximum values are used, the risk is overestimated. The work revealed that the structure configuration strongly influences the time-dependent values—even though the same average overtopping discharge, the risk is different by the flow depth and flow velocity combination. On top, it indicates a risk of return flows when a vertical structure leeward the dike. Such an investigation was achieved owing to the characteristics of SWASH: wave transformation and overtopping characteristics are reproduced accurately and are computationally less demanding.

Chang [5] applied a three-dimensional, fully non-linear potential wave model based on a curvilinear grid system to investigate the interaction of a solitary wave with vertical fully/partially submerged circular cylinders with/without a hollow zone. The model was validated with data in the literature and then was applied to study cases of wave-cylinder interaction. In general, 3D simulations are extremely demanding in terms of computational time and resources, and thus, it is not feasible for engineering practices. However, this paper tackled it by using a potential model. The target wave is one (i.e., solitary wave); thus, the computational time is shorter than a real case study, which often requires longer simulation (e.g., investigation of marine platforms). The work gives a good insight into the wave behaviour around a circular cylinder with or without hollow zones, while cases where the vortex effect is dominant cannot be dealt with due to the nature of potential models.

Li et al. [6] studied vertical breakwater stability under extreme waves. Vertical breakwaters are conventional coastal structures; thus, many investigations and knowledge on the stability of breakwaters have been accumulated in the literature. However, dealing with stability using numerical simulation is somewhat limited due to the complexity of the dynamic loading on the structure combined with overtopping waves. The authors investigated the pressure distribution under a condition with wave overtopping using a two-dimensional RANS model and discussed the potential risk of the failure mechanism.

Gruwez et al. [7] conducted an intensive validation of a two-dimensional RANS multiphase solver (i.e., OpenFOAM) for a case with sea dike in shallow foreshores. The authors compared the results of wave properties along the wave flume, wave runup and forcing in great detail with a quantified and objective validation. Based on it, the numerical model conducted in the study gives promising results in terms of wave propagation and force estimation without calibration after a convergence analysis. Evaluation/comparison of time series between the physical and numerical models is not a trivial task, but the authors proposed a useful classification of the relative refined index of agreement and corresponding rating.

Gruwez et al. [8] extended the work of Gruwez et al. [7] further, including the SWASH and DualSPHysics models, conducted through comparisons based on the same physical model validation case. It is an interesting work since each model is becoming very popular in coastal engineering and growing in the range of applications, yet the characteristics of the models are very different. The RANS model is a conventional numerical approach in coastal engineering, and the knowledge has been accumulated through a long history. A number of numerical schemes and techniques have been developed, and often, the model gives highly accurate modelling. DualSPHysics is a mesh-less model based on the smoothed particle hydrodynamics method (SPH). Compared to mesh-based approaches, SPH has a relatively short history (developed in the 70s of the last century for astrophysics and applied to fluid dynamics for the first time about 30 years ago). However, the method is becoming very popular due to its particle-based approach, capable of inherently catching non-linearities and very violent phenomena. SWASH is based on the non-linear shallow water equation with non-hydrostatic pressure; thus, it is suitable for the estimation of wave propagation in the shallow zone. Due to the nature of the equation, the computational time is much smaller than the RANS and SPH models. The three models are compared in terms of model accuracy and computational speed. The results indicate that all three models give a good accuracy for wave transformation and force estimation while OpenFOAM and DualSPHysics models give a slightly better representation in time series than the SWASH model. On the other hand, the SWASH simulation is much less time consuming compared to OpenFOAM and DualSPHysics. Each model has its characteristics with pros and cons; thus, the user needs to choose which type of numerical model is suitable. This work is one of the benchmark papers which contains a good discussion of detailed capabilities.

Hasanpour et al. [9] investigated tsunami-borne large debris flow and impact on coastal structures using a coupling model of SPH and FE (finite elements) in 2D. The developed model was validated with physical modelling in the literature. After validating the non-linear transformation of the tsunami wave, the debris–fluid interaction and the impact on a coastal structure, the model was further applied to debris–structure interaction. The result indicates the necessity of 3D simulation.

Altomare et al. [10] used the DualSPHysics model to understand the failure mechanism of the Pont del Petroli Pier in Spain, which occurred during the so-called "Storm Gloria" in January 2020. A preliminary 2D modelling was carried out to reproduce wave properties at the toe of the structure. Based on the water surface elevation and velocity field near the structure obtained in the 2D model, the 3D model case was carried out with the state-of-the-art inlet-wave-generation technique. The total number of particles of the 3D simulation was 1.7 million, with a physical time of 20 s, which corresponds to two to three waves characterizing the most extreme events, which were simulated (the run time was 9 hours). Such a simulation used to be very challenging due to the nature of the irregular wave and large domain, but this work proved that the DualSPHysics model can be applied to such events using a kind of nesting method (i.e., reproducing the incident wave time series from the 2D simulation). The simulation results have later been used to set up the experimental campaign carried out in a large-scale wave flume and will be exploited by local authorities to reconstruct the pier, a key element of the urban assets in the coastal town of Badalona. As this practice shows, numerical models can also assist the physical models instead of the physical model only feeding the validation data.

Author Contributions: Conceptualization, T.S. and C.A.; writing—original draft preparation, T.S.; writing—review and editing, C.A.; project administration, T.S. All authors have read and agreed to the published version of the manuscript.

Funding: This research received no external funding.

Conflicts of Interest: The authors declare no conflict of interest.

References

1. Maiolo, M.; Mel, R.A.; Sinopoli, S. A Simplified Method for an Evaluation of the Effect of Submerged Breakwaters on Wave Damping: The Case Study of Calabaia Beach. *J. Mar. Sci. Eng.* **2020**, *8*, 510. [CrossRef]
2. Altomare, C.; Gironella, X.; Suzuki, T.; Viccione, G.; Saponieri, A. Overtopping Metrics and Coastal Safety: A Case of Study from the Catalan Coast. *J. Mar. Sci. Eng.* **2020**, *8*, 556. [CrossRef]
3. Dan, S.; Altomare, C.; Suzuki, T.; Spiesschaert, T.; Verwaest, T. Reduction of Wave Overtopping and Force Impact at Harbor Quays Due to Very Oblique Waves. *J. Mar. Sci. Eng.* **2020**, *8*, 598. [CrossRef]
4. Suzuki, T.; Altomare, C.; Yasuda, T.; Verwaest, T. Characterization of Overtopping Waves on Sea Dikes with Gentle and Shallow Foreshores. *J. Mar. Sci. Eng.* **2020**, *8*, 752. [CrossRef]
5. Chang, C.-H. Interaction of a Solitary Wave with Vertical Fully/Partially Submerged Circular Cylinders with/without a Hollow Zone. *J. Mar. Sci. Eng.* **2020**, *8*, 1022. [CrossRef]
6. Li, M.-S.; Hsu, C.-J.; Hsu, H.-C.; Tsai, L.-H. Numerical Analysis of Vertical Breakwater Stability under Extreme Waves. *J. Mar. Sci. Eng.* **2020**, *8*, 986. [CrossRef]
7. Gruwez, V.; Altomare, C.; Suzuki, T.; Streicher, M.; Cappietti, L.; Kortenhaus, A.; Troch, P. Validation of RANS Modelling for Wave Interactions with Sea Dikes on Shallow Foreshores Using a Large-Scale Experimental Dataset. *J. Mar. Sci. Eng.* **2020**, *8*, 650. [CrossRef]
8. Gruwez, V.; Altomare, C.; Suzuki, T.; Streicher, M.; Cappietti, L.; Kortenhaus, A.; Troch, P. An Inter-Model Comparison for Wave Interactions with Sea Dikes on Shallow Foreshores. *J. Mar. Sci. Eng.* **2020**, *8*, 985. [CrossRef]
9. Hasanpour, A.; Istrati, D.; Buckle, I. Coupled SPH–FEM Modeling of Tsunami-Borne Large Debris Flow and Impact on Coastal Structures. *J. Mar. Sci. Eng.* **2021**, *9*, 1068. [CrossRef]
10. Altomare, C.; Tafuni, A.; Domínguez, J.M.; Crespo, A.J.C.; Gironella, X.; Sospedra, J. SPH Simulations of Real Sea Waves Impacting a Large-Scale Structure. *J. Mar. Sci. Eng.* **2020**, *8*, 826. [CrossRef]

Journal of
Marine Science and Engineering

Article

A Simplified Method for an Evaluation of the Effect of Submerged Breakwaters on Wave Damping: The Case Study of Calabaia Beach

Mario Maiolo *, Riccardo Alvise Mel and Salvatore Sinopoli

Department of Environmental Engineering, Capo Tirone Experimental Marine Station, University of Calabria, 87036 Rende (CS), Italy; riccardo_alvise.mel@unical.it (R.A.M.); salvatore.sinopoli@unical.it (S.S.)
* Correspondence: mario.maiolo@unical.it; Tel.: +39-984-496556

Received: 4 June 2020; Accepted: 8 July 2020; Published: 12 July 2020

Abstract: Erosion processes threaten the economy, the environment and the ecosystem of coastal areas. In addition, human action can significantly affect the characteristics of the soil and the landscape of the shoreline. In this context, pursuing environmental sustainability is of paramount importance in solving environmental degradation of coastal areas worldwide, with particular reference to the design of complex engineering structures. Among all the measures conceived to protect the shoreline, environmentally friendly interventions should be supported by the stakeholders and tested by means of mathematical models, in order to evaluate their effectiveness in coastal protection through the evaluation of wave damping and bedload. This study focuses on protected nourishments, as strategic interventions aimed to counteract coastal erosion without affecting the environment. Here, we develop a simplified method to provide a preliminary assessment of the efficiency of submerged breakwaters in reducing wave energy at a relatively low computational cost, if compared to the standard 2D or full 3D mathematical models. The methodology is applied at Calabaia Beach, located in the southern Tyrrhenian Sea (Italy), in the area of the Marine Experimental Station of Capo Tirone. The results show that the simplified method is proven to be an essential tool in assisting researchers and institutions to address the effects of submerged breakwaters on nourishment protection.

Keywords: shallow waters; wave energy; coastal erosion; beach restoration; submerged breakwaters; protected nourishments

1. Introduction

The erosion process threatens urban settlements, the environment and the ecosystem located in the Mediterranean Sea [1,2]. Coastal areas involve strong relationships between the needs of human communities and the environment [3]. The increasing demand of coastal use should aim to conserve the resilience of the environment, allowing the ecosystem to absorb the unavoidable human impacts, keep functioning, and provide goods and services to the population [4–6]. However, coastal areas regularly deal with complex environmental and ecosystem challenges, which are worsened by mean sea level rises driven by climate change, subjecting these fragile lands to large economic and environmental losses and damage produced by flooding and erosion processes [7–14]. Increasing urban pressure will further exacerbate these damages, requiring an integrated approach to conserve the land and the environment [15,16], evolving from sea defence interventions only aimed at protecting human and goods, to coastal protection measures focused on the overall needs of the land [17].

The effectiveness of sea defence interventions depends on the designer's ability to identify the correct strategy to pursue in the area of concern, and to assess their short-term and long-term impacts, which can produce further alterations and new and more complex issues in the coastal environment, ecosystem and landscape [18,19].

A correct evaluation of the causes of the coastal erosion requires a deep knowledge of the land, which should be achieved through a morphometric and granulometric characterization of the soil and specific climate analysis [20]. Although the most important coastal erosion events occur during the major storms, the ordinary action of the waves and anthropic activities can significantly affect the shoreline [21]. Human actions trigger the erosion process, mostly due to an incorrect understanding of long-term effects of their actions on the shoreline, with particular reference to the impact of coastal buildings on the environment.

The main natural causes of the net retreat of the coastline are: sea level rise; the effect of climate change on the reduction in the sediment supply from water courses; the negative balance of offshore sediment transport due to storm waves and surge overwash; wind erosion; longshore sediment transport; loss of fine material moved from the shoreline in a seaward direction [21–23].

The main anthropogenic causes of the net retreat of the coastline are: land subsidence from the suppression of the subsurface resources; interruption of the sediment transport due to buildings and sea defence structures located in the active beach; river works reducing the sediment input from the water courses; changes of the natural configuration of the beach profile; removal or displacement of the material from the active beach [21–23].

A global analysis of these phenomena allows us to provide a realistic assessment of the erosion process and the identification of the most effective sea defence measures [21]. The modeling of these phenomena shows multiple constraints, including a paucity of meteorological and oceanographic gauged data and large computational times. In addition, there is an increasing need to restore the environment avoiding the use of sea defences that, while effective against the erosion process, threaten the ecosystem and the landscape, affecting tourism, fisheries and the overall economy of coastal communities [24–26]. Natural-based and environmentally friendly measures, such as the use of mangroves and *Posidonia oceanica* meadows [27–29], are increasingly supported by the scientific community, due to their positive impact on the land.

The effectiveness of sea defences is often constrained by the local characteristics of the land. In this context, artificial nourishments can support the restoration of the shoreline without affecting the environment and the use of the sea. However, beach nourishments are threatened by erosion and the degradation of the environment. Here, we describe the protection of beach nourishments through submerged breakwaters, and herein we refer to this intervention as "protected nourishment".

Despite the fact that several mathematical models can properly evaluate the long-term effectiveness of sea defence structures, they are often constrained to three-dimensional schemes for simulating processes related to benthic suspensions, flocculation, entrainment, erosion, sedimentation and bed consolidation. Moreover, two (2D) and three-dimensional (3D) coastal modeling usually requires a significant computational cost, limiting most of the analysis to the use of one-dimensional systems or statistic models [30–34].

In this work, we first address the characteristics of protected nourishments on wave damping [35,36] by means of Mike 21/3 Coupled Model FM 2020, an integrated 2D model (hereinafter denoted as MIKE), which computes wave climate, hydrodynamic processes, and bedload (Section 2). We then develop a simplified methodology to assess the effectiveness of the submerged breakwaters on wave damping, reducing the computational cost (Section 3). This approach has been applied at Calabaia Beach (Belvedere Marittimo municipality, Italy, Section 4) to evaluate the long-term shoreline evolution. Our conclusions bring the paper to a close.

2. Materials and Methods

The design of submerged breakwaters as a support for protected nourishments first requires the evaluation of the effect of the barrier on the reduction in the wave height from offshore to the shore by means of the wave transmission coefficient *Kt* [37].

$$K_t = \frac{H_{out}}{H_{in}} \text{ with } 0 < \text{ Kt } < 1 \tag{1}$$

where H_{out} and H_{in} stand for the wave height upstream and downstream of the barrier, respectively [38]. Although this simplified approach does not describe the variability of the wave climate and the possible effect of the anthropic structures, it is consistent with the aim of this study.

2.1. The Mathematical Model

In recent decades, several models have been developed to compute the morphodynamic evolution of coastal environments, characterized by several computational schemes, scales, levels of detail and computational costs [39–44]. The spatial scale of these models ranges from meters to kilometres. The temporal scale typically ranges from hours to decades. Most of the existing large-scale models are limited to an assessment of the evolution of the shoreline, while the smaller space and time scale models are mainly used to reproduce specific forcing events, involving explicitly reductionist methodologies where the conservation of momentum forms the explicit means for the evolution of the system. [45–47].

Here, we use MIKE 21-3 Coupled Model FM, based on an unstructured grid which consists of an integrated system capable of reproducing the coastal processes and dynamics of the shoreline at all scales. MIKE evaluates the effectiveness of the sea defence interventions, such as the optimisation of the beach nourishments and costal protection structures, and the impact of the buildings located in the active beach.

The Hydrodynamic (HD) module of MIKE solves the 2D Navier–Stokes equations of incompressible fluids, under the hypothesis of hydrostatic pressure [48]. The numerical solution of shallow water equations is achieved through the approximate Riemann solver [49], which calculates the convective flows at the interface of the element of the grid. Second-order precision is achieved through the use of a linear gradient reconstruction technique. The average gradient is computed by means of the Jawahar and Kamath [50] approach.

The Spectral Wave (SW) module describes the wave phenomena by means of the wave action conservation equations [39,40], which can be written by using the Cartesian system as:

$$\frac{\partial N}{\partial t} + \nabla \cdot \left(\vec{v} N\right) = \frac{S}{\sigma} \tag{2}$$

where N (θ, σ) is the density of the wave, which is function of θ, wave direction and wave frequency σ. S is the source term accounting for different processes:

$$S = S_{in} + S_{n1} + S_{ds} + S_{bot} + S_{surf} \tag{3}$$

where

S_{in} is the wind action;

S_{nl} is the waves non-linear interaction;

S_{ds} is the white capping effect;

S_{bot} is the bottom dissipative action;

S_{surf} is the dissipation due to wave breaking phenomena.

Among all these parameters, S_{surf} is essential in representing the wave breaking, which occurs when the waves propagate in shallow water and can no longer be supported by the depth. This process is described following the formulation of Battjes and Janssen [51].

If the shoreline is characterized by shallow waters surrounding the shoreline, the maximum wave height can be calculated as:

$$H_{max} = \gamma d \tag{4}$$

where d is the water depth and γ a coefficient which ranges between 0.5 and one, depending on the slope of the bottom and the wave characteristics [52].

The formulation of γ can be chosen according to Nelson's formulations [53,54], or by means of the more recent Ruessink approach [55], valid both for steep seabeds without bars and flat bottoms. The Ruessink approach calculates γ at each cell of the domain as:

$$\gamma = 0.76kd + 0.29 \tag{5}$$

where k is the local wave number and d the local depth.

The bedload is solved by means of the Sand TransPort Quasi-3-Dimensional module (STPQ3D) [56], which computes the transport of the sediments along two horizontal directions by processing the input data provided by the hydrodynamic model by time-averaging the results.

Waves affect the motion of sediments in shallow waters due to wave breaking, with particular reference to the transport of sediments transversal to the coast, which is mostly involved in the erosion process [57,58]. Morphological changes due to erosion and deposition are accounted in terms of bed elevation, expressed with respect to the centre of each triangular element of the grid as $\frac{\delta z}{\delta t}$. This parameter can be computed by means of the continuity equation of the sediment proposed by Exner [59], where n is the porosity:

$$z_{new} = z_{old} + \frac{1}{1-n}\frac{\delta z}{\delta t}\Delta t_{\ HD} \tag{6}$$

The bedload is computed as the divergence of the sediment flow with respect to the boundary of the elements and it is equal to the sum of all the flows that cross the boundaries, identifying whether the element is receiving or losing sediment.

$$\frac{\delta S_x}{\delta x} + \frac{\delta S_y}{\delta y} = \sum_{i=1}^{m} S_{in} ds_i \tag{7}$$

where

S_{in} is the sediment flux normalized to the face of the element;
ds_i is the lenght of the face of the element;
m is the index of the element;

The continuity equation of sediments affects the bed elevation of each element, based on the total sediment flux crossing the boundaries of each element.

The coupling of the models is based on the establishment of an overall time step of the system, which is necessary to match the instant in which the models exchange the information. Each model has its own time step, which is a multiple of the overall time step. The exchange of information occurs when the time of each model match with an overall time step (e.g., the HD module acquires a radiation stress field from the SW module).

Since the information sharing between each module of MIKE is very complex, we refer to the manuals for a more detailed description [48,56,60]. A summary of the main steps is provided in Figure 1.

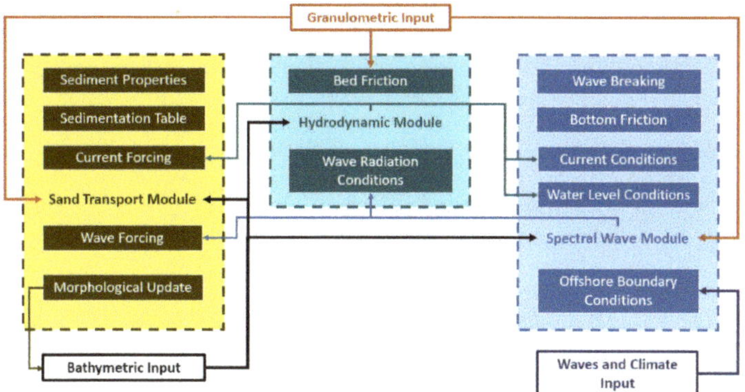

Figure 1. Coupling scheme representing the information flux among the modules of Mike 21/3 Coupled Model FM 2020 (MIKE).

2.2. Submerged Breakwaters

The effects of submerged breakwaters on wave damping is typically evaluated through the formulation of Goda [61,62], which computes the transmission coefficient Kt (1) as:

$$
\begin{gathered}
K_t = K_{t,\,max} \text{ if } \frac{f}{H_i} < \left(\frac{f}{H_i}\right)_{min} \text{ with } K_{t,\,max} < 1 \\
K_t = \frac{1}{2}\left(1 - sin\left(\frac{\pi}{2}\cdot\frac{\frac{f}{H_i}+\beta}{\alpha}\right)\right) if \left(\frac{f}{H_i}\right)_{min} < \frac{f}{H_i} < \left(\frac{f}{H_i}\right)_{max} \\
K_t = K_{t,\,min} \text{ if } \frac{f}{H_i} < \left(\frac{f}{H_i}\right)_{max} \text{ with } K_{t,\,min} > 0
\end{gathered}
\tag{8}
$$

The values of $K_{t,\,max}$ and $K_{t,\,min}$ are set according to the boundary conditions; the parameter f (i.e., the free board) represents the height of the breakwater crest minus the surface level; H_i the value of the incoming wave height, i.e., the input wave height. If the breakwater is submerged, the wave damping effect is negligible when the wave height is equal to or smaller than the depth of the crest, computed with respect to the mean sea level. For parameter α, Goda proposed a value equal to 2.2, while for parameter β, which depends on the inclination of the breakwater, a value ranging between 0.15 and 0.8 [61,62]. Goda's formula allows to evaluate the performance of the breakwater at different water levels, and it is not affected by the angle of attack of the waves with respect to the breakwater.

2D and 3D modeling of long and narrow structures such as the submerged barriers significantly increases the computational effort to avoid model instability. In this study, we propose a simplified method based on the implementation of Goda's formula directly on the domain, avoiding the 2D modeling of the barrier but keeping the high accuracy of the results. Similar approaches are widely implemented into mathematical models available in the literature by developing a set of specific links to reproduce the presence of long and narrow structures, such as sills and levees [63–66].

Here, we link the nodes of the elements located upstream and downstream of the submerged barrier by implementing Goda's formula. We calibrated parameter β of Goda's formula, by comparing the results computed by means of the full 2D model, using different depths of its ridge (z) and different wave climates (H_{m0}).

In the simulations performed by means of the full 2D model, we reproduced the rigid submerged breakwater by means of 2D elements (Figure 2). The hydrodynamic of the barrier is solved by the HD and SW modules, without any change in bottom elevation, by disabling the use of the sedimentological model.

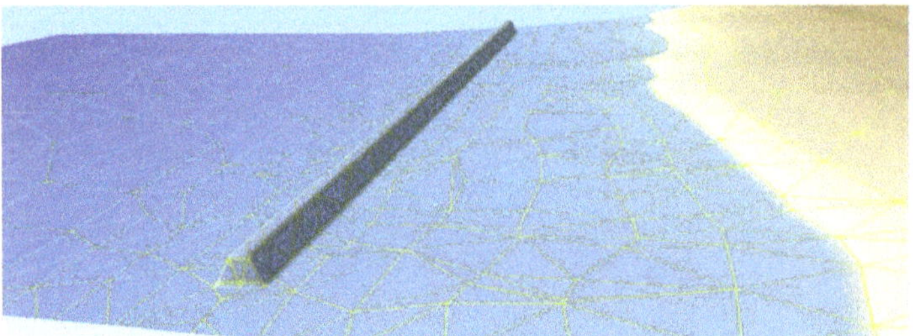

Figure 2. Full two-dimensional (2D) modeling of the submerged breakwater.

The simplified approach we develop allows us to evaluate the effect of the submerged breakwater by reproducing the structure by linking the elements of the grid located before and after the barrier, which works as a weir on whose edges we apply Goda's equation (Figure 3).

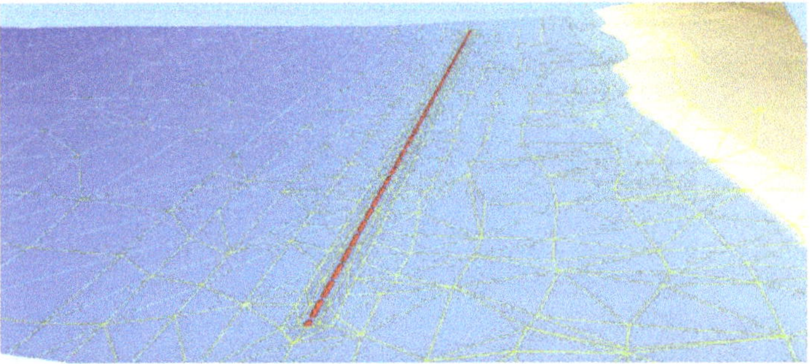

Figure 3. Simplified 2D modeling of the submerged breakwater.

The comparison between the simulations performed by means of full 2D modeling and the simplified scheme allows us to identify the optimum value of β of Goda's formula that better reproduces the effects of the breakwaters on the wave damping. The results of the simplified methodology will be synthetized in two different formulations, which aim to relate the wave damping only to the ridge depth of the barrier and to the wave climate, without performing any simulation by means of the models.

2.3. The Study Area: Calabaia Beach

Calabaia Beach is located 1 km south of Belvedere Marittimo, into the Lao region, which is extended 25 km from Capo Scalea to Capo Bonifati. The coast is characterized by sandy and pebbly beaches bounded by narrow dune belts and sedimentary rocks with deposits of conglomerates of pebbles, sand and clay beds [67] (Figure 4).

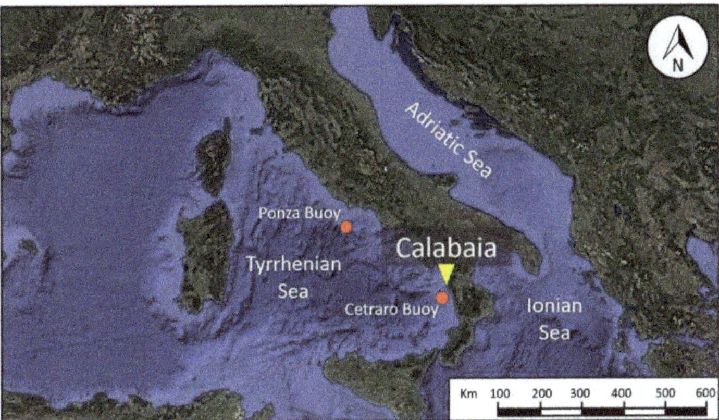

Figure 4. Localization of the Calabaia Beach and the two buoys used to characterize the wave climate.

A granulometric characterization of the site has been carried out by means of several surveys along the shoreline performed in the years 2003, 2006 and 2008, using the ASTM 200 procedure [68].

For each year of monitoring, 24 surveys were carried out, six on each section located at different depths, allowing us to evaluate the composition of the soil (Figure 5). The sample analysis shows that the materials that constitute the bottom have similar features, with an average diameter d_{50} = 0.22 mm, a specific weight of the soil γ_s/γ_w = 2.65, and a porosity Φ = 0.4. The results of the surveys, together with the reduced extension of the coastal stretch (700 m), allow us to assign averaged values to the whole domain.

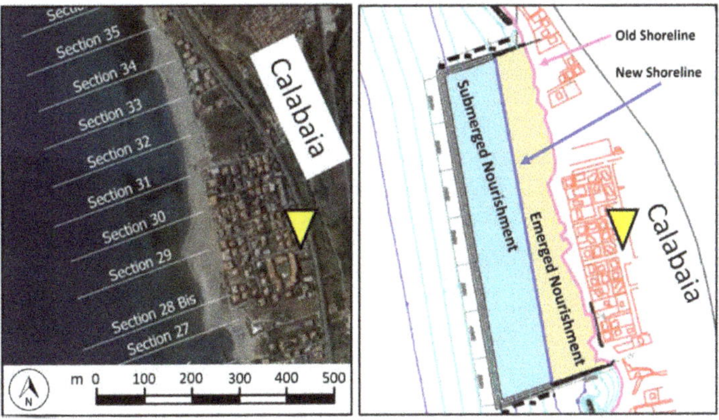

Figure 5. Calabaia Beach. Some survey sections at Calabaia Beach (**left side**); bathymetric characteristics of Calabaia used in the three-dimensional (3D) model (**right side**).

Bathymetric surveys ware performed in the years 2005, 2006 and 2008 up to a depth of −10 m (Figure 5, left side), allowing us to build a digital bathymetric model of the area (Figure 5, right side).

During the year 2002, a protected nourishment intervention was built at Calabaia Beach. The submerged breakwater is located parallel to the shoreline, 700-m long, 250-m seaward and with a ridge 2.5 m below the mean sea level (Figure 6). Two perpendicular semi-submerged groynes complete the structure, delimiting a submerged closed cell aimed at supporting the nourishment.

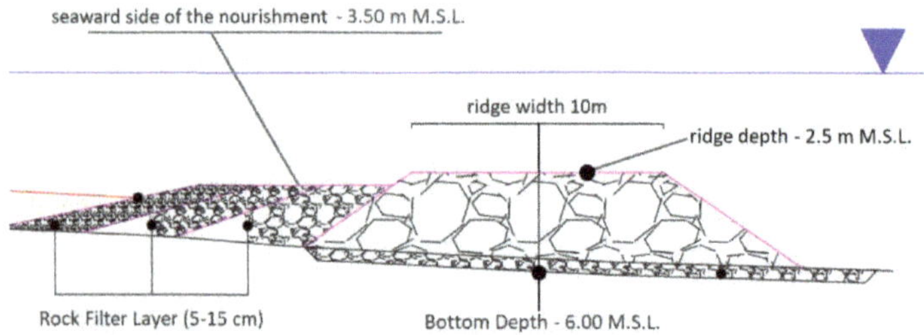

Figure 6. Schematization of the of the submerged breakwater built in the year 2002 at Calabaia Beach.

The characterization of the local wave climate at Calabaia Beach has been carried out by using the hindcast dataset based on the data of the Ponza buoy (40°52′00″ N, 12°57′00″ E), collected from 1st July 1989 to 31th December 1999, and the Cetraro buoy (39°27′01″ N, 15°55′01″ E), collected in the year 1999 (Figure 7). Both the buoys are anchored at depth of 100 m. The Ponza buoy is a Wavec MKI directional wave meter; the Cetraro buoy is a Waverider directional wave meter. The hindcast dataset have been reproduced on the entire Tyrrhenian coast, supported by the WAM wave model [40] based on the wind fields provided by the European Center for Medium-Range Weather Forecasts. Calabaia Beach shows a predominant 270° north direction for the waves, with a significant wave height of 7.2 m for a return period of 200 years [69].

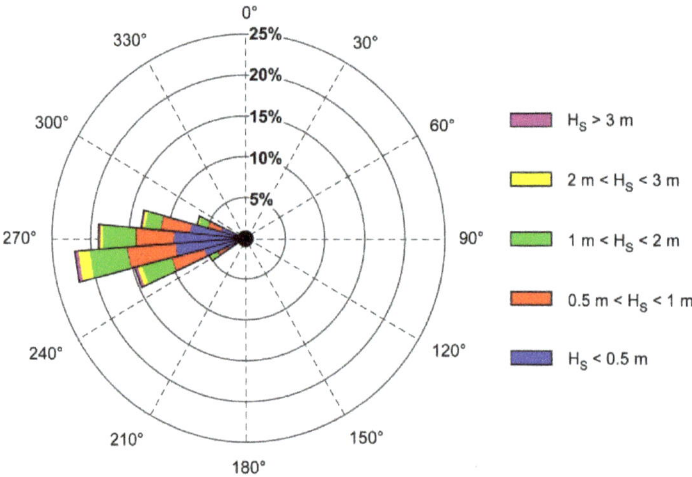

Figure 7. Wave Rose at Calabaia Beach based on the hindcast data of the Ponza and Cetraro buoys.

Figure 8 relates the significant height of the wave with the return period. The statistical analysis conducted led to a Weibull distribution characterized by a λ value = 0.922 and a k value of 0.454.

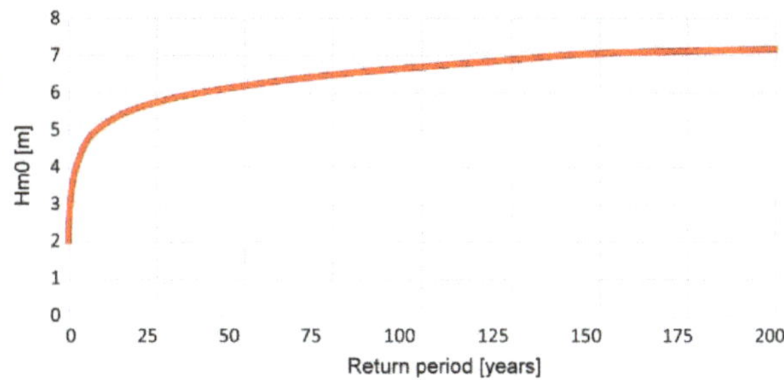

Figure 8. Relationship between significant wave height and return period.

2.4. Model Set-Up

The domain of the model is an unstructured mesh, with triangles of different sizes, which are larger offshore and smaller near the barrier and the shoreline, in order to represent the area of concern in more detail (Figure 9).

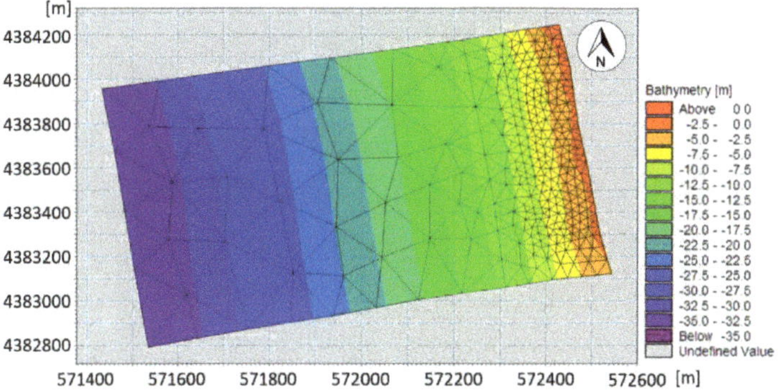

Figure 9. Grid of the hydrodynamic model used in this study. Different colours refer to different bed elevation. The metric scale refers to WGS84 UTM 33N, EPSG-32633.

Wave forcing has been imposed on the west side of the domain. The north and the south side of the grid are characterized by lateral boundary conditions [70–72], stated in the set-up manual as good approximations when the boundary line is almost straight and when the depth contours are almost perpendicular to the line [60]. The east side of the mesh, on the other hand, has the condition of land imposed in the MIKE mesh generator. For each run, we set a condition for gradual offshore wave formation, obtaining a linear growth in the significant wave height up to the value we simulated, in order to avoid wave energy peaks that would generate instability in the model. When the targeted wave height has been achieved, the simulation begins, with a duration of 24 h and a print step of 600 s, for a total of 144 pieces of output data. For each run, we consider the forcing as a constant wave (in time and space), setting the wave model with a quasi-stationary formulation, which is suitable for shorelines not longer than 10 km [60].

3. Results

We simulate the efficiency of the submerged barrier for five different offshore waves (Table 1), in order to cover the significant return periods illustrated in Figure 8.

Table 1. Simulated waves heights and periods.

H_{m0} [m]	3	4	5	6	7
T_p [s]	9.18	10.60	9.18	11.85	14.02

A first set of simulations was carried out by representing the submerged breakwater by means of full 2D modeling (Figure 2). The effects of the breakwater were further evaluated by means of a second set of simulations performed by applying the simplified approach through Goda's formula instead of the 2D modeling of the barrier. Using the same boundary conditions, five values of β have been tested with the aim being to achieve the same results computed through full 2D modeling (Figure 10).

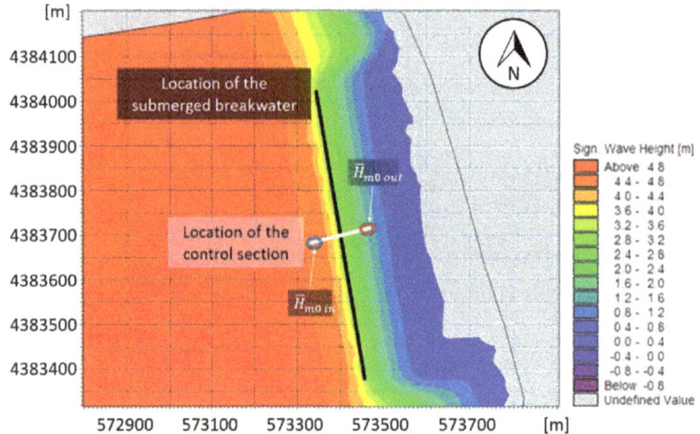

Figure 10. Wave climate computing by means of the simplified simulation with H_{m0} = 5 m. The white line represents the location of the control section, highlighting the two sections used to compute the decay efficiency (9). The wave direction is 270° N. The metric scale refers to WGS84 UTM 33N, EPSG-32633.

We compared the results in a control section 1200-m long, perpendicular to the shoreline and located at the mid-point of the length of the barrier, in order to avoid boundary effects (Figure 10, white line).

Figure 11 compares the damping of the wave height produced by the submerged breakwater along the control section, computed by means of full 2D modelling and the simplified 2D approach, with entering waves with H_{m0} values of 3 m, 5 m and 7 m (Table 1) as boundary conditions (Figure 11).

The results show that simplified 2D modeling is capable of reproducing wave height damping. With the aim of identifying the value of β that better reproduces the effect of the barrier, we compared the wave height reduction due to the breakwater for all the input waves reported in Table 1 (Figure 11d). We compared the height decay efficiency E_h (9), setting the height of the offshore significant wave as the average of the first 25 m of the control section ($\overline{H}_{m0\ in}$), and the height of the outcoming significant wave ($\overline{H}_{m0\ out}$) as the average of the last 25 m of the control section (Figure 10, white line).

$$E_h = \frac{\overline{H}_{m0\ in} - \overline{H}_{m0\ out}}{\overline{H}_{m0\ in}} \tag{9}$$

The β coefficient that better represents the effects of the breakwater is 0.30. Table 2 compares the root mean square error (RMS) between the results of the two different sets of simulations, confirming the optimum value for parameter β as equal to 0.30.

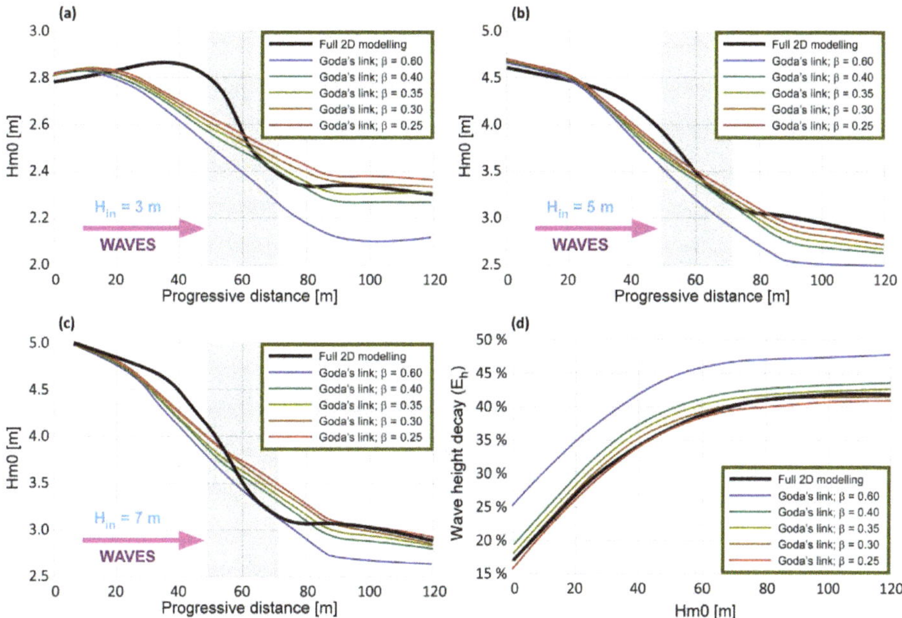

Figure 11. Wave height decay (height efficiency E_h) along the cross section of the submerged breakwater. (**a**) H_{m0} = 3 m, (**b**) H_{m0} = 5 m, (**c**) H_{m0} = 7m. (**d**) Decay efficiency for different values of β.

Table 2. RMS error of the wave height damping computed through simplified 2D modeling compared to the results obtained by means of the full 2D model.

H_{m0}	Full 2D Model	Beta Values				
		0.25	0.3	0.35	0.4	0.6
3	17.06	15.81	16.98	18.10	19.45	25.29
4	31.22	30.71	31.98	33.11	34.29	38.88
5	38.96	38.29	39.35	40.36	41.26	46.00
6	41.25	40.17	40.98	41.89	42.82	47.05
7	41.75	40.89	41.70	42.56	43.46	47.55
RMS Error		**0.908**	**0.404**	**1.23**	**2.27**	**6.97**

We further investigated the damping efficiency using different ridge depths for the breakwater z, ranging between −1 and −2.5 m, in order to provide a more general result [61]. Figure 12 illustrates the logarithmic relation between the height and power damping efficiency and the wave height normalized to the ridge depth of the breakwater $\frac{H_{m0}}{z}$. We selected the range $\frac{Hm0}{z}$, varying from 0.5 (i.e., the breakwater begins to affect the waves) to 3.0, which is the limit constrained by the depth of the breakwater.

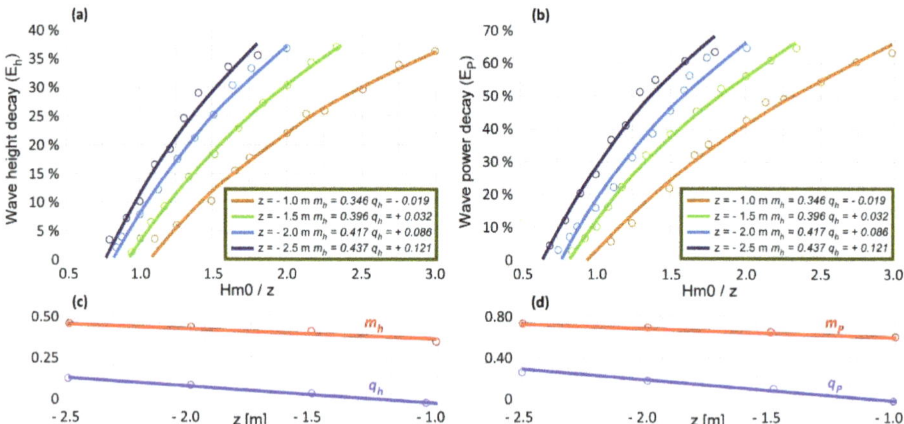

Figure 12. Wave damping efficiency for different values of the ratio $\frac{Hm0}{z}$. (**a**) Wave height decay, (**b**) wave power decay; (**c**) and (**d**) show the relationship between the value of the parameters of the logarithmic relation and the depth of the ridge of the submerged breakwater.

The regressive parameters m and q can be further related to the ridge depth z, showing a linear relationship (Figure 12c,d) and allowing us to relate the damping efficiency only to the wave climate and the depth of the ridge of the breakwater, without performing any simulation (10, wave height damping; 11, wave power damping):

$$E_h = m_h \cdot \ln(\frac{H_{m0}}{z}) + q_h \quad (10)$$

$$m_h = -0.599 \cdot z + 0.294$$

$$q_h = -0.0958 \cdot z - 0.112$$

$$E_p = m_p \cdot \ln(\frac{H_{m0}}{z}) + q_p \quad (11)$$

$$m_p = -0.0866{\cdot}z + 0.513$$

$$q_p = -0.1918{\cdot}z - 0.180$$

Our findings show that the simplified formulation we propose can successfully support the stakeholders in providing an initial assessment of the effectiveness of the submerged breakwater, with particular reference to its design and the further maintenance measures. Although both Equations (10) and (11) have been computed as site specific depending on the surrounding bathymetry, the prevailing environmental conditions and the configuration of the breakwater, the results can have a general validity since, in the area of concern, there are no other structures that can significantly affect the outcomes. The ranges of wave damping we simulated fall within the range of variability provided in the literature [73,74], both in reference to the wave height and to the wave power, showing that the breakwater efficiency is a function of the relative submergence of the crest.

4. Application at Calabaia Beach

We finally applied simplified 2D modeling to assess the effectiveness of the submerged breakwater located at Calabaia Beach through the evaluation of the erosion process affecting the shoreline. Using the same boundary conditions illustrated in Section 2.4, we used the model to reproduce the erosion process affecting the nourishment at Calabaia Beach (Figure 13). Since the area has been subject to a large volume of nourishment, we simulated only one erodible layer, which was characterized by the homogeneous granulometry of the material.

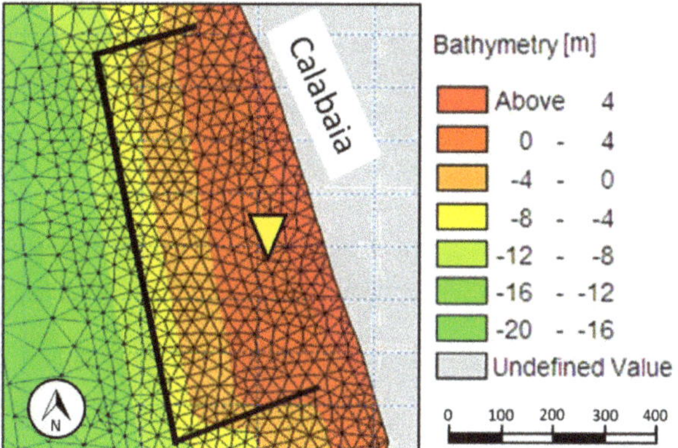

Figure 13. Detail of the domain of the model used to evaluate the erosion process at Calabaia Beach.

We simulated input waves with a return period of 200 and 50 years, respectively, and two ordinary events with monthly and daily frequencies (Table 3):

The analysis shows a significant loss of material from the nourishment, with a bedload from the protected cell to an offshore direction (Figure 14).

Table 3. Simulated event characteristic for material loss analysis.

Simulated H_{m0} [m]	Event Duration [h]	Peak Period [s]	Wave Direction [deg]
7.20	8	14.02	270
5.00	16	11.85	270
2.00	32	7.49	270
0.25	6 months	2.65	270

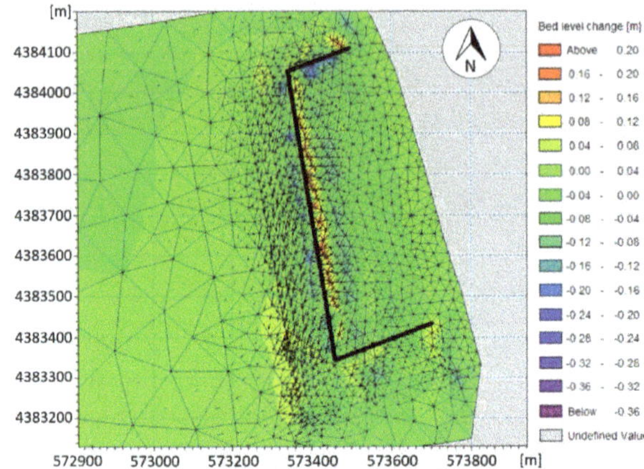

Figure 14. Bed level change and sediment flux for an event with a return period of 50 years, simulated with the presence of the breakwater. The metric scale refers to WGS84 UTM 33N, EPSG-32633.

In order to evaluate the magnitude of the erosion, we performed a quantitative analysis of the material eroded based on the change in the elevation of each element of the grid located within the nourishment zone, multiplied by the value of its area (12).

$$Material\ Loss = \sum_{i=1}^{n} Area_{i-element} \cdot \Delta Elevation_{i-element} \quad (12)$$

The aggregated data account for the erosion process within the event (Table 4).

Table 4. Summary of material loss data.

H_{m0}	Event Duration [hours]	Return Time [years]	With Breakwater		Without Breakwater		Ratio of Loss
			Volume Loss [m^3]	Mean Elevation Loss [m]	Volume Loss [m^3]	Mean Elevation Loss [m]	
7.20	8	200	−629.40	−0.00058	−25245.93	−0.114	40.11
5.00	16	50	−563.94	−0.00045	−21671.71	−0.101	38.43
2.00	32	0.1	−1681.02	−0.0059	−10910.4	−0.0502	6.49
0.25	6 months	0.03	−700.29	−0.00798	−13307.92	−0.06552	19.00

The results indicate that, with the submerged breakwater, 6 months of ordinary wave action produces a volume loss of 700 m^3, comparable to a single event with a return time of 200 years, which produces a loss of 629 m^3. Submerged barriers are generally not suitable to counteract the erosive process produced by ordinary events, requiring complementary measures. Depending on the duration and extent of the event, nourishment is largely eroded without the protection of breakwater, with a loss of material six to 40 times larger than in the protected configuration.

5. Conclusions

This study focuses on protected nourishment, an environmentally friendly sea defence intervention. Protected nourishments prove to be effective against extreme phenomena, without providing a valid support to face the regular action of waves, which can be very detrimental to the shoreline, subtracting huge amounts of material where the nourishment has been designed. In this work, we develop a simplified methodology to address the effects of submerged breakwaters on the reduction in energy and power of the entering wave, at a significantly smaller computational cost if compared to full 2D modeling. We apply this method to evaluate the shoreline evolution at Calabaia Beach, comparing the effectiveness of the local submerged barrier with the different characteristics of the waves.

The need for such simplified methods arises from the large number of coastal interventions that are planned worldwide and from the need to protect them from the consequences of mean sea level rises and changes in the frequency driven by climate change. The proposed methodology can be used for analysing and predicting the shoreline development over a period of decades, proving to be an essential tool in the modeling and design of different submerged sea defences. Moreover, it can be used to achieve an initial assessment of the long-term efficiency of former submerged breakwaters and the effectiveness of further interventions, e.g., the use of a larger diameter of nourishment sand or the implementation of nature-based solutions, such as the planting of native seagrass meadows (e.g., *Posidonia oceanica*). However, this method cannot replace the use of the full 2D or 3D mathematical models, since several phenomena, such as local turbulence, flocculation, benthic suspensions and entrainment of fluid mud by shear flow, usually require a two or three-dimensional approach to assess their effect on bedload and erosion. In conclusion, the application of the presented methodology provides the first evaluation of the effectiveness of the former sea defences built at Calabaia Beach, and proves to be promising in assisting engineers and/or environmentalists in designing/evaluating the efficiency of coastal interventions not only in this specific area, but also in worldwide coastal areas where protected nourishments have been designed or already implemented.

Author Contributions: Conceptualization, M.M.; methodology, M.M., R.A.M. and S.S.; software, S.S.; validation, M.M., R.A.M. and S.S.; formal analysis, M.M., R.A.M. and S.S.; investigation, M.M., R.M.A. and S.S.; resources, M.M. and S.S.; data curation, M.M., R.A.M. and S.S.; writing—original draft preparation, M.M., R.A.M. and S.S.; writing—review and editing, M.M., R.A.M. and S.S.; visualization, R.A.M. and S.S.; supervision, M.M.; project administration, M.M.; funding acquisition, M.M. All authors have read and agreed to the published version of the manuscript.

Funding: This research was funded by Governo Italiano, Piano di interventi infrastrutturali di emergenza in Calabria—eventi alluvionali settembre 2000, Ordinanza di Protezione Civile n°3081/2000", by Regione Calabria, Progetto Integrato Strategico—Rete Ecologica Regionale, POR Calabria—Asse I—Misura 1.10, Convenzione n°1562, and by Provincia di Cosenza, Progetto esecutivo per la difesa e riqualificazione del litorale nei Comuni di Bonifati, Sangineto e Belvedere M.—Tratto da nord Capo Bonifati a nord Capo Tirone, PEG—Bilancio 2004.

Acknowledgments: The authors thank the participants in this study for their time and useful insights. We also thank the Province of Cosenza for their support with the technical documentation provided. The authors sincerely thank the *"Danish Hydraulic Institute—DHI"* for providing the *"MIKE-21 2020 powered by DHI"* license in the *"Agreement for the development of PhD thesis"* between the Danish Hydraulic Institute (DHI) and 'Capo Tirone Marine Experimental Station' (UNICAL).

Conflicts of Interest: The authors declare no conflict of interest. The funders had no role in the design of the study; in the collection, analyses, or interpretation of data; in the writing of the manuscript, or in the decision to publish the results.

References

1. Alexandrakis, G.; Manasakis, C.; Kampanis, N.A. Valuating the effects of beach erosion to tourism revenue. A management perspective. *Ocean Coast. Manag.* **2015**, *111*, 1–11. [CrossRef]
2. Roebeling, P.C.; Costa, L.; Magalhães-Filho, L.; Tekken, V. Ecosystem service value losses from coastal erosion in Europe: Historical trends and future projections. *J. Coast. Conserv.* **2013**, *17*, 389–395. [CrossRef]
3. Folke, C.; Biggs, R.; Norström, A.V.; Reyers, B.; Rockström, J. Social-ecological resilience and sustainability. *Ecol. Soc.* **2016**, *21*, 41. [CrossRef]
4. Leslie, H.M.; Basurto, X.; Nenadovic, M.; Sievanen, L.; Cavanaugh, K.C.; Cota-Nieto, J.J.; Erisman, B.E.; Finkbeiner, E.; Hinojosa-Arango, G.; Moreno-Báez, M.; et al. Operationalizing the social-ecological systems framework to assess sustainability. *Proc. Natl. Acad. Sci. USA* **2015**, *112*, 5979–5984. [CrossRef]
5. Sun, C.; Zhang, K.; Zou, W.; Li, B.; Qin, X. Assessment and Evolution of the Sustainable Development Ability of Human–Ocean Systems in Coastal Regions of China. *Sustainability* **2015**, *7*, 1–29. [CrossRef]
6. Mele, B.H.; Russo, L.; D'Alelio, D. Combining marine ecology and economy to roadmap the integrated coastal management: A systematic literature review. *Sustainability* **2019**, *11*, 4393. [CrossRef]
7. Nicholls, R.J. Coastal Megacities and Climate Change. *GeoJournal* **1995**, *37*, 369–379. [CrossRef]
8. De Zolt, S.; Lionello, P.; Nuhu, A.; Tomasin, A. The disastrous storm of 4 November 1966 on Italy. *Nat. Hazards Earth Syst. Sci.* **2006**, *6*, 861–879. [CrossRef]
9. Mel, R.; Sterl, A.; Lionello, P. High resolution climate projection of storm surge at the Venetian coast. *Nat. Hazards Earth Syst. Sci.* **2013**, *13*, 1135–1142. [CrossRef]
10. Mel, R.; Lionello, P. Verification of an ensemble prediction system for storm surge forecast in the Adriatic Sea. *Ocean Dyn.* **2014**, *64*, 1803–1814. [CrossRef]
11. Mel, R.; Lionello, P. Probabilistic dressing of a storm surge prediction in the Adriatic Sea. *Adv. Meteorol.* **2016**, *2016*, 3764519. [CrossRef]
12. Pranzini, E.; Wetzel, A.; Williams, A. Aspects of coastal erosion and protection in Europe. *J. Coast. Conserv.* **2015**, *19*. [CrossRef]
13. Bendoni, M.; Mel, R.; Solari, L.; Lanzoni, S.; Francalanci, S.; Oumeraci, H. Insights into lateral marsh retreat mechanism through localized field measurements. *Water Resour. Res.* **2016**, *52*, 1446–1464. [CrossRef]
14. Finotello, A.; Marani, M.; Carniello, L.; Pivato, M.; Roner, M.; Tommasini, L.; D'alpaos, A. Control of wind-wave power on morphological shape of salt marsh margins. *Water Sci. Eng.* **2020**. [CrossRef]
15. Huang, W. The Influence of Cruise Tourism Dining Waste on the Process of Self-recovery of Natural Ecological Environment. *Ekoloji* **2019**, *28*, 49–54.
16. Lamine, I.; Alla, A.A.; Bourouache, M.; Moukrim, A. Monitoring of physicochemical and microbiological quality of Taghazout seawater (Southwest of Morocco): Impact of the new tourist resort "Taghazout bay". *J. Ecol. Eng.* **2019**, *20*, 79–89. [CrossRef]
17. McKenna, J.; Cooper, A.; O'Hagan, A.M. Managing by principle: A critical analysis of the European principles of Integrated Coastal Zone Management (ICZM). *Mar. Policy* **2008**, *32*, 941–955. [CrossRef]
18. Pope, J. Responding to coastal erosion and flooding damages. *J. Coast. Res.* **1997**, *13*, 704–710.
19. Sakamoto, R.; Seino, S.; Suzaki, H. Coastal Alteration and Changes in Shoreline Morphology Due To Artificial Structures in Miiraku Town on Fukue Is. in the Goto Archipelago. In Proceedings of the EMECS 11-Sea Coasts XXVI. Joint Conference. Managing Risks to Coastal Regions and Communities in a Changing World, Saint Petersburg, Russia, 22–27 August 2016.
20. Reeve, D.E.; Spivack, M. Stochastic prediction of long-term coastal evolution. *Environ. Stud.* **2001**, *58*, 55–64.
21. EUROSION. Living with Coastal Erosion in Europe: Sediment and Space for Sustainability—PART I—Major Findings and Policy Recommendations of the EUROSION Project. In *Results from the Eurosion study*; European Commission: Luxembourg, 2004; p. 54.
22. Post, J.C.; Lundin, C.G. *Guidelines for Integrated Coastal Zone Management*; The World Bank: Washington, WA, USA, 1996; Volume 9, ISBN 0821337351.
23. OECD. *The Ocean Economy in 2030*; OECD Publishing: Paris, France, 2016.

24. Seixas, C.S.; Davidson-hunt, I.; Kalikoski, D.C.; Davy, B.; Berkes, F.; de Castro, F.; Medeiros, R.P.; Minte-vera, C.V.; Araujo, L.G. Collaborative Coastal Management in Brazil: Advancements, Challenges, and Opportunities. In *Viability and Sustainability of Small-Scale Fisheries in Latin America and The Caribbean*; Springer: Cham, Switzerland, 2019; pp. 421–451. ISBN 9783319760780.
25. Crain, C.M.; Halpern, B.S.; Beck, M.W.; Kappel, C.V. Understanding and managing human threats to the coastal marine environment. *Ann. N. Y. Acad. Sci.* **2009**, *1162*, 39–62. [CrossRef]
26. Maiolo, M.; Pantusa, D. Sustainable Water Management Index, SWaM_Index. *Cogent Eng.* **2019**, *6*, 1–14. [CrossRef]
27. Ondiviela, B.; Losada, I.J.; Lara, J.L.; Maza, M.; Galván, C.; Bouma, T.J.; van Belzen, J. The role of seagrasses in coastal protection in a changing climate. *Coast. Eng.* **2014**, *87*, 158–168. [CrossRef]
28. Spalding, M.D.; Ruffo, S.; Lacambra, C.; Meliane, I.; Hale, L.Z.; Shepard, C.C.; Beck, M.W. The role of ecosystems in coastal protection: Adapting to climate change and coastal hazards. *Ocean Coast. Manag.* **2014**, *90*, 50–57. [CrossRef]
29. Cantasano, N. Posidonia Oceanica per la Difesa Degli Ambienti. In Proceedings of the SOS Dune, Roma, Italy, 11–18 May 2014; pp. 170–183.
30. Callaghan, D.; Ranasinghe, R.; Nielsen, P.; Larson, M.; Short, A. Process-determined coastal erosion hazards. *Proc. Coast. Eng. Conf.* **2009**, *5*, 4227–4236.
31. Li, X.; Zhang, W. 3D numerical simulation of wave transmission for low-crested and submerged breakwaters. *Coast. Eng.* **2019**, *152*, 103517. [CrossRef]
32. Hur, D.S.; Lee, W.D.; Cho, W.C. Three-dimensional flow characteristics around permeable submerged breakwaters with open inlet. *Ocean Eng.* **2012**, *44*, 100–116. [CrossRef]
33. Hur, D.S.; Lee, W.D.; Cho, W.C. Characteristics of wave run-up height on a sandy beach behind dual-submerged breakwaters. *Ocean Eng.* **2012**, *45*, 38–55. [CrossRef]
34. Sharifahmadian, A.; Simons, R.R. A 3D numerical model of nearshore wave field behind submerged breakwaters. *Coast. Eng.* **2014**, *83*, 190–204. [CrossRef]
35. Sorensen, R.M.; Beil, N.J. Perched beach profile response to wave action. In Proceedings of the 21st International Conference on Coastal Engineering, Malaga, Spain, 20–25 June 1988; pp. 1482–1492.
36. González, M.; Medina, R.; Losada, M.A. Equilibrium beach profile model for perched beaches. *Coast. Eng.* **1999**, *36*, 343–357. [CrossRef]
37. Sollitt, C.K.; Cross, R.H.; Engineer, C. Wave Transmission Through Permeable Breakwaters. *Coast. Eng.* **1972**, 1827–1846.
38. Rahimzadeh, A.; Ghadimi, P.; Feizi Chekab, M.A.; Jabbari, M.H. Determining transmission coefficient of propagating solitary wave over trapezoidal breakwater and parametric studies on different influential factors. *ISRN Mech. Eng.* **2014**, *2014*, 841327. [CrossRef]
39. Anastasiau, K.; Chan, C.T. Solution of the 2D shallow water equations using the finite-volume method on unstructured triangular meshes. *Int. J. Numer. Methods Fluids* **1997**, *24*, 1225–1245. [CrossRef]
40. Komen, G.J.; Cavaleri, L.; Doneland, M.; Hasselmann, K.; Hasselmann, S.; Janssen, P.A.E.M. *Dynamics and Modelling of Ocean Waves*; Cambridge University Press: New York, NY, USA, 1994; ISBN 0-521-47047-1.
41. Young, I.R. *Wind Generated Ocean Waves*; Elsevier: Amsterdam, The Netherlands, 1999; Volume 2, ISBN 9780080433172.
42. Ekebjærg, L.; Justesenu, P. An explicit scheme for advection-diffusion modelling in two dimensions. *Comput. Methods Appl. Mech. Eng.* **1991**, *88*, 287–297. [CrossRef]
43. Mel, R.; Carniello, L.; D'Alpaos, L. Addressing the effect of the Mo.S.E. barriers closure on wind setup within the Venice lagoon. *Estuar. Coast. Shelf Sci.* **2019**, *225*, 106249. [CrossRef]
44. Mel, R.; Carniello, L.; D'Alpaos, L. Dataset of wind setup in a regulated Venice lagoon. *Data Br.* **2019**, *26*, 104386. [CrossRef] [PubMed]
45. Syvitski, J.P.M.; Slingerland, R.L.; Burgess, P.; Murray, A.B.; Wiberg, P.; Tucker, G.; Voinov, A. Morphodynamic models: An overview. In *River, Coastal and Estuarine Morphodynamics: RCEM 2009*; Vionnet, C.A., Garcia, M.H., Latrubesse, E.M., Perillo, G.M.E., Eds.; Taylor & Francis: London, UK, 2010; pp. 3–20.

46. Short, A.D.; Jackson, D.W.T. Beach Morphodynamics. In *Treatise on Geomorphology*; Shroder, J.F., Ed.; Academic Press: San Diego, CA, USA, 2013; Volume 10, pp. 106–129.
47. Hamza, W.; Tomasicchio, G.R.; Ligorio, F.; Lusito, L.; Francone, A. A Nourishment Performance Index for Beach Erosion/Accretion at Saadiyat Island in Abu Dhabi. *J. Mar. Sci. Eng.* **2019**, *7*, 173. [CrossRef]
48. Danish Hydraulics Institute (DHI). *MIKE 21 & MIKE 3 Flow Model FM—Hydrodynamic and Transport Module, Scientific Documentation*; DHI, Ed.; DHI: Hørsholm, Denmark, 2017.
49. Roe, P.L. Approximate Riemann Solvers, Parameter Vectors, and Difference Sschemes. *J. Comput. Phys.* **1997**, *135*, 250–258. [CrossRef]
50. Jawahar, P.; Kamath, H. A High-Resolution Procedure for Euler and Navier-Stokes Computations on Unstructured Grids. *J. Comput. Phys.* **2000**, *164*, 165–203. [CrossRef]
51. Battjes, J.A.; Janssen, J.P.F.M. Energy oss and Set-up due to breaking of random waves. *Coast. Eng.* **1978**, 569–587.
52. Eldeberky, Y.; Battjes, J.A. Spectral modeling of wave breaking: Application to Boussinesq equations. *J. Geophys. Res.* **1996**, *101*, 1253–1264. [CrossRef]
53. Nelson, R.C. Design wave heights on very mild slopes—An Experimental Study. *Trans. Inst. Eng. Aust. Civ. Eng.* **1987**, *3*, 157–161.
54. Nelson, R.C. Depth limited design wave heights in very flat regions. *Coast. Eng.* **1994**, *23*, 43–59. [CrossRef]
55. Ruessink, B.G.; Walstra, D.J.R.; Southgate, H.N. Calibration and verification of a parametric wave model on barred beaches. *Coast. Eng.* **2003**, *48*, 139–149. [CrossRef]
56. Danish Hydraulics Institute (DHI). *MIKE 21 Sand Transport Module, Scientific Documentation*; DHI, Ed.; DHI: Hørsholm, Denmark, 2017.
57. Aagaard, T.; Nielsen, J.; Jensen, S.G.; Friderichsen, J. Longshore sediment transport and coastal erosion at Skallingen, Denmark. *Geogr. Tidsskr.* **2004**, *104*, 5–14. [CrossRef]
58. Elfrink, B.; Brøker, I.; Deigaard, R.; Asp Hansen, E.; Justesen, P. Modelling of 3D sediment transport in the surf zone. *Coast. Eng.* **1996**, 3805–3817.
59. Murillo, J.; García-Navarro, P. An Exner-based coupled model for two-dimensional transient flow over erodible bed. *J. Comput. Phys.* **2010**, *229*, 8704–8732. [CrossRef]
60. Danish Hydraulics Institute (DHI). *MIKE 21 Spectral Wave Module, Scientific Documentation*; DHI, Ed.; DHI: Hørsholm, Denmark, 2017.
61. Goda, Y.; Takeda, H. *Laboratory Investigation on Wave Transmission over Breakwater*; The Port and Harbour Research Institute: Yokosuka, Japan, 1967; Volume 8.
62. Goda, Y. *Re-analysis of Laboratory Data on Wave Transmission over Breakwaters*; The Port and Harbour Research Institute: Yokosuka, Japan, 1969; Volume 8.
63. Viero, D.P.; D'Alpaos, A.; Carniello, L.; Defina, A. Mathematical modeling of flooding due to river bank failure. *Adv. Water Resour.* **2013**, *59*, 82–94. [CrossRef]
64. Viero, D.P.; Defina, A. Multiple states in the flow through a sluice gate. *J. Hydraul. Res.* **2019**, *57*, 39–50. [CrossRef]
65. Mel, R.; Viero, D.P.; Carniello, L.; D'Alpaos, L. Optimal floodgate operation for river flood management: The case study of Padova (Italy). *J. Hydrol. Reg.* **2020**, *30*, 100702. [CrossRef]
66. Mel, R.A.; Viero, D.P.; Carniello, L.; D'Alpaos, L. Multipurpose use of artificial channel networks for flood risk reduction: The case of the waterway Padova–Venice (Italy). *Water* **2020**, *12*, 1609. [CrossRef]
67. Maiolo, M.; Versace, P.; Natale, L.; Irish, J.; Pope, J.; Frega, F. A comprehensive study of the tyrrhenian shoreline of the Province of Cosenza. In Proceedings of the AIPCN 2000—Giornate Italiane di Ingegneria Costiera V Edizione, Reggio Calabria, Italy, 11–13 October 2000.
68. Rubio, C.M. A laboratory procedure to determine the thermal properties of silt loam soils based on ASTM D 5334. *Appl. Ecol. Environ. Sci.* **2013**, *1*, 45–48.
69. U.S.A.C.E.-C.H.L.; U.S. *Army Corps of Engineers (USACE) Coastal and Hydraulics Laboratory (CHL)—Centro Studi di Ingegneria Ambientale (CSdIA) Analisi Regionale del Litorale della Provincia di Cosenza*, Provincia di Cosenza Ed. 2000.
70. Orlanski, I. A simple boundary condition for unbounded hyperbolic flows. *J. Comput. Phys.* **1976**, *21*, 251–269. [CrossRef]

71. Chapman, D.C. Numerical treatment of cross-shelf open boundaries in a barotropic coastal ocean model. *J. Phys. Oceanogr.* **1985**, *15*, 1060–1075. [CrossRef]
72. Shabangu, P.E. Investigation of a Simplified Open Boundary Condition for Coastal and Shelf Sea Hydrodynamic Models. Ph.D. Thesis, Stellenbosch University, Stellenbosch, South Africa, March 2015.
73. Hur, D.S.; Lee, W.D.; Cho, W.C.; Jeong, Y.H.; Jeong, Y.M. Rip current reduction at the open inlet between double submerged breakwaters by installing a drainage channel. *Ocean. Eng.* **2019**, *193*, 106580. [CrossRef]
74. Goda, Y.; Ahrens, J.P. New formulation of wave transmission over and through low-crested structures. *Coast. Eng.* **2009**, *5*, 3530–3541.

Journal of
Marine Science and Engineering

Article

Overtopping Metrics and Coastal Safety: A Case of Study from the Catalan Coast

Corrado Altomare [1,2,*], Xavi Gironella [1], Tomohiro Suzuki [3,4] Giacomo Viccione [5] and Alessandra Saponieri [6]

1 Laboratori d'Enginyeria Maritima, Universitat Politècnica de Catalunya—BarcelonaTech, Calle Jordi Girona 1-3, 08034 Barcelona, Spain; xavi.gironella@upc.edu
2 Department of Civil Engineering, Ghent University, Technologiepark 60, 9052 Gent, Belgium
3 Flanders Hydraulics Research, Berchemlei 115, 2140 Antwerp, Belgium; tomohiro.suzuki@mow.vlaanderen.be
4 Department of Hydraulic Engineering, Delft University of Technology, Stevinweg 1, 2628 CN Delft, The Netherlands
5 Environmental and Maritime Hydraulics Laboratory (LIDAM), University of Salerno, Via Giovanni Paolo II, 132, 84084 Fisciano SA, Italy; gviccion@unisa.it
6 Department of Engineering for Innovation, University of Salento, Ecotekne, 73047 Lecce, Italy; alessandra.saponieri@unisalento.it
* Correspondence: corrado.altomare@upc.edu; Tel.: +34-93-401-70-17

Received: 5 June 2020; Accepted: 21 July 2020; Published: 24 July 2020

Abstract: Design criteria for coastal defenses exposed to wave overtopping are usually assessed by mean overtopping discharges and maximum individual overtopping volumes. However, it is often difficult to give clear and precise limits of tolerable overtopping for all kinds of layouts. A few studies analyzed the relationship between wave overtopping flows and hazard levels for people on sea dikes, confirming that one single value of admissible mean discharge or individual overtopping volume is not a sufficient indicator of the hazard, but detailed characterization of flow velocities and depths is required. This work presents the results of an experimental campaign aiming at analyzing the validity of the safety limits and design criteria for overtopping discharge applied to an urbanized stretch of the Catalan coast, exposed to significant overtopping events every stormy season. The work compares different safety criteria for pedestrians. The results prove that the safety of pedestrians on a sea dike can be still guaranteed, even for overtopping volumes larger than 1,000 L/m. Sea storms characterized by deep-water wave height between 3.6 and 4.5 m lead to overtopping flow depth values larger than 1 m and flow velocities up to 20 m/s. However, pedestrian hazard is proved to be linked to the combination of overtopping flow velocity and flow depth rather than to single maximum values of one of these parameters. The use of stability curves to assess people's stability under overtopping waves is therefore advised.

Keywords: wave overtopping; coastal safety; flow velocity; flow depth; sea dikes

1. Introduction

Wave overtopping assessment is a key procedure within the general design of any coastal defense. The groundwork for the assessment of wave overtopping was laid by [1], who chose the average discharge as a design value, stating that "there is no such thing as an absolute discharge: because the wave heights and periods exhibit a random distribution about a given mean, the discharge will also vary randomly". Since [1], several experimental campaigns have been carried out worldwide, leading to semi-empirical models, which are nowadays largely used for wave overtopping assessment (e.g., [2–4]). Nevertheless, during the last few years, the increased storminess caused by climate change has raised concerns among researchers, engineers and decision-makers: is the average overtopping an appropriate design criterion? Or are the biggest waves the ones that cause the major damages and casualties? Allsop et al. [5] highlighted that tests on the effects of overtopping flows on people indicate that the assessment of mean discharge is not enough to evaluate people's safety; the authors proposed maximum individual overtopping volumes as more suitable hazard indicators. Limits for individual overtopping volumes together with tolerable mean discharges were proposed in [5], related to overtopping velocities below 10 m/s. However, the authors suggested that lower volumes may be required for violent overtopping processes with higher velocities. Therefore, it was emphasized but not discussed further that other flow parameters might play an important role and be linked to overtopping hazards, namely overtopping flow velocity and overtopping flow depth. Generally speaking, "the character of overtopping flow hazards depends on geometries of the defence, the hinterland and the form of overtopping" [5]. Several studies have been gathered and presented in EurOtop [6], where finally a tolerable limit for maximum individual volumes equal to 600 L/m was suggested, over which a single event cannot be tolerated by people (Table 1).

Table 1. Tolerable overtopping limits for people proposed in EurOtop (2018).

Hazard Type and Reason	Mean Discharge, q (L/s/m)	Maximum Individual Volume V_{max} (L/m)
People at structures with possible violent overtopping, mostly vertical structures	No access for any predicted overtopping	No access for any predicted overtopping
People at seawall/dike crest. Clear view of the sea.	-	-
H_{m0} = 3 m	0.3	600
H_{m0} = 2 m	1	600
H_{m0} = 1 m	10–20	600
H_{m0} < 0.5 m	No limit	No limit

Despite many efforts having been done to assess overtopping flow velocities and overtopping flow depths on the seaward side, crest and landward side of coastal defenses [7–10], no direct link has been clearly established between tolerable overtopping individual volumes and discharges with tolerable flow velocities and flow depths.

Few studies considered overtopping flow depths and velocities in order to estimate the overtopping hazard for pedestrians [11,12]. In some cases, the referred limits were obtained from physical experiments using anthropomorphic dummies, and in other cases [12], a video analysis of real flooding events was performed. In addition, any formula for overtopping flow depth and velocity does not necessarily refer to the same specific overtopping event, since the 2% exceedance probability of the two quantities is considered. These quantities are proved to not be correlated [10]. Nevertheless, when looking at the maximum individual overtopping volumes, the aim is to characterize the flow depth and velocity associated to that volume, since it is the one reported as threshold value for safety [6].

The present work analyses the overtopping discharges, volumes, overtopping flow depth and velocities that can lead to risk scenarios for people. This work focuses on coastal defenses in highly urbanized areas. For this purpose, the case study of Premià de Mar is analyzed, which schematically represents the coastline north of Barcelona, Spain. The main purpose of this study is to discuss the validity of the present safety limits and design criteria for the Catalan coast. To reach this objective, physical model tests were carried out, modelling a layout that resembles the case study for different wave conditions corresponding to events with different return periods. The acquired experimental data were collected, analyzed and compared with the state-of-the-art semi-empirical formulas; the results, finally, were compared with the safety criteria from [6] and with the stability curves for people combining flow velocities and depths [12–14].

2. Overtopping Flow Parameters and People Safety

Wave overtopping can be distinguished between smooth "green" overtopping flows and a highly turbulent "white" overtopping flows. In [6], "green water" was defined as a wave overtopping which runs up the face of the seawall and over the crest in a (relatively) continuous sheet of water.

In contrast, 'white water' or spray overtopping tends to occur in cases of significant splashes, due to heavy waves breaking on the seaward face of defense structures which produces non-continuous overtopping and/or significant volumes of spray. The main parameters that are commonly employed to characterize overtopping flows are the mean overtopping discharge, q, and the individual maximum overtopping volume, V_{max}. Plenty of semi-empirical models have been derived that allow quantifying mean discharge values and depend on the kind of structure and hydraulic boundary conditions. Dikes with gentle slope, namely between 1:7 and 1:2, are studied in [3,4,6,15,16]. Steeper slopes up to vertical walls were analyzed in [17–20], where the last one includes cases of structures with emergent toe. Goda [21] proposed a set of unified formulas for smooth impermeable sea dikes where new coefficients were derived as a function of local water depth, foreshore slope and dike slope. The influence of the foreshore slope for very and extremely shallow water conditions [22] was taken into account in [15] by means of the equivalent slope concept. Similar to this, [20] employed an imaginary slope for wave run-up and overtopping calculations. Other authors focused their attention on the characterization of overtopping flow depths and velocities, such as [9,10,23,24].

Even though the scientific literature available on wave overtopping prediction is very extensive, there are few works dealing with the stability of people under the effect of wave overtopping flows; that is limited to a modest number of studies. In this area of investigation, some studies have tested human subjects in controlled flow conditions which have generated a quantification of the critical flow parameters and mechanisms, possibly leading to a person losing stability and falling in the surrounding flow. The first study with the purpose to analyze the human stability in wave overtopping flows was promoted by the Japanese Port and Harbor Research Institute (PHRI), the results of which were published in [11]. The authors proposed two models of human instability: "slipping" and "tumbling". The first one occurs when the flow force against the body (F_f) is bigger than the maximum available bottom friction resistance of the subject (F_r). The second mechanism models the falling process arising when the unbalancing moment produced by the flow around the feet of the subject is bigger than the restoring moment produced by the weight of the person (M_r). Sandoval and Bruce [12] revisited the model of [11], accounting for the buoyancy of the subject, as well as its related position respect to the incoming flows. The analysis for each mechanism of instability can be derived as follows:

1. Friction stability (F_f-F_r)

$$u^2{\cdot}d = \frac{2{\cdot}\mu{\cdot}m_g{\cdot}g}{C_d{\cdot}\rho{\cdot}B1} \tag{1}$$

2. Momentum stability (M_r-M_f)

$$u{\cdot}d = 2\sqrt{\frac{d_1{\cdot}m_g{\cdot}g}{C_d{\cdot}\rho{\cdot}B1}} \tag{2}$$

where

u = flow velocity (m/s);
d = flow depth (m);
ρ = density of water (kg/m^3);
μ = coefficient of friction between shoe sole and ground (-);
B_1 = average diameter of the subject legs (m);
C_d = drag coefficient (-)
d_1 = distance from pivot point to the center gravity (m);
m_g = subject's mass (kg)

The moments generated by the flows (M_f) can be calculated as the drag force applied at the half of the depth. On the other hand, the restoring moment (M_r) is a function of the person's weight and the distance to the pivot point (d_1).

More recently, Arrighi et al. [13,14] compared the experimental evidence for humans and vehicles with the results of a numerical investigation. The authors identified relative submergence and the Froude number as the most relevant parameters to express the vulnerability of pedestrians. They derived a regression curve for human stability, expressed as follows:

$$\frac{H_{crP}}{H_P} = \frac{0.29}{0.24 + Fr} \tag{3}$$

where H_{crP} is the critical flow depth, H_P is the height of the subject, while their ratio represents the relative submergence. If $H/H_{crP} < 1$, the person is stable. The Froude number is calculated as $u/\sqrt{(gd)}$.

3. Overtopping Flow Velocity and Flow Depth Estimation on the Dike Crest

There is already plenty of literature dealing with the estimation of overtopping flow parameters [10,25,26]. The equations for overtopping flow depth and velocity at the dike crest can be expressed in a general form as follows:

$$d_{2\%} = c_{d2\%}(Ru_{2\%} - R_c)exp\left(-c_{c,d}\frac{x_c}{B}\right) \tag{4}$$

$$u_{2\%} = c_{u2\%}(Ru_{2\%} - R_c)exp\left(-c_{c,u}\frac{x_c \mu}{2d_{2\%}}\right) \tag{5}$$

where $d_{2\%}$ is the overtopping flow depth on the dike crest, $u_{2\%}$ is the overtopping flow velocity on the dike crest, x_c is the streamwise coordinate on the dike crest, μ is the bottom friction coefficient, $Ru_{2\%}$ is the wave run-up, R_c is the crest freeboard respect to the still water level (Figure 1). The subscript 2% refers to quantities exceeded by 2% of the number of the incident waves. The coefficients $c_{d2\%}$, $c_{u2\%}$, $c_{c,d}$, and $c_{c,u}$ are empirical coefficients, and the values can vary according to the literature. For run-up assessment, formulas are here omitted for sake of simplicity. The reader can refer, for example, to [6].

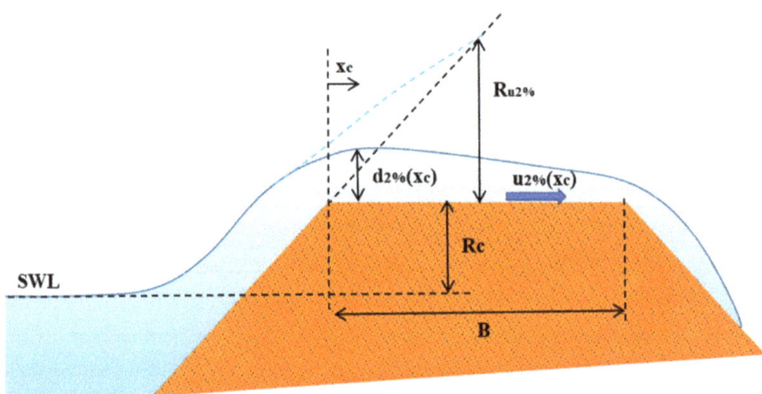

Figure 1. Scheme of overtopping flow on a sea dike.

Instead, Hughes [24] established empirical relationships between individual overtopping wave volumes and the maximum velocity, maximum flow depth and individual discharge, based on data for 1:3 and 1:6 dike slopes:

$$u = 27.67\frac{\sqrt{V_{max}tan\alpha}}{T_{m-1,0}} \tag{6}$$

$$d = 0.324\sqrt{V_{max}} \tag{7}$$

$$q_{max} = 7.405V_{max}\frac{\sqrt{tan\alpha}}{T_{m-1,0}} \tag{8}$$

In this case, u and d refer to the maximum velocity and depth on the dike crest, respectively. The dike slope and the wave period are included in the relationship, but most important there is a direct link with the individual overtopping volume. This was also confirmed by [27], who derived expressions for overtopping flow parameters analytically based on wave momentum flux, in analogy with [28].

4. Case of Study

4.1. Site Description and Model Geometry

Physical model experiments were carried out in the CIEMito wave flume at the Maritime Engineering Laboratory of Universitat Politècnica de Catalunya—BarcelonaTech (LIM/UPC), in Barcelona, Spain. The geometrical layout used for the experimental campaign resembles the beach and coastal protection in the area of Premià de Mar, municipality in the Comarca of the Maresme in Catalonia, Spain (Figure 2). This stretch of the coast is characterised by the presence of both railways and a promenade/bike path which are very exposed to possible sea storms, being located at a few meters from the shoreline. A dike made of natural stone with a relatively steep slope (1:1) characterises this coastal area. In the physical model tests, the effects of the rubble mound have been neglected, considering a smooth slope instead. It is in the interest of the present research to analyse those stretches where nourishment was no longer performed (characteristic in the winter season and storm events). Hence, the beach has been eroded, leaving the dike exposed directly to the sea. A water depth at the dike toe is between 0.5 and 1 m and the crest of the dike is 3–4 m above the mean sea level.

Figure 2. Territorial framework of the study case (source: Google Maps).

In particular, the area near to the railway station was studied. This area has a long history of flooding. This stretch of the coast, in fact, contains a railway and a bike path, and both are very exposed to possible sea storms, being located at a few meters from the shore (see Figure 3). Besides risks related to the safeguarding of human life, the analysis of the overtopping rates for this structure is motivated by the occurrence of violent wave climate events that have been recorded in the last decade. These events have consequent damages to infrastructures located just behind the dike and service interruption of the public transport for a line that is strategic for the zone, connecting it directly to the metropolitan area of Barcelona. Additionally, Premià de Mar can be considered as representative of many other domains of the Mediterranean Sea located in urbanized areas.

Figure 3. Photographs representing the current state of the study area.

The repetition of overtopping episodes shows the inability of the coastal defense to contain the transmitted energy within acceptable limits and therefore the urge to assess the safety of the existing coastal structure and eventually adapt its geometry.

The studied area is mainly composed of low-lying beaches and narrow coarse sand emerged beaches with a relatively gentle slope. It is mainly an urbanized area where several industrial activities are present. The bathymetry data were initially extracted from the EMODnet Digital Terrain Model (DTM). Close to the coastline, these bathymetry data were integrated with more detailed data from a survey carried out by LIM/UPC technicians with a home-built sailing drone (see picture in Figure 4). The drone was equipped with two motors, an echo sounding of 120 KHz with a working range of water depths varying from 0.15 to 13 m. The drone was controlled remotely.

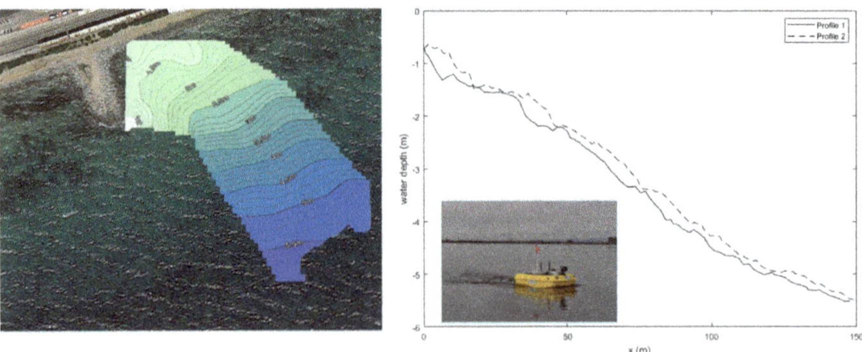

Figure 4. Bathymetry extracted from the LIM/UPC drone survey (**left plot**) and resulting profiles by merging existing data with the carried-out survey (**right plot**). A photo of the drone is in the small picture.

The local water level employed for the drone measurements was corrected based on the information of the tie gauge installed in Barcelona harbor. Example of bathymetric profiles are depicted in Figure 4 (right plot), where x is the distance from the shoreline. An average slope of 1:30 was identified; however, some abrupt bathymetric changes can be noticed in the first 10–20 m, where profile 2 is steeper than profile 1. Hence, to account for local variations of the bathymetry in the studied area, two different foreshore slopes were finally used for the experimental campaign, namely 1:15 and 1:30.

4.2. Experimental Setup and Wave Conditions

The wave flume is 18 m long, 0.38 m wide and 0.56 m high. The wave generation system is a piston-type board. The support structure consists of square metal sections, and both laterals and bottom walls are made of tempered glass which allows a complete view of the tests and clear video camera recording.

The model (scale is 1:50) was built and operated according to Froude's similarity law. A sketch of the layout is proposed in Figure 5, where all dimensions are in model scale. The model consists of a 1:n transition slope followed by a 1:m foreshore slope, where n is equal to 8 for m = 15 and 5 for m = 30, respectively. A 1:1 smooth dike made of polymethyl methacrylate is located at the end of the foreshore. The dike height is 0.09 m. Different widths for the promenade (i.e., crest berm) were modelled, namely 0.12 and 0.24 m, to be representative of the different stretches along the coastline. The freeboard varies between 0.061 and 0.081 m, with toe depths of 0.009–0.029 m (Table 2).

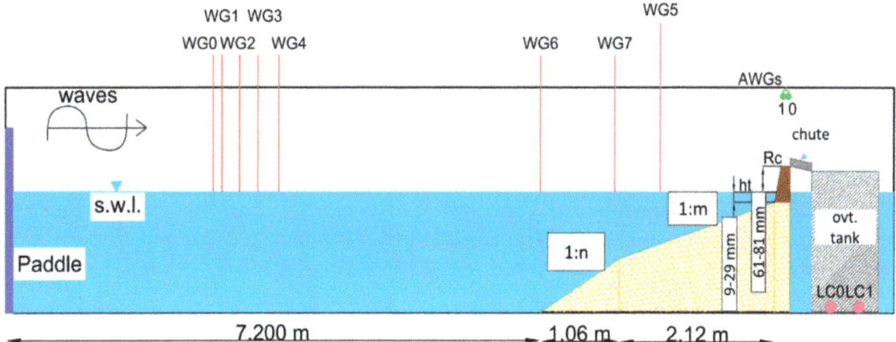

Figure 5. CIEMito wave flume—drawing of the longitudinal section (distorted). Dimensions are in model scale. The value of m is equal to 15 and 30, respectively. Accordingly, n assumes values of 8 and 5, respectively.

Table 2. Geometric characteristics of the tested dike layout.

Scale	h_{toe} (m)	R_c (m)	Promenade Width (m)
Model	0.009–0.019–0.024–0.029	0.81–0.071–0.066–0.061	0.12–0.24
Prototype	0.45–0.905–1.2–1.45	4.05–3.55–3.3–3.05	6–12

Irregular wave tests, employing a JONSWAP spectrum with enhancement factor equal to 3.3, were performed. Each test consisted of approximately 1,000 waves. In total, 420 tests were conducted. We consider 243 tests for the present analysis, having excluded those ones with no or inaccurate overtopping flow property measurements. The tested wave conditions were derived from the extreme wave forecast based on data acquired by the buoy of the Puertos del Estado, located outside Barcelona harbour, for return periods of 1, 2, 5 and >10 years. Deep-water wave height ranges between 2.58 and 4.78 m, with peak periods between 7.96 and 13.16 s. These values correspond to those propagated from offshore to a water depth between 14.5 and 15 m, being the up-scaled water depth values at the wave generation location. These conditions correspond to very shallow and extremely shallow water conditions, based on the definition in [22], $h_{toe}/H_{m0,deep}$ being between 0.1 and 0.38, where h_{toe} is the water depth at the dike toe and $H_{m0,deep}$ is the deep-water wave height.

The experimental tests carried out aimed to measure the following parameters: free surface water elevation along the wave flume, η (m); cumulated overtopping volume, V_{tot} (L/m); mean overtopping discharge, q (L/s/m), obtained as $V_{tot}/T_{mm\text{-}1,0,deep}{}^*N_{w,deep}$, where $T_{mm\text{-}1,0,deep}$ is the spectral wave period close to the wave generation (offshore) and $N_{w,deep}$ is the number of waves offshore; maximum overtopping volume, V_{max} (L/m); flow depth associated to the maximum overtopping volume, d (m); and horizontal velocity associated to the maximum overtopping volume, u (m/s).

4.3. Measurement Setup

Eight resistive sensors were placed along the flume to measure the water surface elevation at different location (WG0-WG8), working at a sample frequency of 80 Hz. The distance of the resistive wave gauges with respect to the mean paddle position is reported in Table 3.

Table 3. Location from paddle of resistive wave gauges (in meters).

Sensor	WG0	WG1	WG2	WG3	WG4	WG6	WG7	WG5
x (m = 15)	2.8	2.96	3.15	3.40	3.69	7.20	8.20	9.20
x (m = 30)	2.8	2.96	3.15	3.40	3.69	7.20	8.20	9.20

Overtopping flow depth and overtopping flow velocity were measured by means of redundant systems, composed of two high-speed cameras and two ultrasonic sensors. The results were compared and averaged. Two ultrasonic sensors were place on the dike top to measure flow depths (AWG0, AWG1), with resolution <0.3 mm. AWG 1 is positioned 2.5 cm away from the dike edge for the crest width of 0.12 m and 6 cm for the crest width of 0.24 m. Distance between AGW1 and AWG0 varied between 5.9 cm and 9.2 cm. The position of the two ultrasonic sensors and the vertical distance from the crest were optimized case per case to avoid interference between the sensors and optimize accuracy. The AWG raw signal was acquired with a sampling rate of 100 Hz and filtered in Matlab environment in order to derive overtopping flow depth and velocity on the dike. The velocity was calculated with the indirect methodology:

$$u_{AWGt_{tip}} = \frac{distance_{AWG}}{\Delta t_{tip}},\ u_{AWGt_{max}} = \frac{distance_{AWG}}{\Delta t_{max}} \tag{9}$$

where the delta values are, respectively, the temporal distance between the two maximum points Δt_{max} and the distance between the two starting points of the event Δt_{tip}. The latter ones correspond to a threshold value of 1mm, which identifies the tip of the overtopping wave. An example of the time series of flow depth for a maximum overtopping event is depicted in Figure 6, where the solid line is the acquisition of the sensor at the end of the promenade (AWG0) and the dashed line is the registration of the sensor at the beginning of the promenade (AWG1). The maximum of each signal is marked with a circle. Threshold values are marked with diamond markers. Here, a clarification is required. The overtopping volumes are measured after the dike promenade. Hence, flow depth and velocity associated to maximum event refers to the same location. For flow depth, the signal of AWG0 is used, being the closest sensor to the end of the promenade. However, as described previously, velocity is measured indirectly, and the calculation employs both ultrasonic sensors. The distance between AWG1 and AWG0 is comparable to the promenade width, and therefore the calculated velocity cannot be considered as the instantaneous velocity at the promenade end. Therefore, the layer velocity associated to the maximum volume and corresponding to the maximum flow depth, d_{AWG0}, has been calculated resolving the following system, considering momentum conservation:

$$\begin{cases} d_{AWG1}u_{AWG1} = d_{AWG0}u_{AWG0} \\ u_{AWGt_{max}} = \frac{u_{AWG1}+u_{AWG0}}{2} \end{cases} \quad (10)$$

where $u_{AWGtmax}$ is the average velocity as previously described, d_{AWG0} and d_{AWG1} are the maximum flow depth values at AWG0 and AWG1 location, respectively, and u_{AWG0} and u_{AWG1} are the instantaneous velocities associated to d_{AWG0} and d_{AWG1}, respectively. For all analyses reported in the next sections, we will refer to d_{AWG0} and d_{AWG1} as maximum flow depth d and overtopping flow velocity u, respectively, for sake of simplicity.

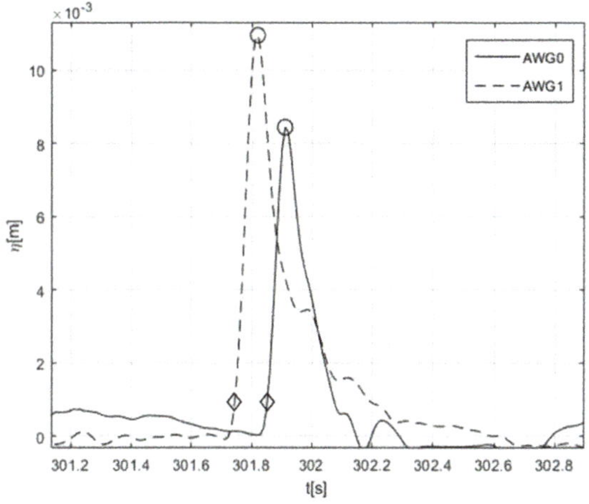

Figure 6. Example of simultaneous time series recorded by ultrasonic sensors.

The high-speed cameras were working at 200 fps. They were placed in order to provide the top and lateral view of the overtopping flows on the dike crest. Camera 1 was placed on the top of the structure. The dike promenade was graduated with equally spaced lines (1 cm). Camera 2 was placed in front of the lateral glass wall, on the side of the sea-dike model, where a transparent paper graduated with 0.5 cm squares was placed. From camera 1, only layer velocity could be measured as $s/\Delta t$, where s corresponds to the distance measured on the reference 1cm frame equal to the distance between the two ultrasonic sensors; Δt is the time needed for the tip of the overtopping flow to run such a distance. From camera 2, flow velocity and depth were measured: the lateral view allowed us to identify both the tip of the overtopping wave and the height of the overtopping layer (Figure 7). The results were compared with the ultrasonic sensors.

The same considerations for instantaneous and average velocities were made for the cameras' measurements. Figure 8 plots the correlation between the measurements of maximum flow depth carried out by means of ultrasonic sensors and high-speed camera (CAM2). Data were divided according to the foreshore slope. A correlation coefficient R^2 between 52% and 73% was calculated.

The individual and mean overtopping volume was collected inside an isolated metallic water box, connected to the edge of the promenade by a 9 cm-wide chute. Two load cells were installed under the overtopping tank located into the box in order to measure the cumulated water weight, with an acquisition frequency of 20 Hz. The volume was obtained by dividing weight by water density.

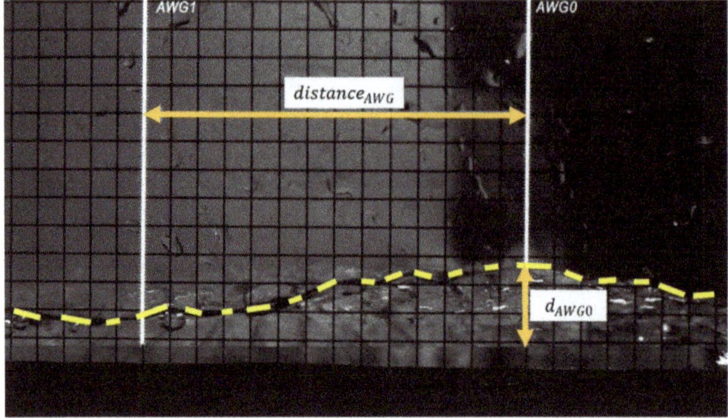

Figure 7. Lateral view with high-speed camera (CAM2).

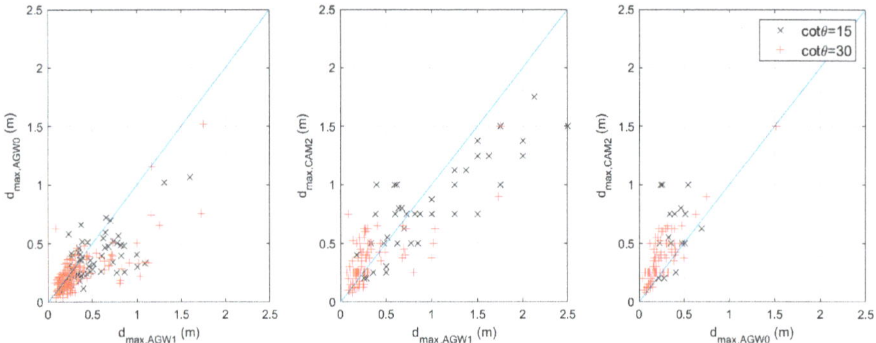

Figure 8. Correlation between overtopping flow depths measured with ultrasonic sensors and high-speed cameras.

4.4. Wave Analysis

Classical linear reflection analysis methods are not suitable in shallow water conditions because non-linear effects are dominant [4], since the generation of low-frequency waves is dominant and affects the value of the mean wave period $T_{m-1,0}$. This wave period is shown to be important for many wave-structure interaction processes, and can be used to assess the response of coastal structures with shallow foreshores [15,20]. Hence, the dike was removed, and horizontal bottom followed by absorption material was placed, instead, to measure incident wave conditions at the dike toe, which are required for the analysis. Sensor WG5 was moved to the dike toe location and used for the scope.

A limitation in the experiments is the use of first-order wave generation and the lack of active absorption of long-wave energy. For spurious long waves in the flume, the lack of second-order wave generation is found to be negligible due to the relatively large water depth at the wave generator and low steepness [29]. The natural frequency of the current flume set-up is around 0.045–0.05 Hz for different test conditions and foreshore slopes, which are found to be outside of the frequency range of the infragravity waves in the flume. Moreover, there is no increasing trend of the observed long-wave energy at the toe of the dike. A passive absorption system was found to be sufficient to reduce the wave reflection, except for the seiching motion. The energy at the seiching frequency band was removed in the analysis of the wave parameters. Despite the energy associated to seiching modes being between 7% and 20% of the total energy at the dike toe, it was decided to neglect it for further analysis. Seiches might have a significant influence on the mean discharge values; however, for individual events, it is very unlikely that the maximum overtopping event occurs when the crest of the seiche is at the dike toe, thus maximizing individual discharge.

An example of typical wave spectrum at the dike toe is depicted in Figure 9, showing an important shift of the wave energy from the JONSWAP typical spectral shape to low frequencies, generated by the heavy wave breaking and consequent release of long waves.

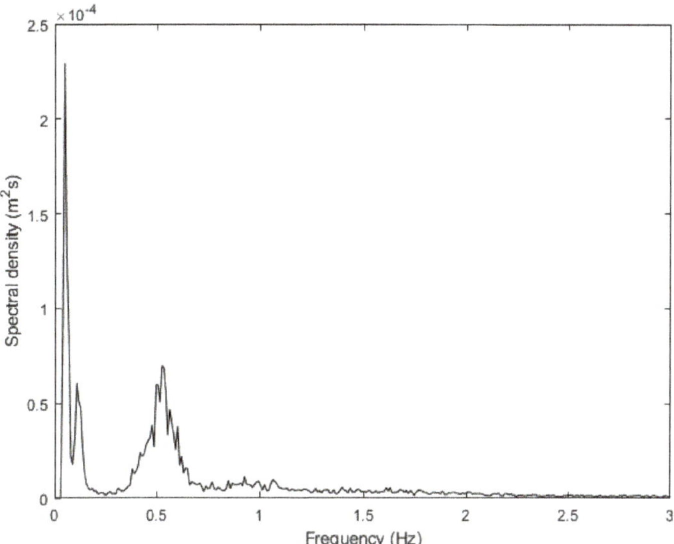

Figure 9. Example of wave spectrum for incident wave conditions at the dike toe (values in model scale).

4.5. Scale Effects

The employed model scale is relatively small and can cause concerns on the possible scale effects that might affect wave overtopping. Heller [30] provided an overview of limited criteria for flow-structure interaction phenomena and showed typical applied scales for the investigation, which are a good compromise between a reasonable model size and moderate scale effects. However, based on this classification, scale effects are not necessarily negligible and need further investigation. Therefore, the influence of viscous forces and surface tension has been analysed, in accordance with what was reported in [6]. The Reynolds and Weber numbers for wave overtopping (R_{eq} and W_{eq}) were calculated. The results were compared versus the proposed critical limits, namely $R_{eq} > 10^3$ and $W_{eq} > 10$. In total, 25 cases showed a $R_{eq} < 10^3$ and $W_{eq} < 10$. Further analysis was carried out to quantify scale effects on those cases following two different methodologies: (1) a correction for the model scale, based on [6] was calculated; (2) an artificial neural network (ANN) proposed by [31] was employed, and the predicted overtopping discharges were compared with the measured ones. Both methodologies prove that scale effects can be neglected. The correction calculated with [6] method ranged between 1 and 2.5. ANN predictions show values in the same order or just smaller than experimental ones. Further details are here omitted for the sake of simplicity.

5. Results

5.1. Relationship between Mean Discharge, Individual Volumes and Overtopping Flow Parameters

Figure 10 shows the relationship between mean overtopping discharge and maximum individual overtopping volume. Data are divided according to the slope of the foreshore. For the same mean discharge, the maximum individual volume is smaller for the 1:30 foreshore slope than for 1:15 foreshore slope: heavy breaking caused by gentle foreshore and shallow water conditions provoke more energy dissipation and reduce the chance of very big individual overtopping events. The steeper slope is characterized by less but more intense overtopping events; meanwhile, the gentle slope experiences more events, but these are smaller in magnitude. This suggests that overtopping processes on milder slopes are somewhat less violent, in agreement with analysis proposed in Chapter 3 of EurOtop [6]. The thresholds proposed by EurOtop for people's safety (Table 1) are drawn in Figure 10, both in terms of mean discharge and maximum volume. It is important to note that while tolerable discharge values vary depending on the local wave conditions at the dike toe, for maximum volume the threshold is fixed at 600 L/m with no further considerations. Almost all experimental results show volumes higher than the proposed limit; however, all values of mean discharge are within the limits of 10–20 L/s/m, a value suggested for wave height at the dike toe equal to 1m (same order of results got during the experimental campaign). This inconsistency will be discussed later.

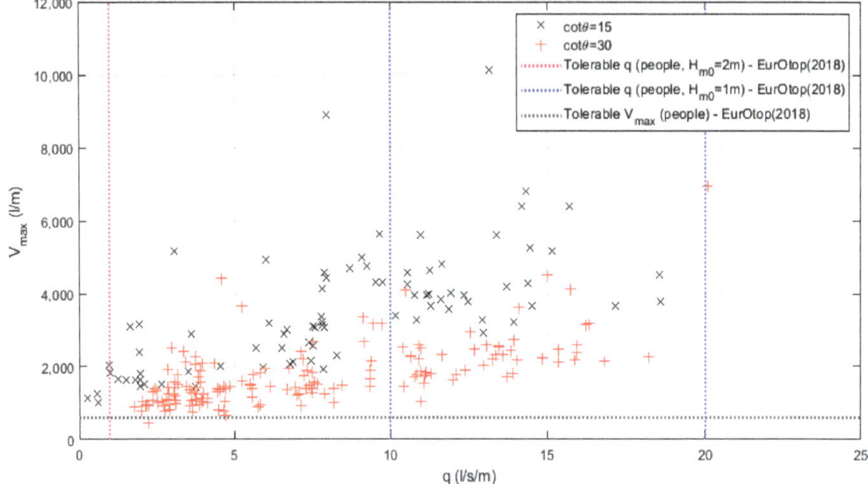

Figure 10. Variation of maximum individual overtopping volume with mean discharge (dimensions in prototype). Limits for safety of pedestrians and assets based on EurOtop (2018) are shown. The two vertical blue lines correspond to tolerable discharges of 10 L/s/m and 20 L/s/m respectively, for H_{m0} = 1 m.

The relationship between q and V_{max} seems pretty linear, whereas more dispersion can be noticed looking at the overtopping flow depth and velocity behavior versus individual overtopping volume (Figure 11). This is especially true for the 1:15 foreshore slope, where more intense overtopping events are far more energetic and violent than the ones for 1:30 slopes.

It must be remarked that the measured overtopping flow velocities are not necessarily the highest velocities recorded during the overtopping process, but they are the ones associated to the overtopping event characterized by the maximum individual volume. In agreement with [24,27], a correlation between overtopping flow parameters and individual maximum volumes is noticeable, differently from mean overtopping discharges (Figure 12), where larger scatters are noticed.

From a practical point of view, it is important to relate the overtopping flows to the sea storm conditions that have generated such flows. The variation of overtopping layer thickness and velocity with deep-wave spectral wave height and peak period is therefore shown in Figure 13. A relatively large scatter is noticed for steeper foreshore slopes, for which overtopping flow depth increases with the deep-water peak period. Low values of flow depth and velocity correspond to wave heights smaller than 3.6 m and wave periods shorter than 10 s. The presence of a gentle slope reduces both the flow depth and velocity, due to the more intense wave breaking that occurs far from the dike toe and generates bore-type flows running on the dike and finally overtopping. These flows can be described as overtopping green water. For the 1:15 slope, especially for the largest values of water depth at the dike toe, the breaking occurs very close or partly on the same dike slope. The overtopping is characterized by very rapid flows running on the seaward edge of the dike, detaching from the crest initially and then splashing on the promenade. For these flows, very large depths are achieved for deep-water wave heights between 3.6 and 4.5 m and wave periods larger than 11.8 s.

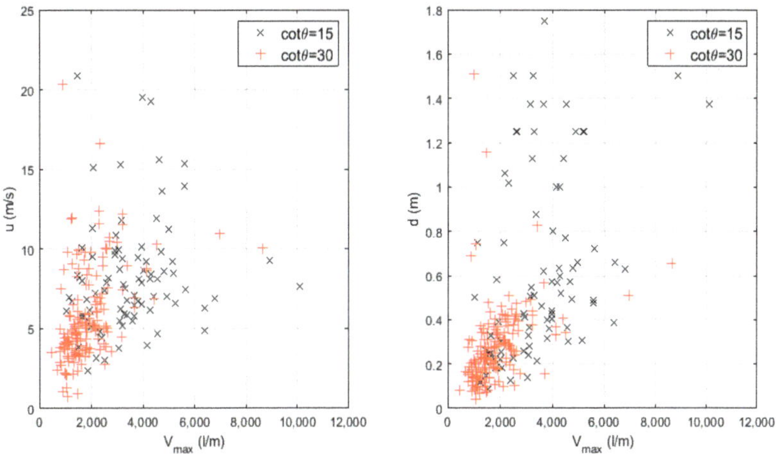

Figure 11. Variation of overtopping flow velocity and depth with maximum individual overtopping volume (dimensions in prototype) for two different foreshore slopes.

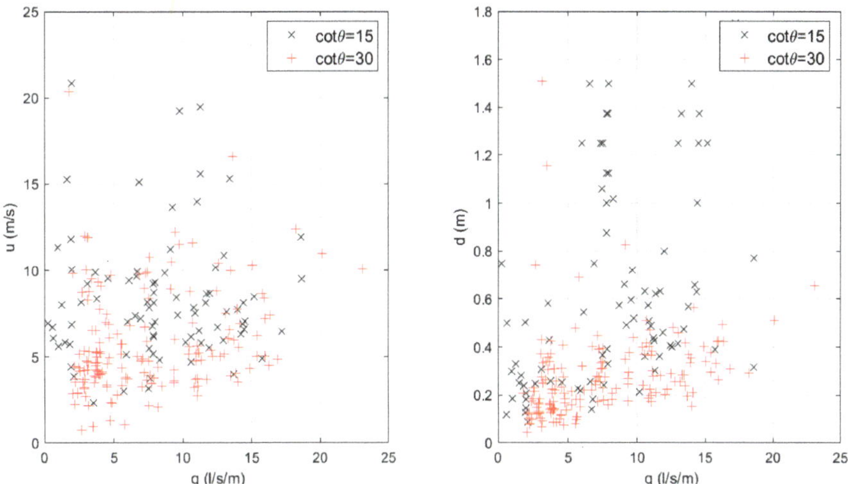

Figure 12. Variation of overtopping layer velocity and depth with average overtopping discharge (dimensions in prototype) for two different foreshore slopes.

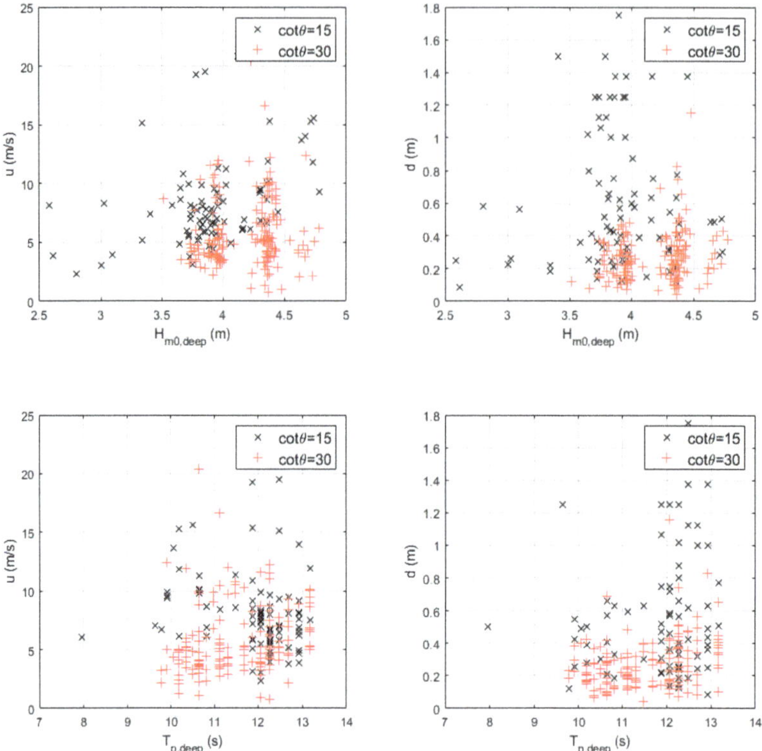

Figure 13. Variation of overtopping layer velocity and depth with deep-water wave characteristics (dimensions in prototype) for two different foreshore slopes.

To better understand the possible correlation among variables, a dimensional analysis was carried out. According to Buckingham's π theorem, if there are n variables in a problem containing k primary dimensions, the equation relating all the variables will have n-m dimensionless groups. In mathematical terms, it is possible to write:

$$f(H_{m0}, T_{m-1,0}, h_{toe}, u, d, q, V_{max}, R_c, g, tan\theta_{eq}, B) = 0 \tag{11}$$

where H_{m0} and $T_{m-1,0}$ are the spectral wave height and period at the dike toe, respectively, h_{toe} is the water depth at the dike toe, B is the width of the dike promenade, g is the gravity acceleration, q is the average overtopping discharge, V_{max} the maximum individual overtopping volume expressed in L per meter of crest width, R_c the crest freeboard, and u and d are the overtopping flow velocity and depth, respectively. The parameter $tan\theta_{eq}$ corresponds to the equivalent slope, calculated starting from the dike and foreshore slope for cases with foreshores in shallow water conditions, as indicated in [14]. Hence, n = 11 and k = 2, leading to nine dimensionless parameters:

$$f\left(\frac{H_{m0}}{h_{toe}}, \frac{T_{m-1,0}}{\sqrt{gh_{toe}}}, \frac{B}{d}, \frac{tan\theta_{eq}}{\sqrt{\frac{H_{m0}}{\frac{g}{2\pi}T^2_{m1-,0}}}}, \frac{u}{\sqrt{gh_{toe}}}, \frac{d}{H_{m0}}, \frac{q}{\sqrt{gH^3_{mo}}}, \frac{V_{max}}{\frac{g}{2\pi}H_{m0}T^2_{m1-,0}}, \frac{R_c}{H_{m0}}\right) = 0 \tag{12}$$

The selection of each dimensionless group is based on the current literature and intends to be physically meaningful. As for example, the wave period and the overtopping flow velocity were divided by the square root of gh_{toe}, corresponding to the wave celerity in shallow waters. Identification of dimensionless groups will help to investigate possible relationships between overtopping flow depth and velocity with other variables at stake. The relationship between overtopping flow depth and velocity with individual maximum overtopping volumes, has been investigated in terms of dimensionless groups as shown in Figure 14. It is possible to distinguish two different trends, one per foreshore slope. Lower values of the dimensionless velocity are shown for a wide range of volumes in case of the 1:30 slope. Opposite to that, a wide variation of velocities is shown within a relatively short range of volumes for the steeper foreshore. A more careful analysis of the results shows that the overtopping flow depth is greatly affected by the promenade width (see Figure 15), which does not show a clear correlation with overtopping flow velocity.

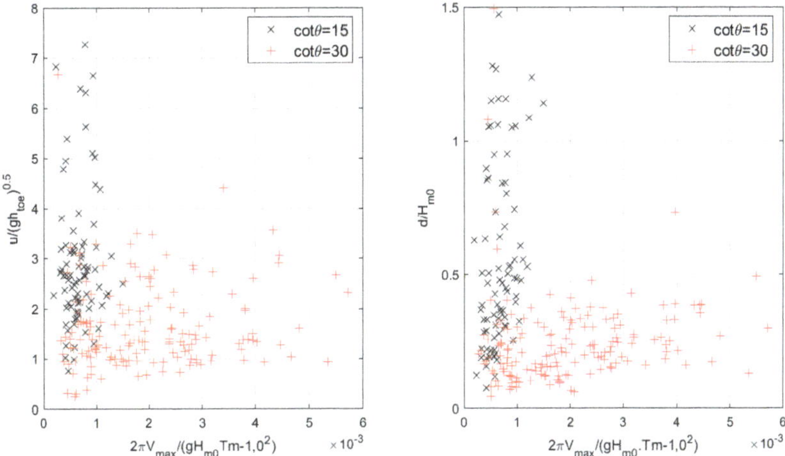

Figure 14. Variation of dimensionless overtopping layer velocity and depth with dimensionless maximum individual overtopping volume for two different foreshore slopes.

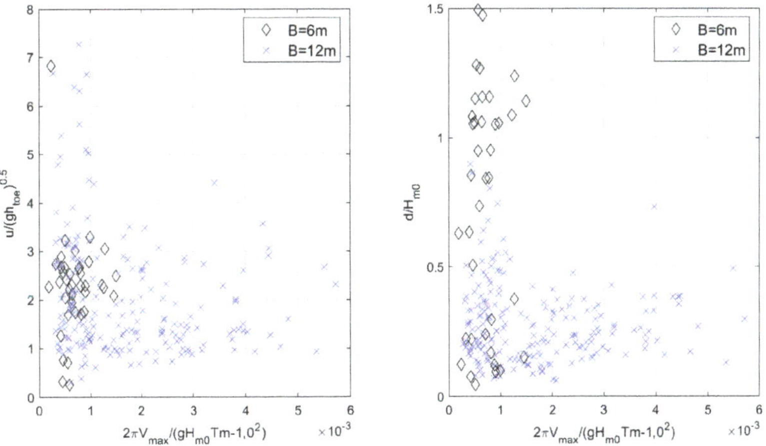

Figure 15. Variation of dimensionless overtopping layer velocity and depth with dimensionless maximum individual overtopping volume for two different promenade width.

The results for overtopping flow depth and velocity were compared with formulas proposed in [6,26,27,32], who finally expressed both overtopping flow parameters as a function of wave run-up and crest freeboard. The application of the aforementioned methods requires an estimation of the run-up. Notwithstanding this, wave run-up assessment for very and extremely shallow waters and relative steep dikes cannot avoid inaccuracies, due to the fact that experimental data fall outside the range of application of any known semi-empirical formula to calculate $R_{2\%}$ or R_{max}. The results are depicted in Figure 16, where it is clear that poor agreement is found between the experimental results and calculated values.

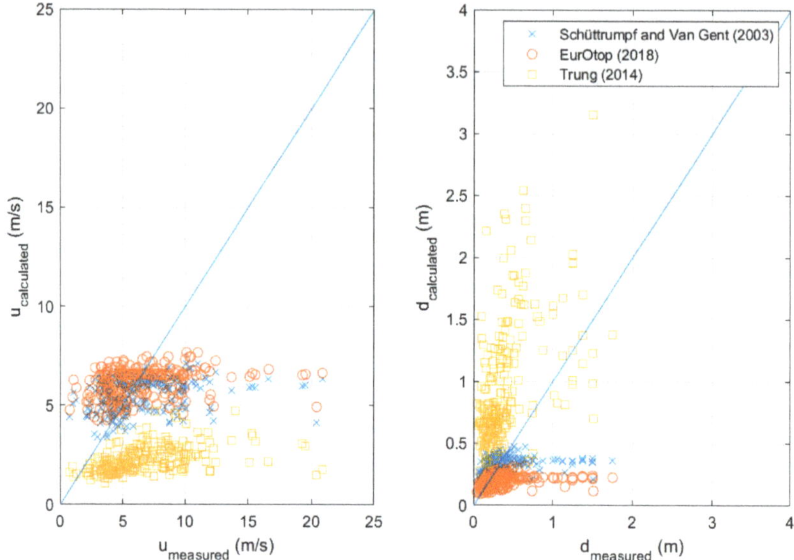

Figure 16. Estimation of overtopping flow depth and velocity employing existing formulas from Schüttrumpf and Van Gent (2003), Trung (2014) and EurOtop (2018).

5.2. Overtopping Flow Parameters Expressed in Terms of Individual Maximum Overtopping Volume

The application of the formulas proposed by Hughes [24], being a function of the maximum individual overtopping volume, are considered more adequate to be applied to the presented results, without further sources of uncertainty. It must be noticed that [24] analyzed overtopping flow parameters only at the seaward edge of the dike crest and not along it. The effects of the crest width were therefore neglected. Comparisons are reported in Figure 17: unlike overtopping flow depth, flow velocities and discharges are considerably under predicted. A possible explanation is the influence of the spectral wave period: $T_{m-1,0}$ can be found at the denominator of Equations (6)–(8); however, the measured wave periods are far larger than the one tested in [24], as deeply affected by heavy breaking and release and shift of the spectral energy to very low frequency as result of the release of infra-gravity waves for very and extremely shallow water conditions.

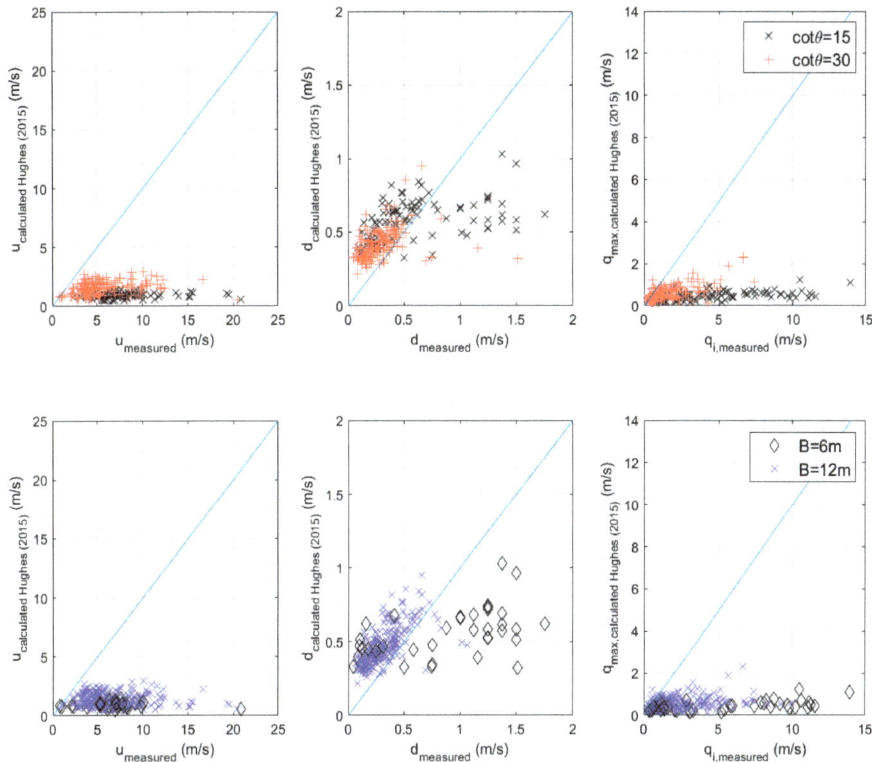

Figure 17. Estimation of overtopping flow depth and velocity employing formulas by Hughes (2015).

Mares-Nasarre et al. [10] demonstrated that overtopping flow velocity and flow depth are correlated if we look at them in statistical terms; they appear to be independent when related to the same individual overtopping event. Notwithstanding this, we aimed to further investigate possible correlations among dimensionless valuable. For the scope, the evolutionary polynomial regression (EPR) technique by [33] was employed. EPR implements a multi-modelling approach with multi-objective genetic algorithm. This technique was already successfully applied to coastal engineering problems [34,35] to find a simple and easily interpretable mathematical models that express the reflection coefficient variation for irregular waves for a particular low-reflective caisson breakwater. After several iterations, the following expressions have been found for overtopping flow depth, overtopping flow velocity and individual overtopping discharge:

$$u_{calculated\ EPR} = 0.254\sqrt{V_{max}\frac{gR_c}{H_{m0}}} \tag{13}$$

$$d_{calculated\ EPR} = 0.343\sqrt{V_{max}\frac{H_{m0}\xi_{m-1,0}}{R_c x_c}} \tag{14}$$

$$q_{max,calculated\ EPR} = 0.25V_{max}\sqrt{\frac{g\xi_{m-1,0}}{x_c}} \tag{15}$$

Similar dependence on the individual overtopping volume is found by employing EPR as by [24,27]. In addition, a certain influence exists of the location, x_c, where flow parameters are measured, of the surf similarity parameter $\xi_{m-1,0}$ and dimensionless freeboard R_c/H_{m0}. The surf similarity parameter is here calculated by employing the equivalent slope as defined in [15]. The influence of the freeboard might be explained as follows: for the same volume, a higher freeboard will lead to lower flow depths (run-up is lower). In addition, the surf similarity parameter will be bigger for longer periods (i.e., bigger wavelengths), leading to bigger individual discharges.

Measured overtopping flow parameters are plotted against the calculated ones in Figure 18. Large scatter is noticed especially for those cases with steeper foreshore slope and narrower crest width: the more violent overtopping events, often characterized by splashes and jets, make it more difficult to measure overtopping flow properties without errors (R^2 is about 50%). In any case, it is important to emphasize here that it is not intended to find new relationships for the overtopping flow parameters to overcome or upgrade the ones already proposed in the literature. The EPR analysis is carried out to provide a general overview of the possible correlations among variables and help to interpret the results. Larger databases are required to optimize any regression for the flow properties on the dike crest, but this is out of scope of the present work.

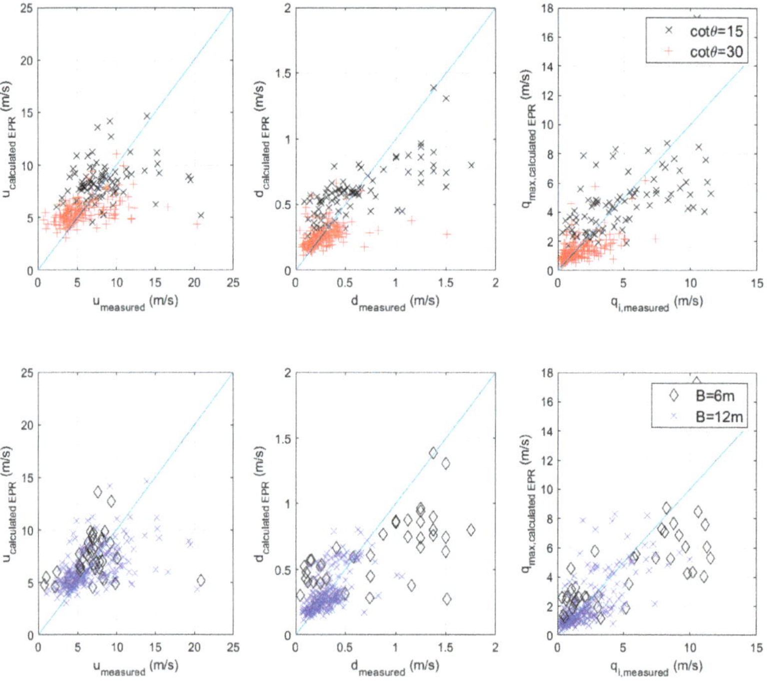

Figure 18. EPR results for overtopping flow properties on the dike crest versus measured ones.

6. Analysis and Comparison of Safety Criteria and Limits of Wave Overtopping for Design of Sea Dikes

In the previous section, a possible correlation among the aforementioned variables was shown. Overtopping flow velocity and flow depth show a clear dependence on the individual volume, as also confirmed by [24,27]. Poor correlation is noticed with the mean overtopping discharge instead. Usually, the only overtopping variables considered during the design of any coastal defense are the mean discharge and the maximum volume. According to [6], people standing and walking on or behind a coastal defense can be considered safe when the volume and discharges values are within certain limits: a preliminary comparison of the obtained experimental data with the tolerable limits of Table 1 is shown in Figure 10. The results indicate that: (1) the measured values of volumes are bigger than the proposed threshold, except for a very few cases; (2) mean overtopping discharge is always within the proposed thresholds.

Considering the limit on individual volume stronger condition than the one on discharge, it can be concluded that the limits presented in [6] are not verified for this particular case. However, a deeper look at overtopping flow properties leads to different conclusions. The safety conditions are evaluated through the stability curves proposed by Sandoval and Bruce [12] and Arrighi et al. [14]. The results are plotted in Figures 19 and 20, respectively. Data are categorized in terms of crest width and mean overtopping discharges, based on [6] (q < 5; 5 < q < 15; q > 15 L/s/m). Moreover, Figures 21 and 22 show the same comparison but data are gathered in three different groups depending on the maximum individual overtopping volume (V_{max} < 1,000; 1,000 < V_{max} < 5000; V_{max} > 5000 L/m). Stability curves calculated for a male adult person and a 10 year old child are plotted in Figures 19 and 21, respectively. All data above the safety curves lie in an unsafe region, whereas all data below the lines correspond to safe flow conditions.

Looking at overtopping flow properties, it becomes clear that overtopping safety criteria based on mean discharge and maximum volume are not sufficient and might lead to overpredicting the hazards related to overtopping events. Although almost all cases present individual maximum volumes higher than 600 L/m, at least 20% of them can be considered safe if looking at combination of flow velocity and depth. The influence of the berm width is remarkable: for shorter berms, almost all cases with V_{max} > 1,000 L/m fall within the unsafe region (above the curves in the figures), while the same volume can be not so critical for longer berms, for which the results are more scattered. Cases with V_{max} > 1,000 L/m that lead to unsafe flows are those with mean discharge q > 5/L/s/m, apart from a few exceptions. In general, very high discharges and volumes, respectively greater than 10 L/s/m and 5000 L/m, lead to unsafe flows for both considered promenade widths.

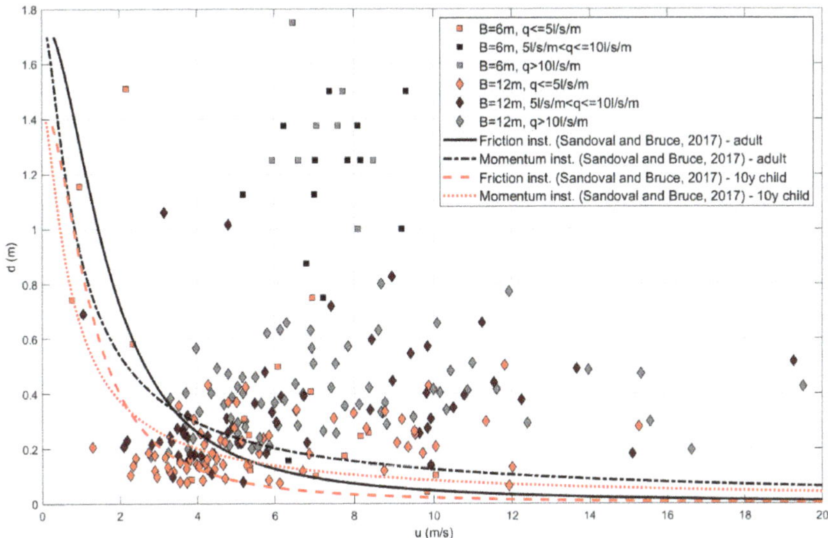

Figure 19. Flow depth versus velocity, comparison with Sandoval and Bruce (2017) curves, the discharges are divided for different promenades and measured average discharge.

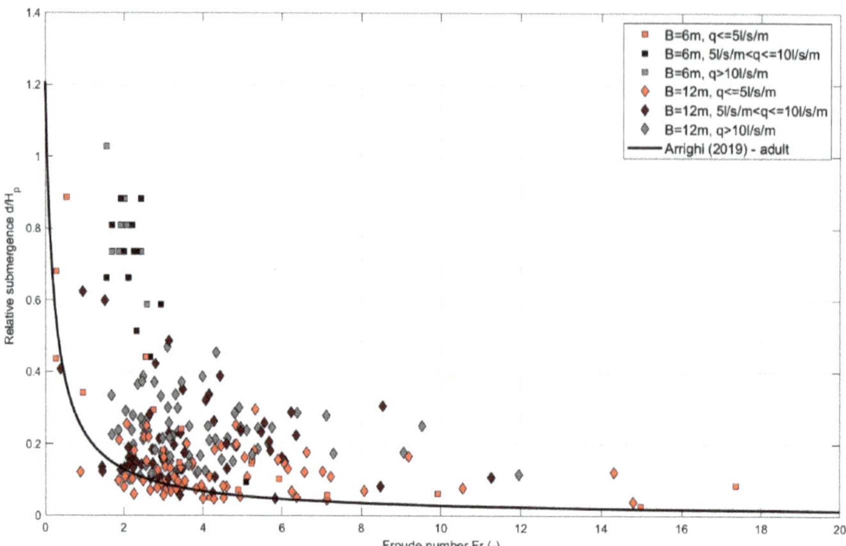

Figure 20. Froude number versus relative submergence, comparison with Arrighi (2017) curve, the discharges are divided for different promenades and measured average discharge.

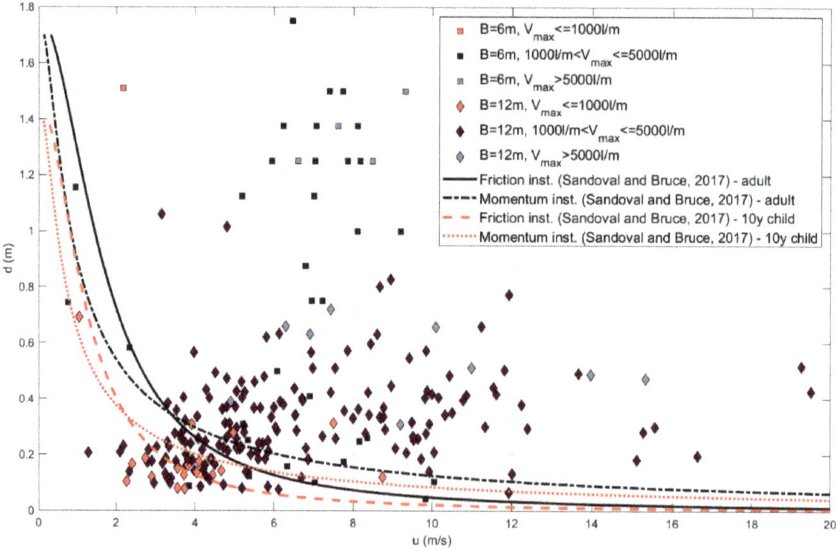

Figure 21. Flow depth versus velocity, comparison with Sandoval and Bruce (2017) curves, the discharges are divided for different promenades and maximum volumes.

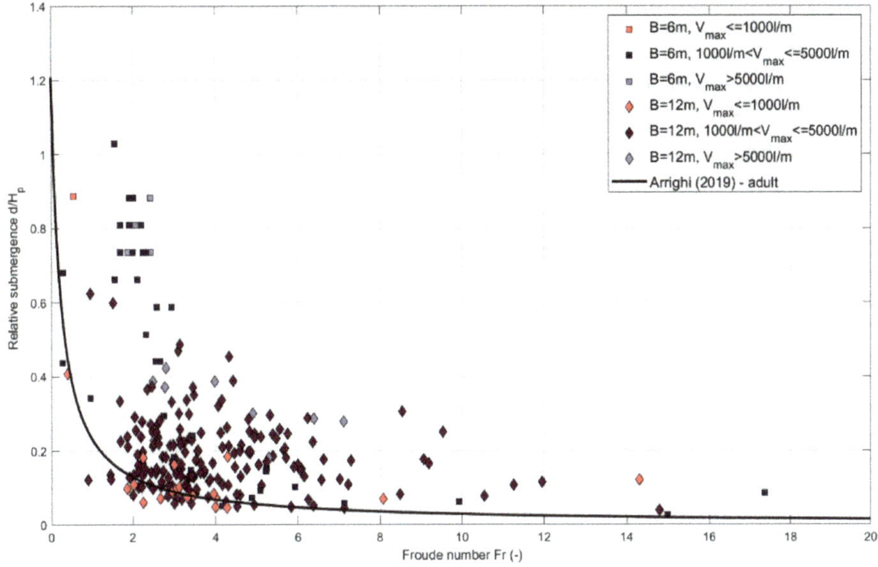

Figure 22. Froude number versus relative submergence, comparison with Arrighi (2017) curve, the discharges are divided for different promenades and maximum volumes.

The results show that the design of costal defenses must fulfil certain conditions that aim to guarantee the safety of people on or behind coastal defenses. These conditions cannot be reduced to the overtopping assessment of mean discharges and maximum volumes. Assessment of overtopping flow velocity and flow depth is required, in order to upgrade the current design criteria. To evaluate the flow velocity and depth associated to the maximum overtopping event is not an easy task and any semi-empirical model that can be derived would be limited to specific hydrodynamic conditions and geometrical layouts. Instead, experimental or, as an alternative, numerical modelling provide a further insight into the processes occurring on the dike crest.

7. Conclusions

In this study, 243 physical model tests of wave overtopping on smooth sea dike in very and extremely shallow water conditions have been carried out in the CIEMito wave flume at LIM/UPC. Two different foreshore slopes have been tested, 1:15 and 1:30, respectively. The dike has a crest or promenade at the end of which the velocity and flow depth are measured. Overtopping volumes are collected right after the promenade. The experimental campaign aimed at modelling sea states that characterize an urbanized stretch of a town along the Catalan coast located a few kilometers north of Barcelona, where a promenade/bike path and a railway running along the coastline are exposed to significant overtopping waves every stormy season.

Overtopping flow properties were measured by means of a redundant system that consists of two ultrasonic sensors and two high-speed cameras. Velocity and depth of the maximum overtopping event characterizing each test were measured. They proved to be dependent on the associated individual volume, crest freeboard and promenade width; however, correlation is weak, especially for the steeper foreshore cases. No clear correlation with mean discharge is found.

Two different methods to evaluate people safety have been used: tolerable limits for mean discharges and volumes, as proposed in [6] from one side, and the vulnerability of pedestrians expressed as a critical combination of flow velocity and depth [12–14]. The outcomes of the two criteria have been compared and discussed.

The results of the experimental campaign are specific for the studied area. It must be considered that all experimental tests have been conducted for the worst-case scenario of smooth dike, where, instead, the actual dike presents a rough and permeable seaward slope. The first evidence emerging from the study is related to the role of the foreshore on the overtopping flows and coastal safety. A gentle foreshore induces intense wave breaking which dissipates part of the incoming wave energy. On gentle foreshores (e.g., 1:30), a bore-type wave develops that will run on the dike and crest with small splashes and relatively small flow depths. When the foreshore is steeper (i.e., 1:15), the waves partly break on the dike (especially for local water depths of 1 m), leading to splashing flows which are very irregular and turbulent that generate, especially for narrow promenades, very high flow depths.

Critical range of deep-water wave height is 3.6–4.5 m with periods larger than 11.8 s. For these conditions, the most dangerous combinations of overtopping flow velocity and thickness is achieved, also related to very large individual overtopping volumes. In the studied area, there are stretches where nourishment is no longer carried out. However, the outcomes of this study show that, especially when a relative narrow promenade is present, a soft protection-like nourishment would be advisable, attaining a higher level of protection by means of more gentle and shallow foreshore slopes.

Besides specific considerations for the studied area, some general conclusions can be drawn:

- Tolerable discharge values proposed by [6] vary depending on the local wave height at the toe of the coastal structure. On the contrary, a fixed value corresponding to 600 L/m is reported as a threshold for individual overtopping volume. It is not clear from [6] whether this value corresponds to some specific value of overtopping flow velocity as, for example, in [5]. If one criterion is fulfilled, it can happen that the other criteria appear stronger. For the case study, average discharges were always within the proposed limits, whereas individual volumes were above the tolerable value.
- Overtopping flow velocities and depths are plotted along with the corresponding maximum volumes and average discharges. What emerges is a not clear two-way relationship between maximum overtopping volumes and velocities or flow depth: a dependence does exist, as also confirmed by applying the EPR technique [33] to the present dataset. Nevertheless, the data scatter is big, and therefore a larger dataset is required to performed more detailed regression analysis on the data.
- Experimental values of overtopping flow velocities and flow depth have been compared with stability curves for pedestrians (adults and children) placed on the sea dike and subjected to overtopping waves. The results show a clear influence of the dike crest width, where for mean discharges lower than 5 L/s/m and volumes lower than 1,000 L/m, a shorter crest does not necessarily lead to safe conditions, where the longer crest shows a combination of values of overtopping flow parameters lower than the thresholds calculated using [12,13].
- Volumes bigger than 600 L/m do not always determine unsafe conditions for pedestrians. At least 20% of all analyzed data are in the safe region, for the specific case of study.
- EurOtop [6] tolerable limits and stability curves lead to discordant results. In fact, due to the non-two-way relationship between volumes and corresponding flow parameters, it can be observed that flow parameters related to 1,000 L/m maximum volumes can be located in the unsafe area, while the same parameters related to bigger volumes can even be included in the safety range for a large enough crest.

Concluding, the experimental campaign suggests that further research is needed in terms of design criteria for wave overtopping, if related to people's safety. The proposed tolerable discharge and volume values from [6] are still valid, but not sufficient to clearly identify a safe or unsafe scenario. Overtopping flow depth and velocity provide further insight and are advised to be employed, together with [6]'s criteria, for coastal safety assessment.

Author Contributions: Conceptualization, C.A. and X.G.; methodology, C.A.; validation, C.A., T.S.; formal analysis, C.A., G.V. and A.S.; investigation, C.A., T.S.; resources, X.G.; data curation, C.A.; writing—original draft preparation, C.A.; writing—review and editing, C.A, X.G., T.S., G.V., A.S.; supervision, X.G., G.V.; project administration, C.A.; funding acquisition, X.G. All authors have read and agreed to the published version of the manuscript.

Funding: This research was funded by European Union's Horizon 2020 research and innovation programme under the Marie Sklodowska-Curie grant agreement No.: 792370.

Acknowledgments: The authors would like to acknowledge the students Mauro Campagnola and Maria Luigia Robustelli, who carried out the experimental campaign under the guidance and supervision of C.A, and the technical/research staff of LIM/UPC, in particular Joaquim Sospedra, Oscar Galego and José M. Alsina for their support and assistance for the model constructions and setup of measurement systems.

Conflicts of Interest: The authors declare no conflict of interest.

Disclaimer: The presented results reflect only the authors' view and Research Executive Agency (REA) is not responsible for any use that may be made of the information it contains.

References

1. Owen, M.W. *Design of Seawalls Allowing for Wave Overtopping*; Report No. EX 924; Hydraulics Research: Wallingford, UK, 1980.
2. Bruce, T.; Van Der Meer, J.; Franco, L.; Pearson, J. Overtopping performance of different armour units for rubble mound breakwaters. *Coast. Eng.* **2009**, *56*, 166–179. [CrossRef]
3. Van Der Meer, J.; Bruce, T. New Physical Insights and Design Formulas on Wave Overtopping at Sloping and Vertical Structures. *J. Waterw. Port Coast. Ocean Eng.* **2014**, *140*, 04014025. [CrossRef]
4. Van Gent, M.R.A. *Physical Model Investigations on Coastal Structures with Shallow Foreshores: 2D Model Tests with Single and Double-Peaked Wave Energy Spectra*; Deltares (WL): Delft, The Netherlands, 1999.
5. Allsop, N.W.H.; Bruce, T.; Pullen, T.; van der Meer, J. Direct Hazards from Wave Overtopping—The Forgotten Aspect of Coastal Flood Risk Assessment? In Proceedings of the 43rd Defra Flood and Coastal Management Conference, Manchester University, Manchester, UK, 1–3 July 2008; pp. 1–11.
6. Van der Meer, J.W.; Allsop, N.W.H.; Bruce, T.; De Rouck, J.; Kortenhaus, A.; Pullen, T.; Schüttrumpf, H. EurOtop: Manual on Wave Overtopping of Sea Defences and Related Structures: An Overtopping Manual Largely Based on European Research, but for Worldwide Application. 2018. Available online: www.overtopping-manual.com (accessed on 16 June 2020).
7. Cappietti, L.; Simonetti, I.; Esposito, A.; Streicher, M.; Kortenhaus, A.; Scheres, B.; Schuettrumpf, H.; Hirt, M.; Hofland, B.; Chen, X. Large-Scale Experiments of Wave-Overtopping Loads on Walls: Layer Thicknesses and Velocities. In Proceedings of the ASME 2018 37th International Conference on Ocean, Offshore and Arctic Engineering, Madrid, Spain, 17–22 June 2018; Volume 7. [CrossRef]
8. Schüttrumpf, H.; Oumeraci, H. Scale and Model Effects in Crest Level Design. In Proceedings of the 2nd Coastal Symposium, Höfn, Iceland, 5–8 June 2005; pp. 1–12.
9. Nørgaard, J.Q.H.; Andersen, T.L.; Burcharth, H.F.; Steendam, G.J. Analysis of overtopping flow on sea dikes in oblique and short-crested waves. *Coast. Eng.* **2013**, *76*, 43–54. [CrossRef]
10. Mares-Nasarre, P.; Argente, G.; Gómez-Martín, M.E.; Medina, J.R. Overtopping layer thickness and overtopping flow velocity on mound breakwaters. *Coast. Eng.* **2019**, *154*, 103561. [CrossRef]
11. Endoh, K.; Takahashi, S. Numerically Modeling Personnel Danger on a Promenade Breakwater Due to Overtopping Waves. *Coast. Eng.* **1994**, 1016–1029. [CrossRef]
12. Sandoval, C.; Bruce, T. Wave Overtopping Hazard to Pedestrians: Video Evidence from Real Accidents. In *Coasts, Marine Structures and Breakwaters 2017*; Thomas Telford Ltd.: Liverpool, UK, 2018; pp. 501–512.
13. Arrighi, C.; Pregnolato, M.; Dawson, R.; Castelli, F. Preparedness against mobility disruption by floods. *Sci. Total Environ.* **2018**, *654*, 1010–1022. [CrossRef]
14. Arrighi, C.; Oumeraci, H.; Castelli, F. Hydrodynamics of pedestrians' instability in floodwaters. *Hydrol. Earth Syst. Sci.* **2017**, *21*, 515–531. [CrossRef]
15. Altomare, C.; Suzuki, T.; Chen, X.; Verwaest, T.; Kortenhaus, A. Wave overtopping of sea dikes with very shallow foreshores. *Coast. Eng.* **2016**, *116*, 236–257. [CrossRef]
16. Van Doorslaer, K.; De Rouck, J.; Audenaert, S.; Duquet, V. Crest modifications to reduce wave overtopping of non-breaking waves over a smooth dike slope. *Coast. Eng.* **2015**, *101*, 69–88. [CrossRef]

17. Bruce, T.; Van Der Meer, J.; Pullen, T.; Allsop, W.; Kim, Y.C. Wave Overtopping at Vertical and Steep Structures. In *Handbook of Coastal and Ocean Engineering*; California State University: Long Beach, CA, USA, 2009; pp. 411–439.
18. Gallach Sanchez, D. Experimental Study of Wave Overtopping Performance of Steep Low-Crested Structures. Ph.D. Thesis, Ghent University, Belgium, The Netherlands, 2018.
19. Victor, L.; Troch, P. Wave Overtopping at Smooth Impermeable Steep Slopes with Low Crest Freeboards. *J. Waterw. Port Coast. Ocean Eng.* **2012**, *138*, 372–385. [CrossRef]
20. Mase, H.; Tamada, T.; Yasuda, T.; Hedges, T.S.; Reis, M.T. Wave Runup and Overtopping at Seawalls Built on Land and in Very Shallow Water. *J. Waterw. Port Coast. Ocean Eng.* **2013**, *139*, 346–357. [CrossRef]
21. Goda, Y. Derivation of unified wave overtopping formulas for seawalls with smooth, impermeable surfaces based on selected CLASH datasets. *Coast. Eng.* **2009**, *56*, 385–399. [CrossRef]
22. Hofland, B.; Chen, X.; Altomare, C.; Oosterlo, P. Prediction formula for the spectral wave period T m-1,0 on mildly sloping shallow foreshores. *Coast. Eng.* **2017**, *123*, 21–28. [CrossRef]
23. Schüttrumpf, H.; Oumeraci, H. Layer thicknesses and velocities of wave overtopping flow at seadikes. *Coast. Eng.* **2005**, *52*, 473–495. [CrossRef]
24. Hughes, S.A. Hydraulic Parameters of Overtopping Wave Volumes. In *Coastal Structures and Solutions to Coastal Disasters*; American Society of Civil Engineers: Reston, VA, USA, 2015; p. 15.
25. Van Bergeijk, V.; Warmink, J.; Van Gent, M.R.; Hulscher, S.J.M.H. An analytical model of wave overtopping flow velocities on dike crests and landward slopes. *Coast. Eng.* **2019**, *149*, 28–38. [CrossRef]
26. Schüttrumpf, H.; Möller, J.; Oumeraci, H.; Smith, J.M. Overtopping Flow Parameters on the Inner Slope of Seadikes. In *Coastal Engineering*; World Scientific: Cardiff, Wales, 2002; pp. 2116–2127. [CrossRef]
27. Trung, L.H. Velocity and Water-Layer Thickness of Overtopping Flows on Sea Dikes. In *Communications on Hydraulic and Geotechnical Engineering 2014-02*; Delft University of Technology: Delft, The Netherlands, 2014; ISSN 0169-6548.
28. Hughes, S.A. Wave momentum flux parameter: A descriptor for nearshore waves. *Coast. Eng.* **2004**, *51*, 1067–1084. [CrossRef]
29. Hansen, N.-E.O.; Sand, S.E.; Lundgren, H.; Sorensen, T.; Gravesen, H. Correct Reproduction of Group-Induced Long Waves. In *Coastal Engineering*; ASCE: Sydney, Australia, 1980; pp. 784–800.
30. Heller, V. Scale effects in physical hydraulic engineering models. *J. Hydraul. Res.* **2011**, *49*, 293–306. [CrossRef]
31. Formentin, S.M.; Zanuttigh, B.; Van Der Meer, J.W.; Van Der Meer, J.W. A Neural Network Tool for Predicting Wave Reflection, Overtopping and Transmission. *Coast. Eng. J.* **2017**, *59*, 1750006–1750031. [CrossRef]
32. Schüttrumpf, H.; Van Gent, M.R. Wave Overtopping at Sea Dikes. In *Coastal Structures 2003*; American Society of Civil Engineers: Portland, OR, USA, 2003; pp. 431–443.
33. Giustolisi, O.; Savic, D. A symbolic data-driven technique based on evolutionary polynomial regression. *J. Hydroinform.* **2006**, *8*, 207–222. [CrossRef]
34. Altomare, C.; Gironella, X.; Laucelli, D.B. Evolutionary data-modelling of an innovative low reflective vertical quay. *J. Hydroinform.* **2012**, *15*, 763–779. [CrossRef]
35. Altomare, C.; Gironella, X. An experimental study on scale effects in wave reflection of low-reflective quay walls with internal rubble mound for regular and random waves. *Coast. Eng.* **2014**, *90*, 51–63. [CrossRef]

Journal of
Marine Science and Engineering

Article

Reduction of Wave Overtopping and Force Impact at Harbor Quays Due to Very Oblique Waves

Sebastian Dan [1,*], Corrado Altomare [2], Tomohiro Suzuki [1,3], Tim Spiesschaert [1] and Toon Verwaest [1]

1 Flanders Hydraulics Research, Berchemlei 115, 2140 Antwerp, Belgium; tomohiro.suzuki@mow.vlaanderen.be (T.S.); tim.spiesschaert@mow.vlaanderen.be (T.S.); toon.verwaest@mow.vlaanderen.be (T.V.)

2 Laboratori d'Enginyeria Maritima, Universitat Politècnica de Catalunya - BarcelonaTech (UPC), 08034 Barcelona, Spain; corrado.altomare@upc.edu

3 Faculty of Civil Engineering and Geosciences, Delft University of Technology, Stevinweg 1, 2628 CN Delft, The Netherlands

* Correspondence: sebastian.dan@mow.vlaanderen.be

Received: 10 July 2020; Accepted: 7 August 2020; Published: 11 August 2020

Abstract: Physical model experiments were conducted in a wave tank at Flanders Hydraulics Research, Antwerp, Belgium, to characterize the wave overtopping and impact force on vertical quay walls and sloping sea dike (1:2.5) under very oblique wave attack (angle between 45° and 80°). This study was triggered by the scarce scientific literature on the overtopping and force reduction due to very oblique waves since large reduction is expected for both when compared with the perpendicular wave attack. The study aimed to compare the results from the experimental tests with formulas derived from previous experiments and applicable to a Belgian harbor generic case. The influence of storm return walls and crest berm width on top of the dikes has been analyzed in combination with the wave obliqueness. The results indicate significant reduction of the overtopping due to very oblique waves and new reduction coefficients were proposed. When compared with formulas from previous studies the proposed coefficients indicate the best fit for the overtopping reduction. Position of the storm return wall respect to the quay edge rather than its height was found to be more important for preventing wave induced overtopping. The force reduction is up to approximately 50% for the oblique waves with respect to the perpendicular wave impact and reduction coefficients were proposed for two different configurations a sea dike and vertical quay wall, respectively.

Keywords: overtopping reduction; force reduction; oblique waves; storm return wall; EurOtop manual

---

1. Introduction

Densely populated coastal zones with very low freeboards are common worldwide (e.g., Belgium, The Netherlands, Vietnam). Often the flood protection is provided in these zones by the sandy beaches, but when it is insufficient or in the case of harbors, the most common solution is storm wall construction. A storm wall is located on top of the crest of a quay or a dike at a certain distance from the seaward edge of the crest, providing additional protection against the overtopping waves. During each overtopping event, the waves runup in form of a bore along its crests before reaching the wall. Usually, this flow is turbulent, and its velocity is decreased along the crest width. Consequently, the distance between the edge of the structure and the storm wall is important because it characterizes the wave impact on the storm wall and overtopping over the storm wall.

Typically, the wave's angle is assumed to be perpendicular or at an angle lower than 45° with respect to the quay's normal. However, when the harbor opening is orientated against the main wave direction very oblique waves can approach some of the harbor quays and dikes. There are several formulas proposed for the overtopping computation under oblique wave attack. One of the most widely used is the European Overtopping Manual [1,2] which provides validated formulas to calculate the overtopping discharge for classical configurations (wave angles smaller than 45°). The overtopping is maximum for the perpendicular wave attack on a storm return wall, but for larger wave angles, a reduction factor is applied to account for the overtopping discharge decrease. However, the EurOtop formula suggests keeping the obliqueness reduction factor constant for vertical structures and wave angles larger than 45°. Obviously, the overtopping discharge reduces with the increasing wave angle with respect to the structure normal, but the reduction for very large wave angles has not been fully investigated yet. A similar situation is for the case of the impact force reduction due to the large wave angle, but studies comprehensively analyzing this reduction are not currently available.

The mean wave overtopping is mainly a function of the relative freeboard and the relationship between the overtopping discharge and the freeboard is expressed through, in most of the cases, an exponential formula. Several reduction coefficients are used to account for effects induced by the presence of a berm, a storm return wall, the surface roughness, and the wave obliqueness.

To investigate the overtopping reduction and impact force reduction for oblique waves a physical model was set-up at Flanders Hydraulics Research in Antwerp, Belgium. The present study has three main objectives. Firstly, to investigate overtopping induced by very oblique waves at quay harbors and to propose reliable reduction coefficients for the overtopping calculation. Secondly, to identify the influence on overtopping of a storm return wall placed on the quay at different positions and having variable heights. Thirdly, to evaluate the impact force reduction due wave obliqueness. Although, it was expected that overtopping and impact forces will be reduced at large wave angles (respect to the structure normal), quantification of these reductions is still unclearly defined in previous studies; hence, the necessity of the present study.

2. Overtopping and Force Reduction

2.1. Vertical Quay

A series of formulations describe the overtopping reduction with the large incident wave angle. Most of the formulas used are presented in [1], but significant contributions are given also in other previous studies [3–5]. The overtopping reduction due to very obliques wave angles is usually limited to angles of 45° and for larger wave angles a constant value is proposed.

The most used approach is based on the equations and reduction factors included in the European Overtopping Manual [1] for non-impulsive conditions:

$$\frac{q}{\sqrt{gH_{m0}^3}} = 0.04\exp\left(-2.6\frac{R_c}{H_{m0}\gamma_\beta}\right) \tag{1}$$

where q is the overtopping discharge per meter of width of the structure ($m^3/s/m$), H_{m0} is the significant incident wave height, measured at the toe of the structure (m), R_c is the crest freeboard (m), and γ_β is the reduction coefficient that considers the effects of the obliqueness (-).

The coefficient γ_β is expressed in EurOtop as

$$\gamma_\beta = 1 - 0.0062|\beta| \text{ for : } 0° \le \beta \le 45° \tag{2}$$

For wave angles larger than 45° a constant value of 0.72 is proposed in EurOtop [1].

The formulations contained in the EurOtop manual [1] assume that different regimes of non-breaking, impulsive breaking and broken waves may produce differences in the overtopping.

Although the wave loadings on vertical walls due to individual waves are certainly affected by the wave regime, it is not clear if the overtopping is affected in the same way. The overtopping discharge is a mean value where many non-breaking, breaking, and broken waves can contribute in the same wave train. Therefore Goda [3] proposed an equation that is valid for both non-impulsive and impulsive wave conditions:

$$\frac{q}{\sqrt{gH_{m0}^3}} = \exp\left[-\left(A + B\frac{R_c}{H_{m0}}\frac{1}{\gamma_f\gamma_\beta^*}\right)\right] \tag{3}$$

The constants A and B can be estimated by:

$$\begin{array}{l} A = A_0\ tanh\left[b_1\left(\frac{h_t}{H_{s,\ toe}} + c_1\right)\right] \\ B = B_0\ tanh\left[b_2\left(\frac{h_t}{H_{s,\ toe}} + c_2\right)\right] \end{array} \tag{4}$$

where h_t is the water depth at the toe of the dike and $H_{s,toe}$ is the incident wave height at the toe of the dike. The coefficients b_1, c_1, b_2, and c_2 depend on the foreshore slope as summarized in Table 1.

Table 1. Optimum coefficient values of empirical formulas for intercept A and gradient coefficient B (after Goda, 2009).

Seabed Slope	**Coefficient A**			**Coefficient B**		
tan θ	A0	b1	c2	B0	b2	c2
1/10	3.6	1.4	0.1	2.3	0.6	0.8
1/20–1/1000	3.6	1.0	0.6	2.3	0.8	0.6

The coefficients A_0 and B_0 are calculated as a function of the dike slope, $\cot \alpha_s$, and their value ranges between 0 and 7. The expression for the reduction factor for wave obliqueness has been estimated by Goda [3] as

$$\gamma_\beta = 1 - 0.0096|\beta| + 0.000054\beta 2 \quad \text{for} \quad 0^\circ \leq \beta \leq 80^\circ \tag{5}$$

2.2. Sloping Dike

Several studies investigate the reduction in overtopping due to oblique waves [6,7], but two formulas from literature were considered due to the similarity with the tests from the present study: EurOtop [1] (6) for non-breaking waves and van der Meer and Bruce [8] (8) which is an adaptation of the EurOtop formula.

$$\frac{q}{\sqrt{gH_{m0}^3}} = 0.2\exp\left(-2.6\frac{R_c}{H_{m0}\gamma_f\gamma_\beta}\right) \tag{6}$$

in which the coefficient γ_β is expressed as

$$\gamma_\beta = 1 - 0.0033|\beta| \quad \text{for}: \ 0^\circ \leq \beta \leq 80^\circ \tag{7}$$

For wave angles larger than 80° a constant value of 0.736 is proposed.

The formula given by van der Meer and Bruce [8]:

$$\frac{q}{\sqrt{gH_{m0}^3}} = 0.09\exp\left(-1.5\frac{R_c}{H_{m0}\gamma_f\gamma_\beta}\right)^{1.3} \tag{8}$$

in which γ_β is identical as in (7).

In all the cases γ_f has been assumed equal to 1 (smooth slope). Van Doorslaer et al. [9] propose a reduction factor γ_{prom_v} to take into account the presence of a storm return wall on the top of the dike. This coefficient considers both the effect of the wall height and position. The values of γ_{prom_v} are calculated for each case based on the approach described in Van Doorslaer et al. [9].

$$\frac{q}{\sqrt{gH_{m0}^3}} = 0.2\exp\left(-2.3\frac{R_c}{H_{m0}}\frac{1}{\gamma_f\gamma_\beta\gamma_{prom\_v}}\right) \tag{9}$$

$$\gamma_{prom\_v} = 0.87\gamma_{prom}\gamma_v \tag{10}$$

where γ_{prom} and γ_v are the individual reduction factors to consider the effects respectively of the promenade and of the storm wall. The promenade reduction factor γ_{prom} is expressed as

$$\gamma_{prom} = 1 - 0.47B/L_{m-1,0} \tag{11}$$

where B is the width of the promenade and $L_{m-1,0}$ is the spectral wave length calculated using the spectral period in deep waters $T_{m-1,0} = m_{-1}/m_0$.

The reduction factor γ_v for the presence of a storm return wall is expressed in Van Doorslaer et al. [9] in function of the wall height (h_{wall}) and freeboard (R_c) as follows:

$$\gamma_v = \begin{cases} \exp(-0.56h_{wall}/R_c) \\ 0.5 \end{cases} \text{for} \begin{cases} h_{wall}/R_c < 1.24 \\ h_{wall}/R_c \geq 1.24 \end{cases} \tag{12}$$

2.3. Force Reduction

There is scarce information regarding the impact forces on a storm wall in case of wave overtopping by oblique waves. However, the study of Van Doorslaer et al. [10] performed at Polytechnic University of Catalonia, Barcelona (UPC), used configurations similar to those tested in the present study. Two structures were tested in the wave flume (scale 1:6): a vertical quay wall and a dike with a smooth slope. The storm wall was 1.20 m high (prototype value) and located at 10.14 m (prototype value) behind the edge of the crest. Three water levels were used resulting in freeboard R_c from the still water level to the top of the storm wall of 3.18 m, 2.22 m, and 1.20 m (prototype values). The irregular waves had a Jonswap wave spectrum ($\gamma = 3.3$). The significant wave height H_{m0} ranged from 0.78 m to 3.00 m (prototype values); the wave period T_p was either 7.00 s or 10.00 s. The experiments were carried out in two dimensional conditions with perpendicular waves (no wave obliqueness). The authors proposed a new formula to evaluate the wave force on a storm wall, both for quay walls and sea dikes. The formula can be expressed as follows:

$$F_{1/250} = a\rho g R_c^2 \exp\left(-b\frac{R_c}{H_{m0}}\right) \tag{13}$$

where $F_{1/250}$ is the average force of the highest 1/250 waves. The coefficients a and b (Table 2) are derived from a non-linear regression analysis and they are considered as the mean value of normally distributed variables. Under this hypothesis, the relative standard deviation ($\sigma' = \sigma/\mu$) was calculated for each coefficient and is reported in Table 2 between brackets.

Table 2. Coefficients a and b in Equation (13) for different geometries (after Van Doorslaer et al. [10]).

Geometry	a	b
Dike	8.31 (0.22)	2.45 (0.07)
Quay	18.27 (0.23)	3.99 (0.06)
All	5.96 (0.23)	2.42 (0.09)

3. Methods and Instrumentation

Investigation of the overtopping reduction required a physical model sufficiently large to observe the alongshore variation and to accommodate the collection of the overtopped volumes, respectively. However, this structure was not firm enough to prevent vibrations which can severely alter the impacting forces measurements. Therefore, it was decided to build two different structures, first one for the overtopping reduction and second one for force reduction due to the wave obliqueness. The structural layout and hydraulic boundary conditions were assumed based on real conditions from the Belgian harbors. However, the model geometries do not represent one specific quay or dike, but they were selected in such way that the results could be extended to other similar structures. The experiments were carried out in the wave tank at FHR (Flanders Hydraulics Research) (dimensions 17.50 m × 12.20 m × 0.45 m), equipped with a piston-type wave generator. The wave generator has a width of 12 m and generates long-crested waves. Both regular and irregular wave patterns can be generated at different angles of wave incidence ranging between −22.5° and 22.5° with respect to the center line of the wave tank.

Two sets of wave directions were used in the experimental campaign conducted at FHR: the first set contains the wave directions 0° and 45°, used to validate the results of the FHR experiments against previous experiments and existing formulas; the second set contains wave directions 60°, 70°, and 80°, used to investigate larger angles. Similar configuration tests from CLASH database [11] were used to compare and validate the tests from the present study.

3.1. Model Settings of Overtopping Tests

The first physical model was designed to study very oblique wave attacks and overtopping flows onto vertical quays and sloping dikes with storm return walls (Figure 1).

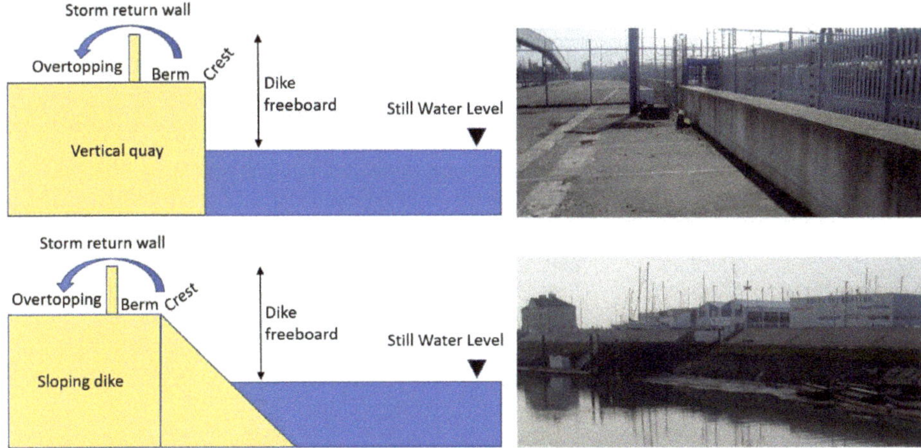

Figure 1. Model set-up based on the real cases from Belgian harbors. Upper part: vertical quay design and an example of storm return wall at Ostend harbor. Lower part: sloping dike design and an example at Blankenberge harbor.

As the wave height can variate along the structure, a smaller scale was necessary to accommodate a model sufficiently long to record the overtopping variations However, the scale cannot be smaller than 1:50 because some wave height scenarios would be smaller than 3 cm and therefore affected by the surface tension, and thus altering the reproduction of the prototype conditions [12]. Following the above mentioned boundary conditions as well as the Froude's law a scale of 1:50 was selected as best fitting for the physical model. For vertical walls, tests in large scale flumes and field measurements have demonstrated that results of overtopping discharge in small scale laboratory studies may be securely scaled up to full scale under impulsive and non-impulsive conditions. Only the wind effects are not considered and may cause a significant difference (for further details see [1]). For dikes the evaluation of scale effects is based on the approach of Schüttrumpf and Oumeraci [13]. Calculation of the Reynolds-number and its comparison with the critical value demonstrated that scale effects are negligible. A minimum distance between the wave maker and the structure equal to two wave lengths was kept for every configuration and wave dampers were placed around the basin to absorb the reflected waves. Considering the limitations, it was decided to build a laminated wooden structure of 8 m long and 1 m wide. Attached to this structure, there are 16 boxes (1.5 m long, 0.48 m wide, and 0.18 m deep), built from the same material, designed to collect the overtopping water during the experiment (Figure 2).

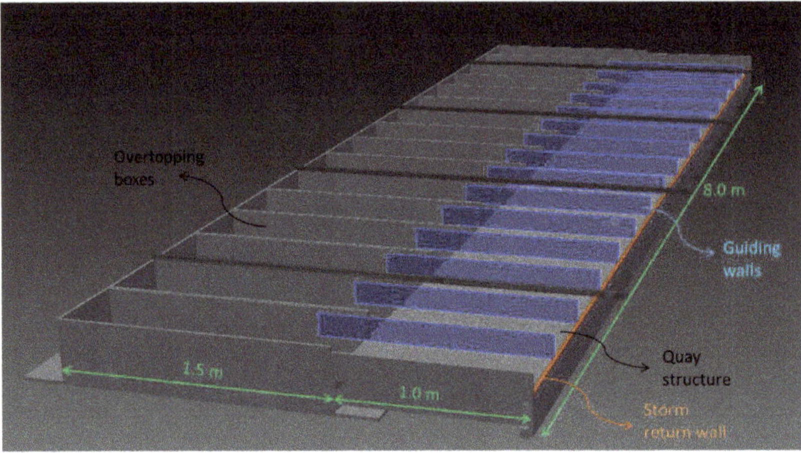

Figure 2. The structure used during the experiment.

Two positions of this structure in the basin were planned. Firstly, the structure was mounted in the central down part of the basin for the 0° wave direction (Figure 3b). Secondly, the structure was moved towards the down left corner to optimize the distance to the wave maker, but also to allow simulation of the wave directions between 45° and 80° just by moving the wave paddle and keep the structure in the same position (Figure 3a,c).

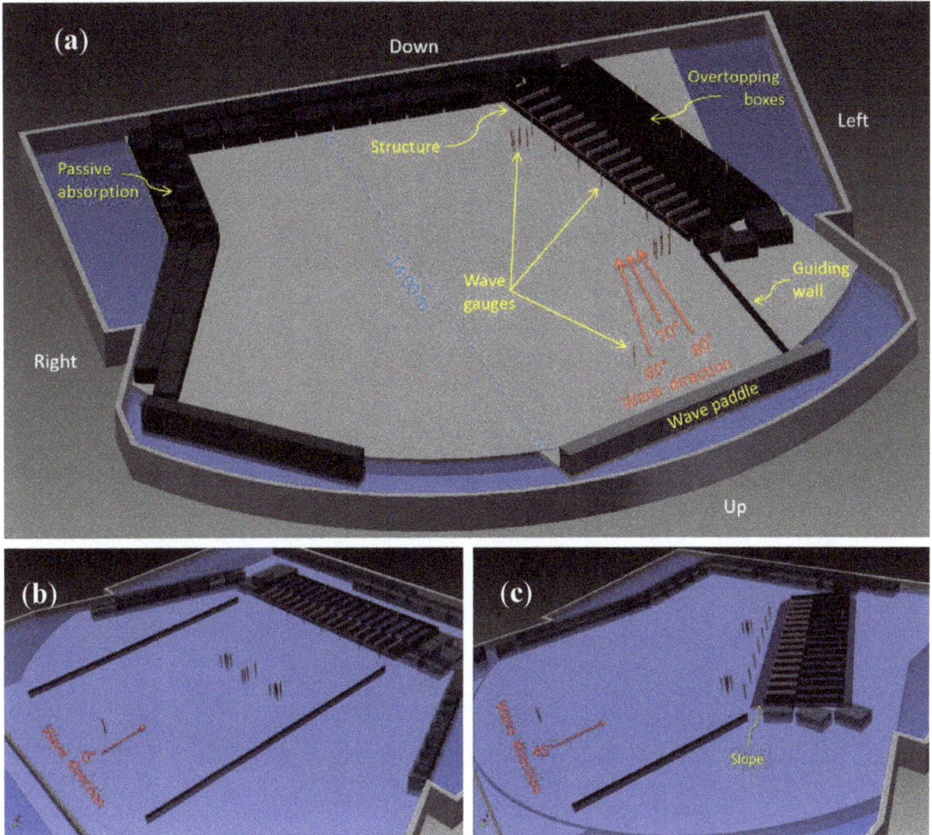

Figure 3. The position of the structure in the wave basin during experiments: (**a**) wave directions 60°, 70° and 80°; (**b**) perpendicular wave attack; (**c**) wave direction 45°.

3.1.1. Instrumentation

Resistance type wave gauges (17 in number) were used to measure the wave characteristics (height, period, and direction). One wave gauge is permanently situated in front of the wave maker to verify the generated waves. Two wave gauges arrays were built, each consisting of five wave gauges; these wave gauges are located in such way that a directional spectral analysis can be performed. The incident wave height has been measured using these two 3D-arrays. The WaveLab software (version 3.39, [14]) which utilizes the Baysian Directional spectrum estimation method (BDM) [15], has been used for the analysis. Using this method, the user generally indicates a circular sector around the expected incident and reflected wave direction, so the analysis will be limited to this sector. It is possible to select a very narrow circular sector, excluding from the wave analysis directions too far from the main one. Alternatively, it also possible to select ±90° around the main direction, so the entire 360° will be covered from the analysis. In this study, the analysis for the perpendicular wave case attack used ±30° around the main expected directions (0° for the incident and 180° for the reflected waves respectively). Hence, spurious transversal effects were removed from the results. Differently, for the oblique wave cases, it has been preferred to extend the analysis to the entire 360°, because in such case the main reflected direction can be assumed, but the effects on wave further reflections on the sides of the basin (even though passive absorption was installed) has to be checked. The rest of the six wave gauges were mounted equidistantly, in the proximity of the structure to provide information about the total wave height variation along the structure. This instrument setup was used for all the wave directions, but some minor changes in distances and positions were made for each wave direction (Figure 4). On every overtopping box a mechanical reader for the water level was installed to measure the accumulated volume.

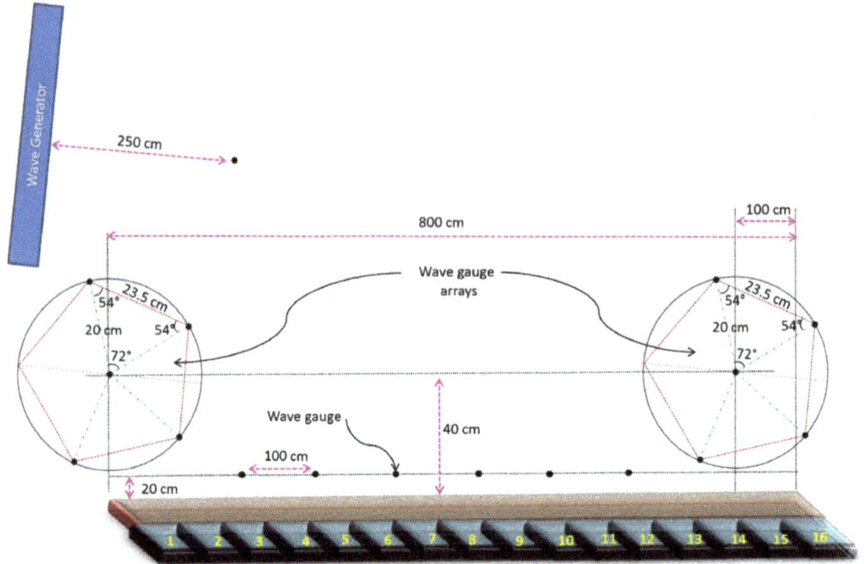

Figure 4. Example of instrument distribution in the basin for the wave angle 80° (not to scale).

3.1.2. Test Programme

A total of 377 tests were run, covering a wide range of wave conditions and structure configurations (Table 3). The wave angle is defined as the angle between the wave direction and the line normal to the quay structure so that 0° defines perpendicular wave attack. For the majority of the wave angles both the vertical and the sloping dike configurations were used. In all the tests long-crested waves were generated. The water level was varied around the crest level: the defined "dike freeboard" (R_c) can assume negative (i.e., Still Water Level, SWL, above the dike crest) and positive values (SWL below the dike crest). Three different wall heights were used (0 m, 1 m, and 2 m in prototype scale). The wall elevation with respect to the SWL defines the dike freeboard R_c. The berm (distance between the storm return wall and the quay or dike crest) lengths used in the experiments were 0, 5, 25, and 50 m in prototype scale. Tests with no reliable measured wave conditions, zero overtopping, and water volumes exceeding the boxes' volume, as well as preliminary tests to set-up the model were excluded from further analyses.

Table 3. Summary of the test conditions for overtopping reduction.

Total no. of Tests	**Used for Analyses**	**Vertical Quay**	**Sloping Dike (1 to 2.5)**	
377	230	191	39	
Wave directions	Wave height (H_{m0})	Wave period (T_p)	Crest freeboard (R_c)	Storm return wall position
0°, 45°, 60°, 70°, 80°	0.96 to 3.39 m	5.1 to 12.6 s	0 to 2.75 m	0 to 50 m

The overtopping discharge per each overtopping box and the measured total wave height along the structure were analysed. In most of the tests, the overtopping boxes of both sides (from 0 to 1 m and from 7 ÷ 7.5 to 8 m) were not included in the calculation to avoid errors generated by model boundary effects. The calculation of the mean overtopping discharge starting from the measured overtopping volume follows geometrical rules as

- For each test the berm length was calculated as a distance between the edge (crest) of the quay (sea dike) and the crown wall.
- For each angle the projection of the berm length was measured on the wave direction; this is the effective berm length that the wave has to run before reaching the wall.
- To calculate the mean overtopping for the entire quay some buffer zones at both edges of the structure were skipped (where possible model effects are noticed). For instance, in the case with no crown wall or crown wall on the quay edge, the entire quay length (8 m) was considered excluding the two overtopping boxes situated at the edges of the structure.
- It was verified on video recordings that the peaks in the overtopping volume were not due to model effects (boundary reflection), but they were due to the wave attack.

3.2. Model Settings of Force Test

The model built to investigate the reduction of the wave impact forces was very similar with the one for the overtopping reduction with the same 1:50 scale reduction. The structure (Figure 5) had a length of 8 m, a width of 0.6 m, and a height of 0.2 m. Based on the distribution of the largest wave heights and largest overtopping volumes along the structure an area of interest was selected approximately in the structure's centre, where four force sensors (Tedea Huntleigh, Tension Compression Load Cells, Model 641) were placed to record the time series of the wave forces acting on the storm return wall (Figures 6 and 7). A minimal distance between the wave maker and the structure of two wave lengths was respected for all tests.

Three positions of the structure in the basin were planned. Firstly, the structure was mounted in the central down part of the basin for the 0° wave direction. Secondly, the structure was moved towards the down left corner to optimize the distance to the wave maker and to obtain the angle of 45° without changing the position of the wave paddle. Thirdly, the structure was moved for the 80° wave angle attack (same positions as in Figure 3).

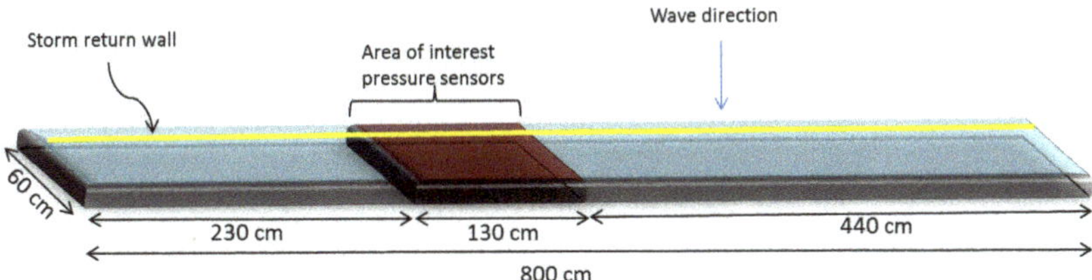

Figure 5. The structure used to investigate the force reduction (posterior view) (not to scale).

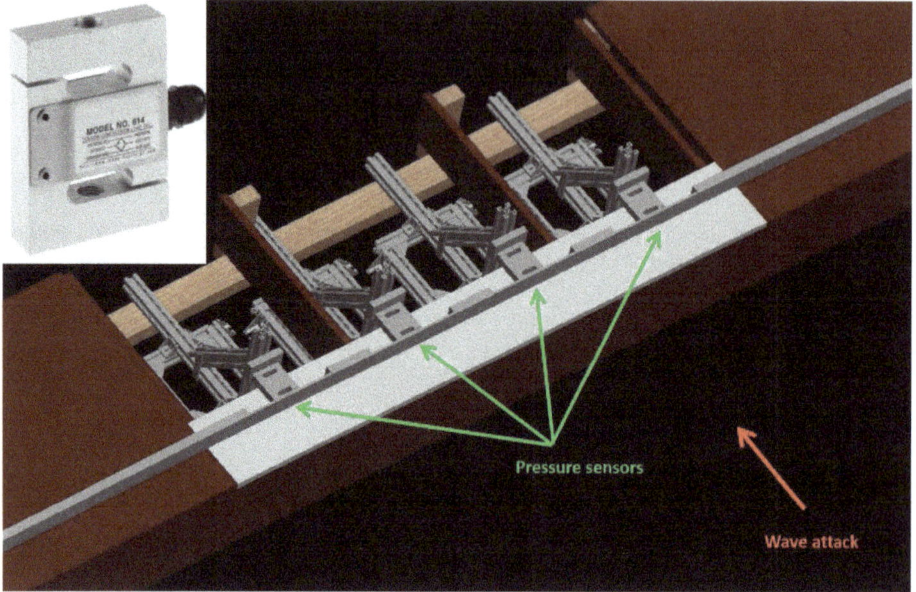

Figure 6. Force sensors were installed location as designed.

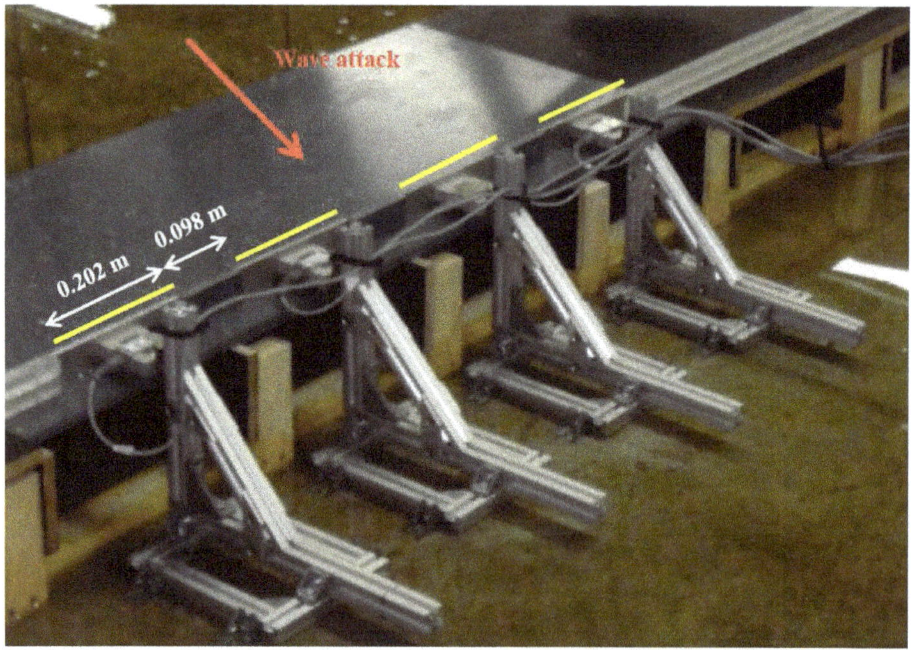

Figure 7. The part of the structure where the force sensors were installed (detailed picture).

Instrumentation

The wave gauges were installed in a similar position as for the overtopping model and four force sensors were installed to measure the forces acting on the storm return wall at a frequency of 1 kHz (Figures 6 and 7). Forty-four successful tests were performed (Table 4).

Table 4. Summary of the test conditions for the force reduction.

Total Number of Tests		**44**		
Wave directions	Wave height (H_{m0})	Wave period (T_p)	Crest freeboard (R_c)	Storm return wall position
0°, 45°, 80°	1.04 to 4.54 m	10.2 to 12.9 s	0 to 3.0 m	0 to 25 m

4. Results

4.1. Overtopping Reduction

The measured average wave overtopping has been compared with the predicted values using the existing formulas. A reduction coefficient for each direction has been assessed using the FHR tests results; both mean value and standard deviation of the reduction coefficient were calculated. The distribution of overtopping along the overtopping boxes was analyzed and correlated to the total wave height measured at the toe of the structure. For the analyses of the overtopping reduction due to the obliqueness, only tests without crest berm or with very short crest berm (5 m in prototype) were considered. The influence of long crest berms has been analyzed afterwards.

To cope with the spatial variation, the calculation of the mean overtopping discharge starting from the collected overtopping volume has been done based on a geometrical rule, summarized as follows:

1. For each test, the berm length was calculated as a distance between the edge of the quay (sea dike) and the crown wall.
2. For each angle, the projection of the berm length along the wave direction was assessed; this represents the effective berm length that the wave has to run before reaching the wall.
3. Starting from the first corner of the dike, the projection of the effective berm along the quay gives the minimum distance before which no wave reaches the wall.
4. The width considered to calculate the mean overtopping for the entire quay is equal to the quay length minus the calculated distance and some buffer zones at the edge (where possible model effects are noticed).

For incident wave height to be used in the formulas, the one coming from the star array after reflection analysis is employed since the information of the wave gauges placed along the dike only corresponds to total wave height (incident + reflected).

Physical model test results included in the CLASH database [11] similar to the test from the present study were used for comparison. In detail:

- Sloping dike: only CLASH data with slope between 1:4 and 1:2 with gentle or no foreshore were considered.
- Vertical quay: only tests with gentle or without foreshore were considered.

4.1.1. Vertical Quay Wall

The results of the tests indicate a clear decrease in the overtopping volumes with the increase of the wave angle. An increase of the overtopping volumes along the structure was observed for all cases, except for the perpendicular waves. In Figure ??, an example is shown, and the horizontal axis represent the quay extension, from 0 to 8.0 m, where the 0 is taken in the corner of the structure closest to the wave paddle. Each line plotted in every figure represents the results from one model test. The distribution of the wave overtopping along the vertical quay is generally consistent with the distribution of the total wave height at the toe.

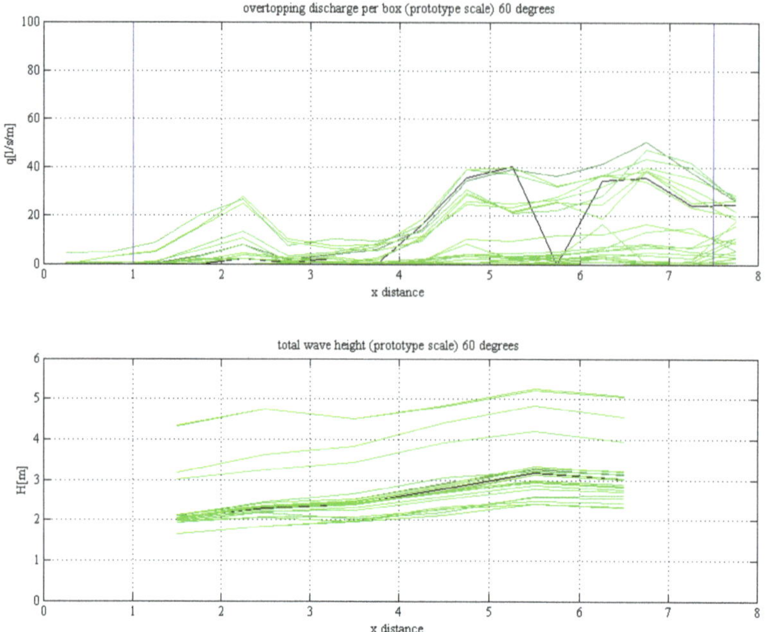

Figure 8. Overtopping discharge per box along the vertical quay and the total wave height for direction 60°.

However, stem wave formation can play a role in increasing wave height and consequently increasing overtopping along the structure. Research on stem waves along vertical wall with different researchers, ref. [16,17] reveal that the normalized significant stem wave height becomes large as the incident angle of wave become large. It was found also that the wave breaking suppresses the growth of the stem waves. These studies were based on wave tank experiments and on various numerical wave models with regular and irregular waves, but the predictions did not match very well the observations. In the present study the effect of the stem waves was not investigated since just one value of mean discharge along the whole quay was considered in the final analysis, a value measured by the instruments array.

Figure 9 shows the results of the FHR tests in a graph with the measured discharges plotted against the predicted ones, expressed in l/s/m (prototype scale). The plotted data include cases without a crest berm (distance of the wall from the edge of the quay, d_w, equal to 0 m) and with a crest berm (d_w larger than 0 m). The dash–dot lines indicate a prediction of 10 times larger and smaller with respect to the central line (ratio predicted/measured equal to 1:1). The formula overestimates the overtopping discharge for the 70° and 80° directions, while for the 0°, 45°, and 60° directions results are in reasonable agreement or within the above mentioned range.

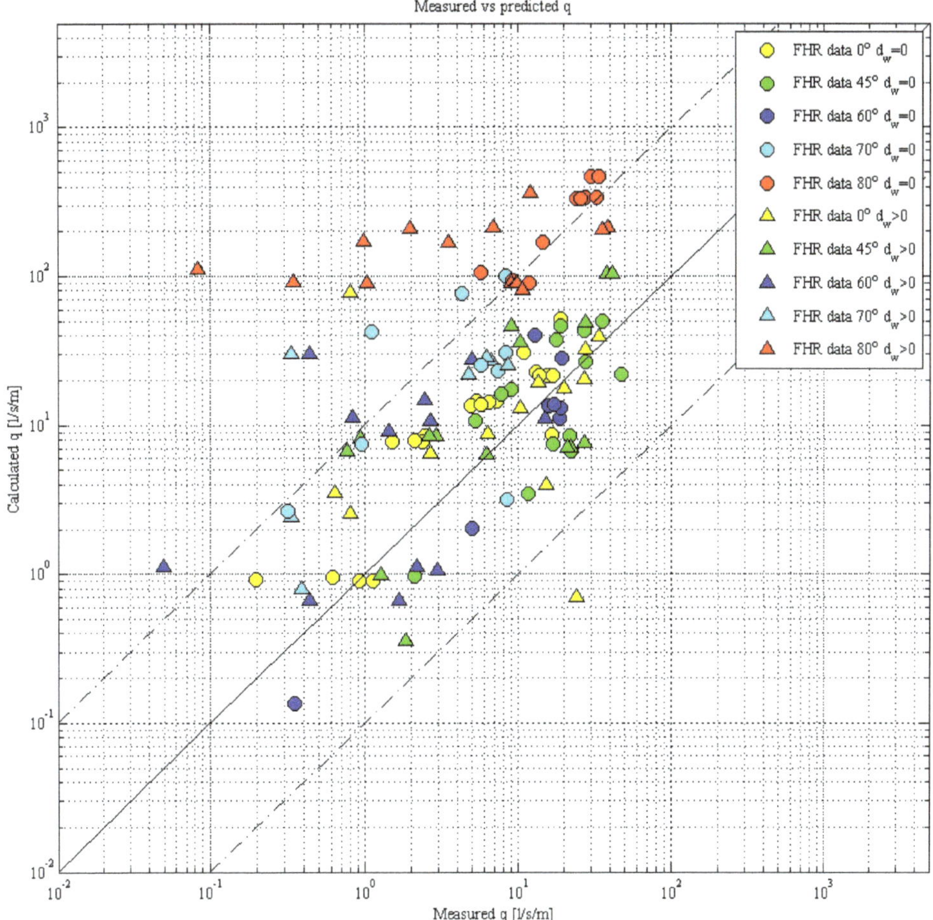

Figure 9. Quay wall: predicted [1] vs. measured overtopping discharges. The circles indicate the cases without berm crest ($d_W = 0$), the triangles indicate the cases where a berm crest is present ($d_W > 0$).

The effects of the obliqueness on the overtopping discharge were evaluated calculating the reduction coefficient of each case, starting from Equation (14), as follows:

$$\gamma_\beta = -2.6\frac{R_c}{H_{m0}}\frac{1}{\ln\left(\frac{q}{0.04\sqrt{gH_{m0}^3}}\right)} \tag{14}$$

The calculation has been performed both for the FHR data and for the selected CLASH data. Figure 10 shows the variation of the reduction coefficient with the wave angle. The existing formulations were analyzed to calculate the reduction coefficient as function of the wave angle. Despite the scattering of the results (similar scatter can also be noticed in Goda, 2009) a certain trend can be identified.

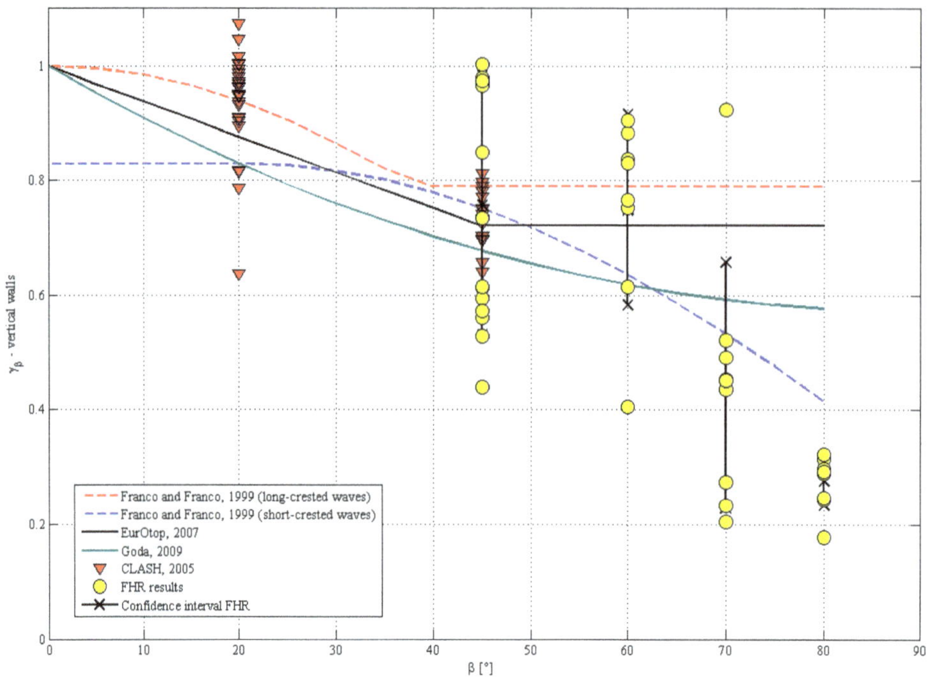

Figure 10. Quay wall: variation of reduction coefficient with wave angle, comparison to existing formulas.

The tests clearly show that the overtopping discharge is inversely proportional to the wave angle: the larger the wave angle, the smaller the wave overtopping. Different formulas propose constant values for the overtopping volumes for waves larger than 37° (long crested waves, [18], or 45° [1]). Franco and Franco formula [18] for short-crested waves seems to be the closest to FHR results, although the FHR tests were conducted using just long-crested waves. However, the differences due to the "short-crestedness" lie within the scattering of the formula, similar to previous studies [19]. Franco and Franco [18] stated that the directional spreading might allow reducing the freeboard with 30% in respect to cases with only long-crested waves.

The results of the experiments indicate that no formula, among those previously proposed predicts accurately the overtopping reduction. However, it is preferable to use the formula proposed by Goda [3] for large angles due to two main reasons:

(a) the correction coefficient represents an upper limit (safe approach) for the present cases with very oblique waves, although not excessively high as EurOtop [1]; and

(b) the expression for γ_β is applicable up to 80°, meanwhile EurOtop [1] indicates a constant value for wave angles larger than 45°.

The mean overtopping discharge is generally expressed by means of an exponential function as follows:

$$\frac{q}{\sqrt{gH_{m0}^3}} = A\exp\left(-B\frac{R_c}{H_{m0}\gamma_\beta}\right) \tag{15}$$

where

- A = 0.040 and B = 2.6 in EurOtop [1],

- A = 0.033 and B = 2.3 in Goda [3], and
- A = 0.116 and B = 3.0 in Franco and Franco [18].

Note that the reduction coefficient γ_β is a function of the A and B coefficients. The differences between Goda [3] and EurOtop [1] can be considered negligible because the values of A and B coefficients are rather similar.

New values for the reduction coefficient are presented here based on the FHR data and it is proposed to be used for similar conditions (Table 3). The resulting values, based on the FHR measurements, including the standard deviation, can be summarized as follows:

- $\gamma_\beta = 0.76$ ($\sigma = 0.23$), for $\beta = 45°$;
- $\gamma_\beta = 0.75$ ($\sigma = 0.17$), for $\beta = 60°$;
- $\gamma_\beta = 0.44$ ($\sigma = 0.21$), for $\beta = 70°$; and
- $\gamma_\beta = 0.28$ ($\sigma = 0.04$), for $\beta = 80°$.

The calculated gamma value is the mean value for each wave angle. The mean values and standard deviation values were calculated for each wave angle starting from the results of γ_β estimated for each single test. The confidence interval represented in Figure 10 is calculated as $\pm\sigma$ with respect to the mean value. As general approach, the mean value of γ_β has to be used for design purposes. It can be noticed that the difference in the reduction coefficient between 0.72 (calculated value using EurOtop [1]) and 0.28 might cause a difference in the calculated discharge of at least 1 order of magnitude (10 times) in the selected data range.

Figure 11 shows the FHR data, the CLASH data and the EurOtop predictions in term of non-dimensional discharge $Q = q/(g \cdot H_{m0}{}^3)^{\hat{}0.5}$. Only the FHR cases with the wall on the edge of the quay are plotted in order to avoid misinterpretations due to the effects of the width of the crest berm. Three different plots are shown in Figure 11:

(a) the values of Q are plotted against the non-dimensional freeboard R_c/H_i;
(b) the values of Q are plotted against the non-dimensional freeboard $R_c/H_i\gamma_\beta$ (EurOtop), where $\gamma\beta$ (EurOtop) is the correction coefficient calculated using the EurOtop (2007) formula; and
(c) (the values of Q are plotted against the non-dimensional freeboard $Rc/Hi\gamma\beta$(Goda), where $\gamma\beta$ (Goda) is the correction coefficient calculated using the Goda [3] formula.

The use of Goda [3] formula is improving the wave overtopping prediction in case of oblique wave attack with respect to the EurOtop [1] formula. In most of the cases, especially for very oblique angles, the EurOtop formula seems to overestimate the overtopping, while using Goda correction factors the results are spread around the formula prediction and only few of them are still overestimated.

The analysis on the berm length effects (distance between the seaward edge of the quay and the storm wall) and on the wall height has been carried out. Figure 12 shows the non-dimensional overtopping discharge in function of two different non-dimensional parameters: (i) the ratio between the wall height and the incident wave height, (ii) the ratio between the berm length and $1.56T_p{}^2$ that can be assumed as the wave length in deep water conditions. The combination of obliqueness, wall height and berm length made it challenging to have a clear view of the phenomena occurring at the structure. Despite the rather wide data scatter, there are clear differences between short or no berm layouts and wide berm layouts. A dependence on the berm length can be detected, the overtopping was reduced when the ratio of the berm length over the wave length was increased and this trend was clearer for larger wave angles. The waves travelled at the dike crest before approaching the storm wall and it was expected that the waves would refract on the berm, and therefore approach the wall with less obliqueness, but still not perpendicular. The distance travelled by the waves to reach the wall was larger for larger angles, so the amount of energy dissipated on the crest might have been larger. The configurations without berm, and with short berm length, 5 m in prototype, show a similar behaviour leading to larger overtopping discharge than the configurations with wider berms (25 m and 50 m in prototype).

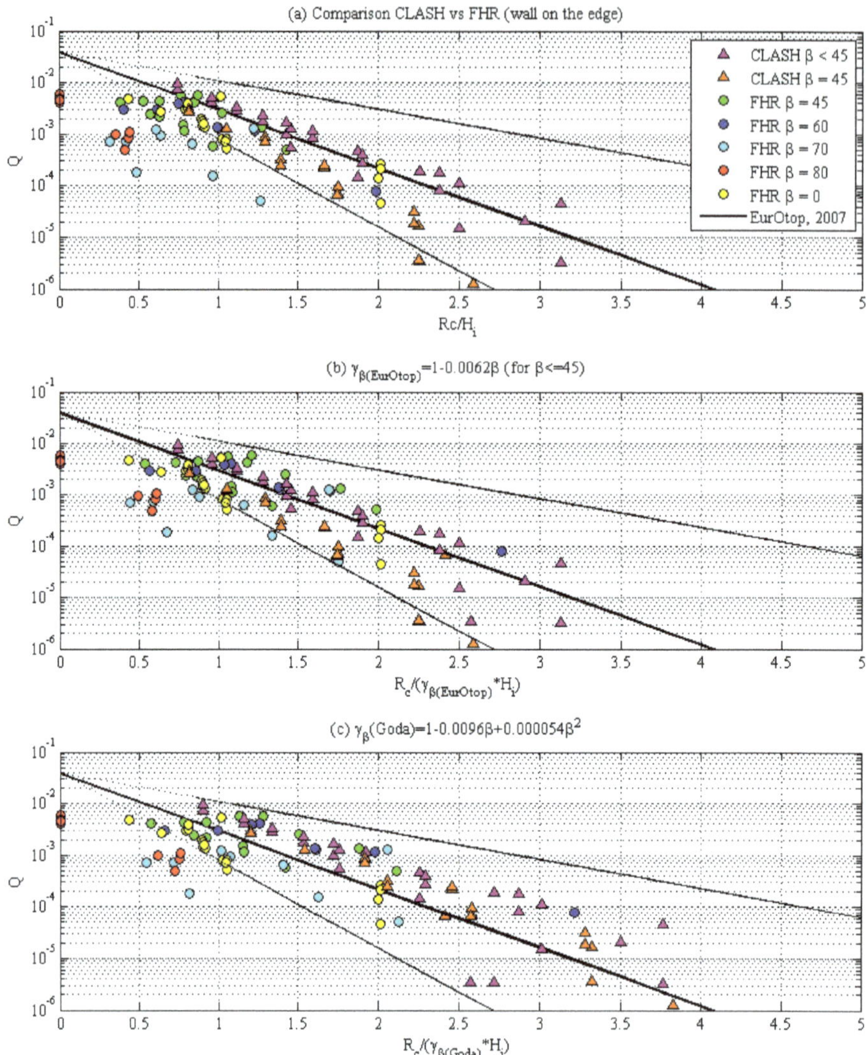

Figure 11. CLASH and FHR (wall on the edge of the quay) data vs. EurOtop predictions: (**a**) overtopping plotted against the non-dimensional freeboard; (**b**) overtopping plotted against the non-dimensional freeboard, with the correction factor from EurOtop (2007) formula; (**c**) overtopping plotted against the non-dimensional freeboard, with the correction factor from Goda (2009) formula.

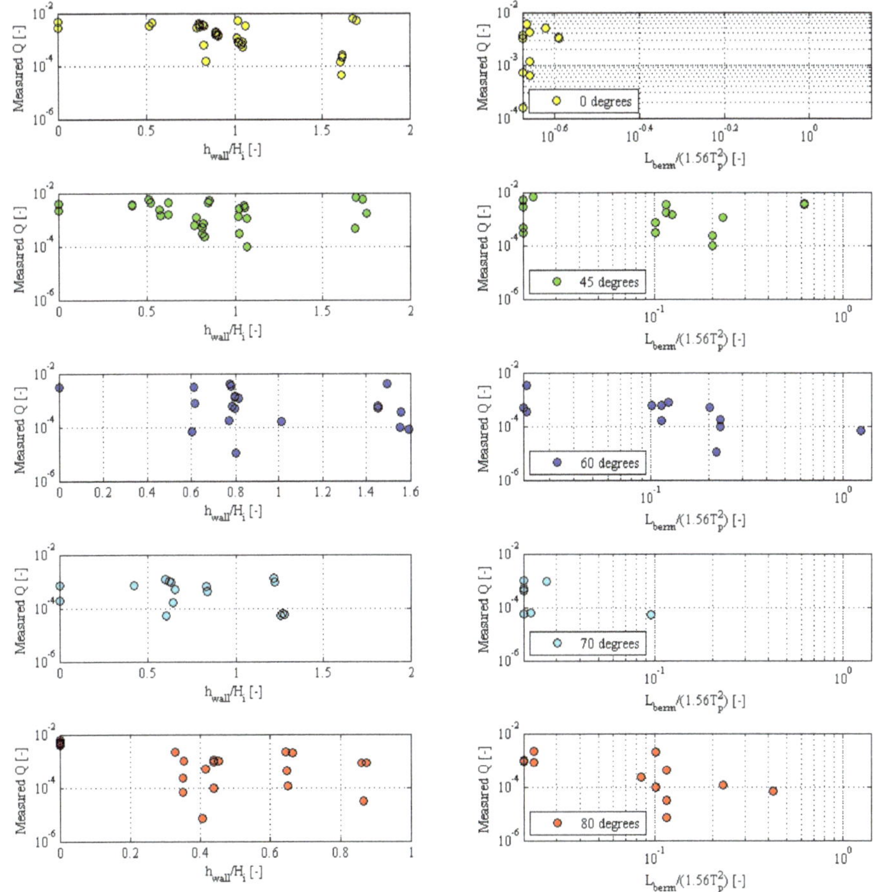

Figure 12. Non-dimensional discharge vs. relative wall height and relative berm length (quay layout).

4.1.2. Sloping Dike

The results of the tests for a sloping dike are similar with those for a vertical quay, indicating the same decrease in the overtopping volumes with the increase of the wave angle. The measured overtopping discharges for FHR data are plotted in Figure 13 against the values predicted using Equations (3) and (4) [3]. As noticed in the previous cases, the formula seems to overestimate the overtopping discharge for very oblique wave attacks.

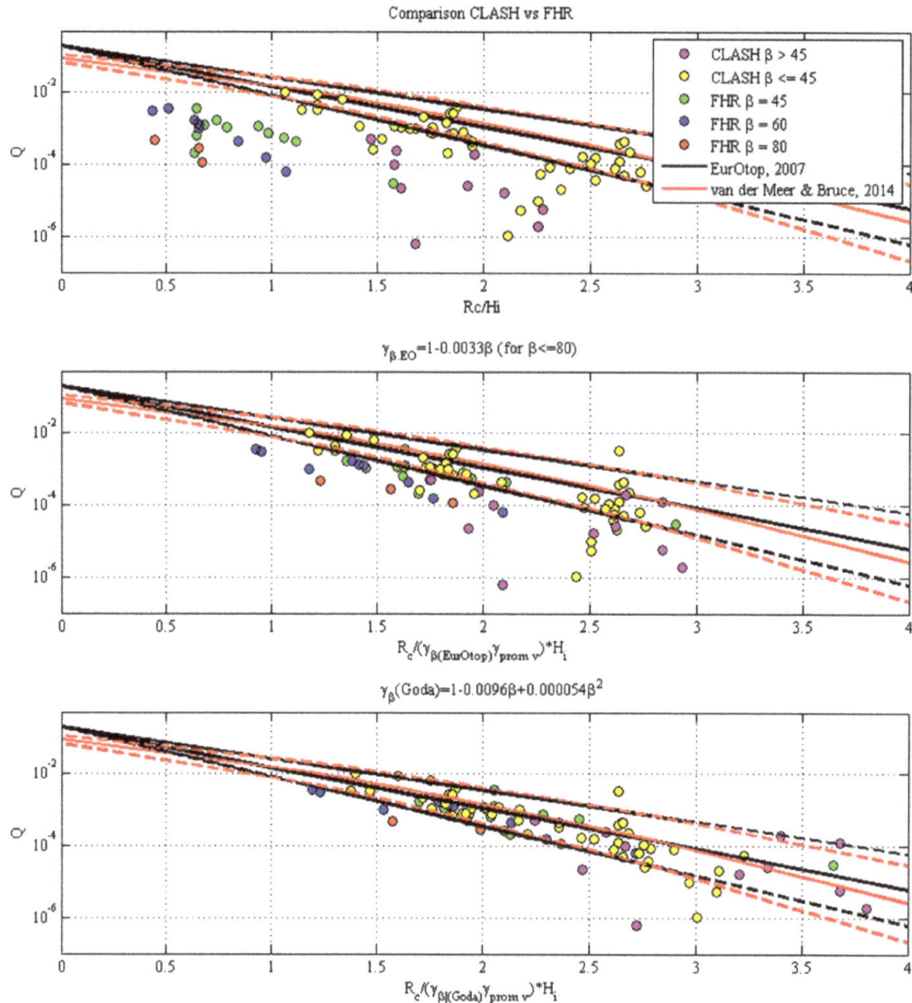

Figure 15. FHR and CLASH data vs. formula predictions.

4.2. Force Reduction

The measured forces are plotted in Figure 16 (in prototype scale both vertical quay and sloping dike) in function of the incident significant wave height. The colors indicate the wave angle, respectively red for 0°, yellow for 45°, and blue for 80°. The different shapes indicate the results from each different load cell; this underlines that, despite the waves are long-crested, the forces exerted along the structure have a certain variability. As expected, the forces increase with the wave height. It is clear that the 0° cases result in larger forces than the 45° case, and the 80° cases have the lowest forces. For the same wave height, the very oblique cases present in average a value of the wave force that is almost half of the perpendicular case.

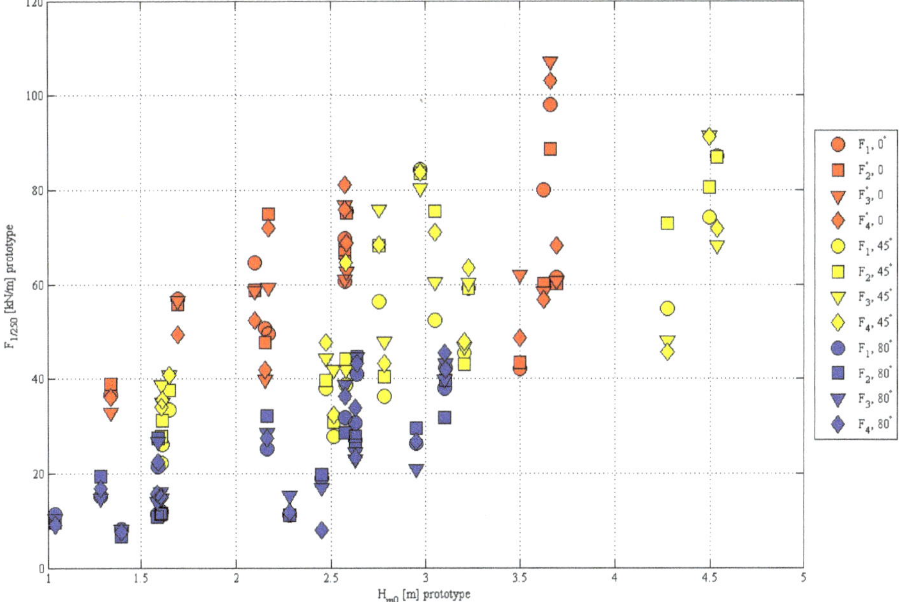

Figure 16. Dependence of the wave forces on the incident wave height for different wave obliqueness.

The data were analyzed to define an analytical expression for the reduction factor. This coefficient is the ratio between the force due to an oblique wave attack over the force in case of perpendicular waves reaching the structure. The reduction factor expresses how much the data from cases with oblique attack should be corrected to be in line with a 0° case that present the same hydraulics boundary conditions (except from the obliqueness).

Two different expressions, respectively for quay walls and dikes, were found:

$$\gamma_{quay} = \frac{F_{quay,\beta>0}}{F_{quay,\beta=0}} = 0.5{\cdot}(1+\cos\beta) \tag{17}$$

$$\gamma_{dike} = \frac{F_{dike,\beta>0}}{F_{dike,\beta=0}} = \exp(-0.007\beta) \tag{18}$$

where β is the wave direction relative to the structure (perpendicular wave direction = 0°, angle expressed in degrees). The variation of the reduction factors as function of the wave angle is depicted in Figure 17. The expression for quay walls is corresponding to the formula for caisson breakwaters proposed by Goda [23].

The two expressions for the reduction factor could certainly be improved if additional data with other wave angles than 45° and 80° were available. However, the new proposed expressions can already be considered as a significant improvement in the prediction. In Figure 18 the measured wave forces are plotted as a function of the relative freeboard. Four figures are reported, two for the dike cases and two for the quay wall cases. In detail:

(a) measured wave force on the storm wall for the quay wall layout;
(b) measured wave force on the storm wall for the sea dike layout;
(c) measured wave force on the storm wall for the quay wall layout, including the correction with the proposed reduction factor for wave obliqueness; and

(d) measured wave force on the storm wall for the sea dike layout, including the correction with the proposed reduction factor for wave obliqueness.

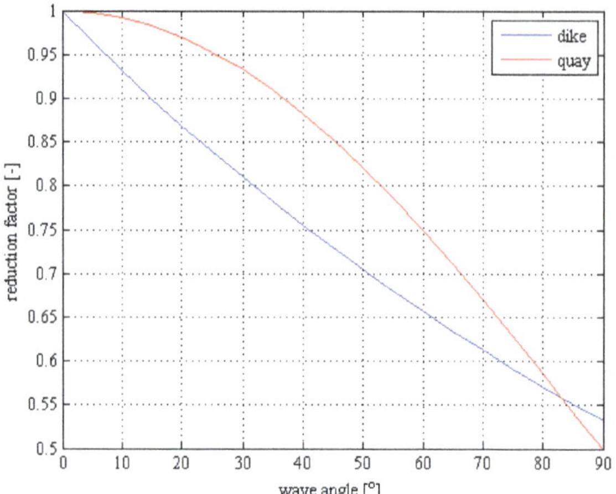

Figure 17. Variation of the reduction factor as function of wave angle.

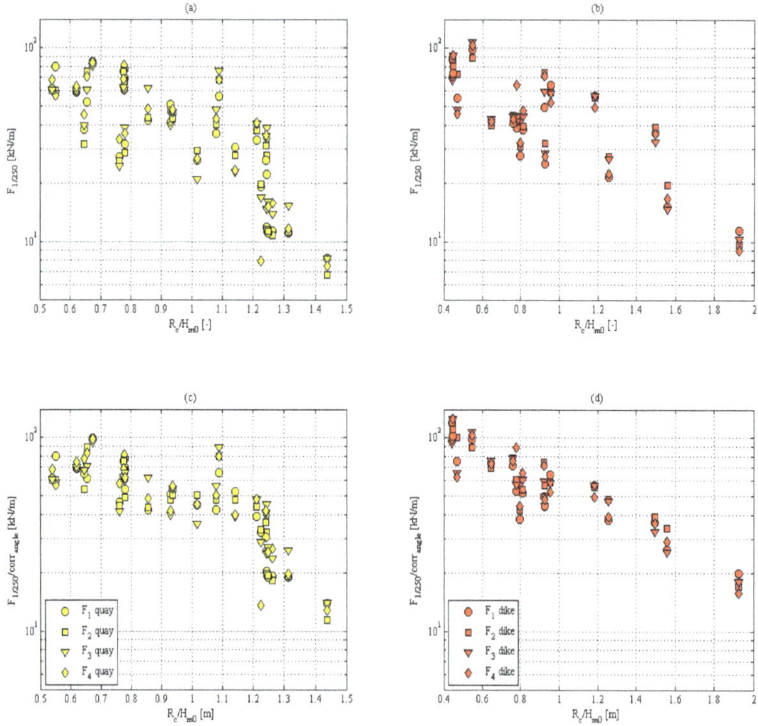

Figure 18. Dependence of the wave force on the relative freeboard with and without reduction factor (**a** and **c**: quay wall; **b** and **d**: dike).

The scatter in the wave forces is significantly reduced if the wave force is corrected using the reduction factor proposed above. This improvement was also quantified by the relative standard deviation for each case ($\mu' = \mu/\sigma$): (a) μ' = 7.9%; (b) μ' = 7.0%; (c) μ' = 8.8%; and (d) μ' = 4.7%.

The analysis of the overall results finally suggests that in case of very oblique wave attack (obliqueness between 70° and 80°), the expected force on the storm wall range between 55% to 65% of the value in case of perpendicular wave attack.

The results of the FHR tests are compared to predictions of the formula proposed by Van Doorslaer et al. [10] for sea dikes, regardless of its range of applicability (e.g., wall position and wall height are different). Figure 19 depicts the variation of the non-dimensional quantity $F_{1/250}/\rho g R_c^2$ as function of the relative freeboard, both for FHR and UPC results. A common trend between the two experimental datasets can be noticed, despite of a certain scatter in the FHR results, mainly due to the different wave angles.

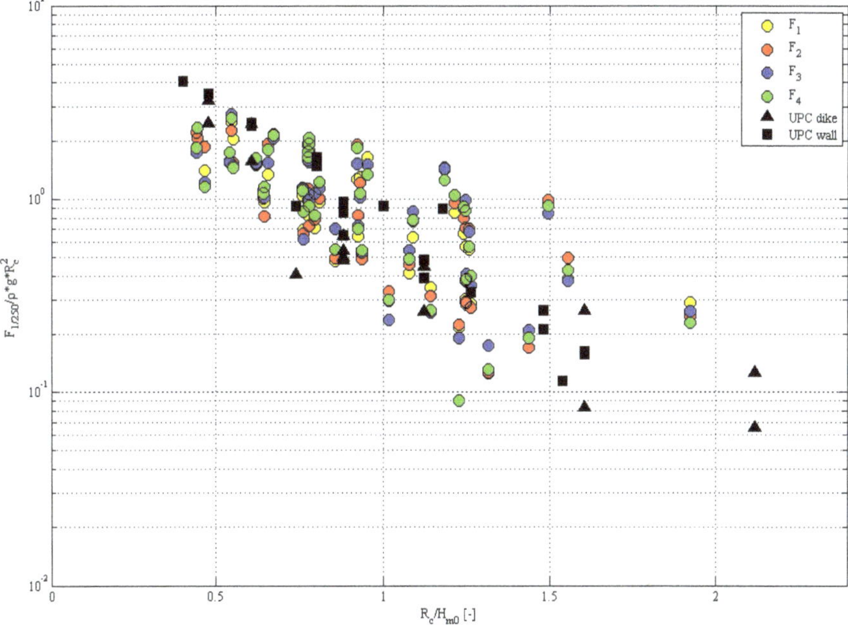

Figure 19. Dependence of the non-dimensional wave forces on the relative freeboard and comparison with data from UPC [10].

Equation (13) was applied to the FHR and UPC data; only FHR cases with 0° were initially considered for comparison, as the UPC data refer to perpendicular wave attack. The results of the are plotted in Figure 20. Generally, Equation (13) underestimates the force for FHR cases probably due to different wall height between UPC data and FHR data, respectively 1.2 m and 2.0 m (in prototype scale). Higher walls would lead to smaller overtopping rates and bigger reflection exerted by the storm wall with consequent higher forces on the same wall.

In the next step, Equation (13) was applied to all FHR data and the results are reported in Figure 21, both without and with, application of the reduction factors (Equations (17) and (18)) for wave obliqueness. Without correction, the prediction showed a large scatter, while the application of the reduction factor reduced significantly the scatter and improved the predictions. Nevertheless, the estimated forces are still slightly smaller than the measured ones and it can be concluded that the correction applied to take into account the wave obliqueness improves the predicted forces.

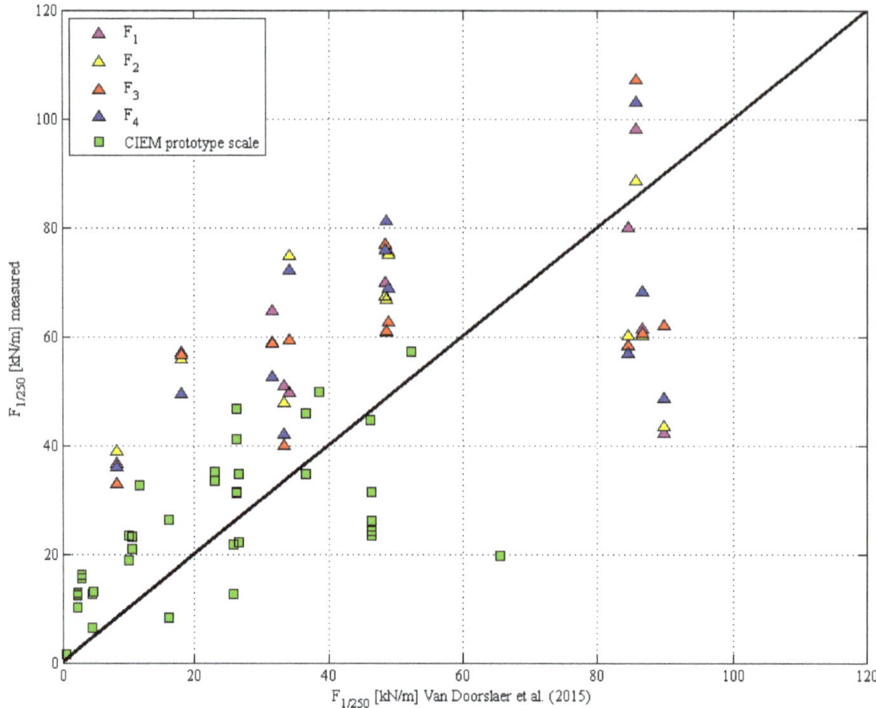

Figure 20. Measured forces versus Van Doorslaer et al. [10] predictions for FHR 0° cases and UPC cases.

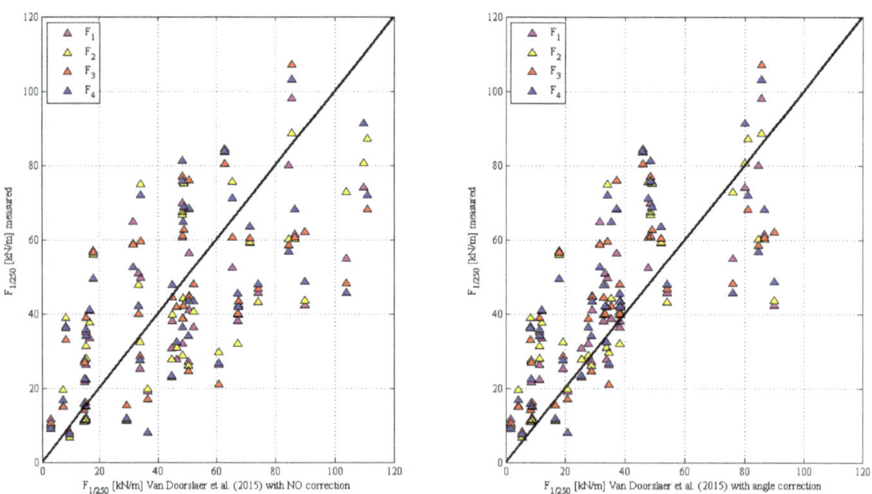

Figure 21. Application of Van Doorslaer et al. [10] formula with and without reduction factor for wave obliqueness.

5. Conclusions

The present study describes the setup and the results of physical model tests carried out at FHR on wave overtopping generated by perpendicular and very oblique waves, based on the generic configurations and conditions at Belgian harbors. A set of different wave angles has been tested: for a vertical quay layout: 0°, 45°, 60°, 70°, and 80° and for the dike layout: 45°, 60°, and 80° were investigated.

The reduction in overtopping discharge has been quantified and the results were compared with similar tests in the CLASH database [11] and with predictions by several semi-empirical formulas and correction factors from literature.

The influences of the storm wall height and the crest berm width were investigated together with the effect of wave obliqueness.

The analysis of the results for the vertical quay and sloping dike layouts leads to the following conclusions:

1. The EurOtop formula [1] generally overestimates the overtopping discharge for large wave obliqueness.
2. The values of the reduction factor γ_β calculated for the vertical quay layout are equal to 0.76, 0.75, 0.44, and 0.28, respectively, for 45°, 60°, 70° and 80°.
3. The values of the reduction factor γ_β calculated for the sloping dike layout are equal to 0.72, 0.54 and 0.44 respectively for 45°, 60° and 80°.
4. A rather large scatter is present in the results similar to the results presented in previous studies [3].
5. The expression of γ_β presented by Goda [3] is finally proposed as a good compromise between accuracy (in comparison with physical model results) and a certain safety in the design of the storm walls.
6. The high obliqueness combined with long berms on the crest (comparable with the wave length) leads to very low or zero overtopping discharge.
7. The berm length (ranging from 0 to 50 m) has a larger influence on the overtopping discharge than the wall height (ranging from 1 to 2 m).

Tests were performed to identify the wave force impact reducing due to wave obliqueness both for sea dike and for quay wall layouts. The wall height was 2.0 m (in prototype scale) and it was located at three different distances from the seaward edge of the main structure (berm).

The results indicate that for high wave obliqueness the force reduction in case of very oblique waves is 0.55–0.65 times the wave forces for similar wave conditions, but for perpendicular wave attack. Two reduction factors were defined, respectively for the sea dike and the vertical quay wall layout as presented in Equations (17) and (18). Finally, the formula of Van Doorslaer et al. [10] was applied, confirming that the use of the above mentioned reduction factors reduces the uncertainties in the wave force predictions due to the effects of the wave obliqueness.

A significant knowledge gap regarding the quantification of overtopping and impact forces reduction due to oblique waves attack was filled by adding valuable data to the scarce existing literature and improving the predicting formulas. These formulas were applied to several specific situations in Belgium [24,25] and they can be used in similar settings worldwide.

Due to the limited amount of available data, the relationship between wave obliqueness and other variables such as wall position has not been analyzed in the present study. Further studies on wave forces on storm walls on top of sea dikes or quay walls should take into account this reduction if the waves are approaching the structure with an angle larger than 45°. Since it was considered limited, the influence of the stem waves formed along the structure on the overtopping were not investigated during this study and further investigation is recommended.

Author Contributions: S.D.: Conceptualization, methodology, resources, data curation and analysis, writing—original draft preparation, supervision, and project administration. C.A.: Conceptualization, methodology, software, validation, data analysis, investigation, data curation, writing—review and editing, and visualization. T.S. (Tomohiro Suzuki): Software, validation, formal analysis, investigation, and writing—review and editing. T.S. (Tim Spiesschaert): Physical test execution, wave basin set up, and data collection and analysis. T.V.: writing—review and editing, project administration, and funding acquisition. All authors have read and agreed to the published version of the manuscript.

Funding: This research was funded by Agentschap voor Maritieme Dienstverlening en Kust (MDK)—Coastal Division, project WL2012_00_050. Corrado Altomare acknowledges funding from the European Union's Horizon 2020 research and innovation programme under the Marie Sklodowska-Curie grant agreement No.:792370.

Acknowledgments: The authors are grateful to Thomas Lykke Andersen (University of Aalborg) and Marc Willems (Flanders Hydraulics Research) for their valuable suggestions regarding wave measuring and processing and the experiment set up, as well as to the four anonymous reviewers who improved the manuscript final version through constructive comments and suggestions.

Conflicts of Interest: The authors declare no conflict of interest. The funders had no role in the design of the study; in the collection, analyses, or interpretation of data; in the writing of the manuscript, or in the decision to publish the results.

References

1. Pullen, T.; Allsop, N.W.H.; Kortenhaus, A.; Schuttrumpf, H.; Van der Meer, J.W. (Eds.) *European Overtopping Manual for the Assessment of Wave Overtopping (EurOtop)*; Die Küste, 73. Kuratorium für Forschung im Küsteningenieurwesen: Heide im Holstein, Germany, 2007; 185p, ISBN 978-3-8042-1064-6. Available online: www.overtopping-manual.com (accessed on 29 April 2020).
2. Van der Meer, J.; Allsop, W.; Bruce, T.; Rouck, J.; Kortenhaus, A.; Pullen, T.; Schüttrumpf, H.; Troch, P.; Zanuttigh, B. *EurOtop: Manual on Wave Overtopping of Sea Defences and Related Structures—An Overtopping Manual Largely Based on European Research, but for Worldwide Application*, 2nd ed.; Environment Agency: Bristol, UK, 2018.
3. Goda, Y. Derivation of unified wave overtopping formulas for seawalls with smooth, impermeable surfaces based on selected CLASH datasets. *Coast. Eng.* **2009**, *56*, 385–399. [CrossRef]
4. De Waal, J.P.; Van der Meer, J.W. Wave run-up and overtopping on coastal structures. In Proceedings of the 23rd International Conference on Coastal Engineering, Venice, Italy, 4–9 October 1992; pp. 1758–1771.
5. Kortenhaus, A.; Geeraerts, J.; Hassan, R. *Wave Run-Up and Overtopping of Sea Dikes with and without Stilling Wave Tank under 3D Wave Attack (DIKE-3D)*; Final Report; Technische Universität Braunschweig: Braunschweig, Germany, 2006.
6. Chen, W.; Van Gent, M.R.A.; Warmink, J.J.; Hulscher, S.J.M.H. The influence of a berm and roughness on the wave overtopping at dikes. *Coast. Eng.* **2020**, *156*, 103613. [CrossRef]
7. Van Gent, M.R.A. Influence of oblique wave attack on wave overtopping at smooth and rough dikes with a berm. *Coast. Eng.* **2020**, *160*, 103734. [CrossRef]
8. Van der Meer, J.W.; Bruce, T. New physical insights and design formulas on wave overtopping at sloping and vertical structures. *J. Waterw. Port Coast. Ocean Eng.* **2014**, *140*, 04014025. [CrossRef]
9. Van Doorslaer, K.; De Rouck, J.; Audenaert, J.; Duquet, V. Crest modifications to reduce wave overtopping of non-breaking waves over a smooth dike slope. *Coast. Eng.* **2015**, *101*, 69–88. [CrossRef]
10. Van Doorslaer, K.; Romano, A.; Bellotti, G.; Altomare, C.; Cáceres, I.; De Rouck, J.; Franco, L.; Van der Meer, J. Force measurements on storm walls due to overtopping waves: A middle-scale model experiment. In Proceedings of the International Conference on Coastal Structures, Boston, MA, USA, 9–11 September 2015.
11. Verhaeghe, H.; Van der Meer, J.W.; Steendam, G.J. *Database on Wave Overtopping at Coastal Structures*; CLASH report, Workpackage 2; Ghent University: Ghent, Belgium, 2004.
12. Hughes, S.A. *Physical Models and Laboratory Techniques in Coastal Engineering*; World Scientific Publishing: Singapore, 1993; Volume 7, p. 568.
13. Schüttrumpf, H.; Oumeraci, H. Layer thicknesses and velocities of wave overtopping flow at seadikes. *Coast. Eng.* **2005**, *52*, 473–495. [CrossRef]
14. WaveLab 3 Homepage Aalborg University. 2012. Available online: http://www.hydrosoft.civil.aau.dk/wavelab/ (accessed on 25 March 2020).

15. Hashimoto, N.; Kobune, K. Directional spectrum estimation from a Bayesian approach. In Proceedings of the 21st Conference on Coastal Engineering, Costa del Sol-Malaga, Spain, 20–25 June 1987; Volume 1, pp. 62–76.
16. Lee, J.I.; Lee, Y.T.; Kim, J.T.; Lee, J.K. Stem Waves along a vertical wall: Comparison between Monochromatic Waves and Random Waves. *J. Coast. Res.* **2009**, 991–994.
17. Mase, H.; Memita, T.; Yuhi, M.; Kitano, T. Stem waves along vertical wall due to random wave incidence. *Coast. Eng.* **2002**, *44*, 339–350. [CrossRef]
18. Franco, C.; Franco, L. Overtopping formulas for caissons breakwaters with nonbreaking 3D waves. *J. Waterw. Port Coast. Ocean Eng.* **1999**, *125*, 98–107. [CrossRef]
19. Vanneste, D.; Verwaest, T.; Mostaert, F. *Overslagberekening Loodswezenplein Nieuwpoort*; Versie 2.0. WL Adviezen, 15_005; Waterbouwkundig Laboratorium (Flanders Hydraulics Research): Antwerpen, België, 2015. (In Dutch)
20. Owen, M.W. Design of seawalls allowing for wave overtopping. In *Hydraulics Research, Wallingford*; Report No. EX 924; HR Wallingford: Wallingford, UK, 1980.
21. Van der Meer, J.W.; En de Waal, J.P. Invloed van scheve golfinval en richtingspreiding op galfoploop en overslag, "Influence of oblique wave attack and directional spreading on wave run-up and overtopping". In *Report on Model Investigation*; H638; Delft Hydraulics: Delft, The Netherlands, 1990. (In Dutch)
22. Bornschein, A.; Pohl, R.; Wolf, V.; Schüttrumpf, H.; Scheres, B.; Troch, P.; Riha, J.; Spano, M.; Van der Meer, J. Wave run-up and wave overtopping under very oblique wave attack (CORNERDIKE-Project). In Proceedings of the HYDRALAB IV Joint User Meeting, Lisbon, Portugal, 2–4 July 2014.
23. Goda, Y. New wave pressure formulae for composite breakwater. In Proceedings of the 14th International Conference Coastal Engineering, Copenhagen, Denmark, 24–28 June 1974; pp. 1702–1720.
24. Dan, S.; Altomare, C.; Spiesschaert, T.; Willems, M.; Verwaest, T.; Mostaert, F. *Overtopping Reduction for the Oblique Waves Attack: Report 1. Wave Overtopping Discharge*; Version 3.0. FHR Reports, 00_050_1; Flanders Hydraulics Research: Antwerp, Belgium, 2016.
25. Dan, S.; Altomare, C.; Spiesschaert, T.; Willems, M.; Verwaest, T.; Mostaert, F. *Oblique Wave Attack on Storm Walls: Report 2. Force Reduction*; Version 3.0. FHR Reports, 00_050_2; Flanders Hydraulics Research: Antwerp, Belgium, 2016.

Journal of
Marine Science and Engineering

Article

Characterization of Overtopping Waves on Sea Dikes with Gentle and Shallow Foreshores

Tomohiro Suzuki [1,2,*], Corrado Altomare [3,4], Tomohiro Yasuda [5] and Toon Verwaest [1]

1 Flanders Hydraulics Research, 2140 Antwerp, Belgium; toon.verwaest@mow.vlaanderen.be
2 Faculty of Civil Engineering and Geosciences, Delft University of Technology, 2628 CN Delft, The Netherlands
3 Maritime Engineering Laboratory, Department of Civil and Environmental Engineering, Universitat Politecnica de Catalunya—BarcelonaTech (UPC), 08034 Barcelona, Spain; corrado.altomare@upc.edu
4 Department of Civil Engineering, Ghent University, Technologiepark 60, 9052 Gent, Belgium
5 Department of Civil, Environmental and Applied System Engineering, Kansai University; Suita, Osaka 564-8680, Japan; yasuda-t@kansai-u.ac.jp
* Correspondence: tomohiro.suzuki@mow.vlaanderen.be; Tel.: +32-3-224-69-34

Received: 15 September 2020; Accepted: 25 September 2020; Published: 27 September 2020

Abstract: Due to ongoing climate change, overtopping risk is increasing. In order to have effective countermeasures, it is useful to understand overtopping processes in details. In this study overtopping flow on a dike with gentle and shallow foreshores are investigated using a non-hydrostatic wave-flow model, SWASH (an acronym of Simulating WAves till SHore). The SWASH model in 2DV (i.e., flume like configuration) is first validated using the data of long crested wave cases with second order wave generation in the physical model test conducted. After that it is used to produce overtopping flow in different wave conditions and bathymetries. The results indicated that the overtopping risk is better characterized by the time dependent h (overtopping flow depth) and u (overtopping flow velocity) instead of h_{max} (maximum overtopping flow depth) and u_{max} (maximum overtopping flow velocity), which led to overestimation of the risk. The time dependent u and h are strongly influenced by the dike configuration, namely by the promenade width and the existence of a vertical wall on the promenade: the simulation shows that the vertical wall induces seaward velocity on the dike which might be an extra risk during extreme events.

Keywords: wave overtopping; average overtopping discharge; individual volume; overtopping flow depth; overtopping flow velocity; promenade; vertical wall; SWASH

1. Introduction

Global climate change has manifold impacts on the ocean and its behavior, which directly translates to the coastal/nearshore region as well as the governing processes. One such climate-induced response is the augmented frequency and intensity of extreme waves, leading to increased overtopping risk for people living in coastal area [1,2]. One of the wave overtopping risks for people is direct wave action, which is not only relevant to pedestrians [3,4] and vehicles on a dike/promenade but also to people in front of and inside dwellings and commercial buildings (e.g., hotels and restaurants on dikes). There is some literature related to the stability of people on the dikes/promenade (e.g., [3–5]), which deals with the relationship between human's stability and flow parameters (i.e., overtopping flow depth and flow velocity). Altomare et al. [6] indicated that the combination of overtopping flow velocity and flow depth rather than single maximum values of one of these parameters is required to understand pedestrians hazard. Arrighi et al. [7,8] conducted a numerical study on human's stability and highlighted the importance of relative submergence and Froude number to interpret the results.

Other works focused on the characterization of overtopping flow depths and/or velocities, see [9–12]. These are important works to understand the basic risk exposed to the flows on dikes, however the present knowledge cannot cover all the risks due to different layouts and hydraulic conditions. In a reality, there are often structures on the dike and thus the overtopping flow characteristics are a bit more complex than the simplified assumptions (e.g., only plain promenade) found in the literature. Overtopping flow can be changed by the interaction with structures. When overtopping is severe, overtopping waves can destroy the facade (i.e., the first defense of the dwellings/apartment buildings, such as windows, see examples in [13]), waves can propagate further even inside buildings. Then, the wave will be reflected to the seaward. The return flow or reflected wave in front of a vertical wall can also influence the human's stability, however a detailed discussion on such different flow directions has not been made explicitly so far.

As such, there are some knowledge gaps in the investigation on the risk of wave overtopping flow on and behind sea dikes together with structures, and therefore it is important to discuss further which physical process is relevant for the risk on people in coastal areas. So far, the safety of pedestrians and vehicles on coastal zones have been evaluated based on average overtopping discharge, maximum individual volume, and associated wave height, see EurOtop [14]. For instance, EurOtop indicates that individual overtopping volume V_{max} of 600 l/m in combination with H_{m0} = 1–3 m is a limit for overtopping for people standing at dikes with clear view of the sea but it does not give further detailed explanation. Those are important indications but the applicability, for example, to the gentle and shallow foreshore cases is still not very clear [15]. Moreover, it is of interest how a fixed criterion (e.g., 1 l/s/m or 10 l/s/m) can be linked to the overtopping characteristics such as V_{max} (maximum individual volume), and time dependent h (overtopping flow depth) and u (overtopping flow velocity). As Altomare et al. [6] indicated, the combination of u and h is linked to the hazard rather than the single maximum values of one of these parameters. According to Suzuki et al. [16], gentle and very shallow foreshore will result in flatter spectrum at the toe of the dike and thus spectral wave period $T_{m-1,0}$ is much longer than ones in deep water conditions due to infragravity waves contribution [17]. In such a situation, the waves have been transformed into bores and therefore overtopping characteristics, namely, flow pattern on dikes /promenades, might be also different from one which toe is at deep water. However, not so many studies have been conducted on the flow characteristics on dikes with gentle and shallow foreshores and discussed the associated risk.

The purpose of this study is to investigate the relationship between the mean discharge, q, and other overtopping parameters (V_{max}, h_{max}, u_{max}, V, h, u) on the dike with and without a vertical structure (i.e., a sea wall or a building) with a gentle and shallow foreshore, and eventually to discuss the proper assessment method for overtopping waves. To this end, a non-hydrostatic wave-flow modelSWASH (an acronym of Simulating WAves till SHore) [18] is employed in this study. The model has been validated for the case of wave overtopping over the impermeable dikes with gentle and shallow foreshore configuration [16]. To ensure the applicability of the model to this study, relevant physical model test results from the Climate Resilient Coast project (CREST, http://www.crestproject.be/) are employed for further validation. Using the validated model, flow characteristics on a wide range of different hydraulic and topographic conditions are further investigated. Note that SWASH can provide not only time dependent wave surface elevation but also velocity field. By post-processing, it is possible to calculate individual overtopping volumes and average overtopping discharge too. Obtaining such outputs, especially the velocity fields on the dike, is not an easy task in physical models since the velocity measurement points are exposed to wet and dry conditions (when overtopping happens the bottom becomes wet while in the other moments the bottom is in general dry) which is often a problem for velocimeters and thus numerical simulation is a good alternative to study overtopping hazard.

2. Methods

2.1. SWASH

SWASH is based on non-linear shallow water equation with non-hydrostatic pressure terms. The model can be run either in depth averaged mode or multi-layer mode. It is possible to maintain frequency dispersion by increasing the number of layers. A model with two or three layers already provides enough accuracy in terms of the frequency dispersion for most of the coastal applications. Combining with HFA (hydrostatic front approximation [19]) the model can deal with wave breaking with enough accuracy even in such a limited number of vertical layers. On top, non-linear wave properties under breaking waves (e.g., asymmetry and skewness) are preserved. See more details in [18].

The features that SWASH offers, namely maintaining a good accuracy of wave transformation and overtopping and is computationally not too demanding, are important factors for this study since the model needs to capture the overtopping process on the sea-dike and at the same time it is necessary to repeat calculations with different bathymetries, different water levels and wave conditions. SWASH is computationally less demanding and thus it is easy to run a long duration (i.e., 1000 waves) and a large number of calculations. One drawback of the SWASH model is that it cannot deal with a complex structure such as a parapet.

2.2. Model Settings

All the simulations are carried out in 2DV (two dimensional vertical). The version of the model applied in this study is version 5.0. The grid size in the horizontal direction is 0.5 m in the prototype scale as recommended in [20] which ensures a good wave propagation and overtopping processes. The threshold water level (DEPMIN) is set 0.001 m for the prototype calculation, which increases the computational stability compared to the default value. Two layers of equidistant layer distribution are employed in this simulation in order to maintain good frequency dispersion and the accuracy of second order wave generation [21], so-called infragravity waves, which plays an important role for wave run-up and overtopping process. Note that it is still possible to use one layer in light of linear dispersion since the kh value of the test is less than 2.9 as indicated in the user manual. However, two layers are better for the accuracy of the wave generation and propagation. Internal wave generation [22] has not been used in this study since the reflection from the structure is very limited in this case.

As for the numeric, the Keller-box scheme is used for the simulation since the number of the vertical layer is two. ILU (incomplete lower–upper factorization) preconditioner is employed for the computational robustness.

The momentum scheme is moment conservative. The standard first order up-wind scheme is used for the discretization of the vertical term for w-momentum equation for the sake of stability of the computation, while other discretization (i.e., the horizontal and vertical terms for u-momentum equation and the horizontal term for w-momentum equation) used MUSCL (monotonic upstream-centered scheme for conservation laws) limiter to achieve second order accuracy. Time integration is explicit and a maximum Courant number of 0.5 is used to cope with high and nonlinear waves used in this study.

The Manning formula with a Manning coefficient of 0.019 $m^{-1/3}s$ is employed to represent bottom friction for the entire domain, both for sandy beach and the dike. Note that 0.019 $m^{-1/3}s$ is the recommended value for wave simulations in the user manual. This must be due to the fact that the Manning's coefficient for sand (e.g., the grain size of 0.3–0.4 mm) is around this value. For the dike it is assumed that the bottom of the promenade is often like unfinished concrete, and which Manning's coefficient is around 0.014–0.020 $m^{-1/3}s$ and thus 0.019 $m^{-1/3}s$ should be an acceptable choice. Standard wave breaking control parameters, alpha = 0.6 and beta = 0.3, are used for wave breaking, and those values are also used in [16].

2.3. Test Matrix

As stated earlier, the purpose of the numerical experiment in this study is to understand the overall overtopping flow characteristics on the dike and in front of / inside buildings for a wide range of the overtopping discharge values. Therefore, the test matrix is designed to be able to obtain 4 different orders of magnitude of wave overtopping discharges, namely 0.1, 1, 10, and 100 l/s/m. The range of average overtopping discharges is achieved by changing the input hydraulic conditions (i.e., water levels, offshore significant wave height) and bathymetries (i.e., toe level, dike crest level, promenade width). The results will be further processed to discuss which physical parameter is relevant to the risk of people at the coast. In total 96 cases for each configuration (Q and W) of the numerical experiment have been conducted. For all the numerical experiments, a fixed seed number is used, and the number of waves is 1000.

Table 1 shows the test matrix. The case name is specified according to the input conditions, e.g., RSK_Q_7_3_12_65_85_00. See the next section for the detailed setting of the bathymetry.

Table 1. Variation of test parameters and the values.

Name [-]	Bathymetry [-]	Water Level [m]	H_{m0} [m]	T_p [m]	Toe Level [m]	Dike Crest Level [m]	Promenade Width [m]
RSK	Q	22 (7) [1]	3	12	21.5 (6.5) [1]	23.5 (8.5) [1]	0
	W	23 (8) [1]	4		21.9 (6.9) [1]	24.0 (9.0) [1]	20
			5			24.5 (9.5) [1]	
						25.0 (10.0) [1]	

[1] The value inside the brackets is based on [m TAW] (Tweede Algemene Waterpassing; Belgian standard datum level, situated near MLLWS) and the value is reflected in the case name.

2.4. Bathymetry

The level of the flat bottom in front of the wave generator is 0 m (−15 m TAW) and the length is 200 m: it is slightly longer than one offshore wave length. The foreshore slope is fixed at 1/35 up to the dike toe at 21.5 and 21.9 m (6.5 and 6.9 m TAW). The slope of the dike is 1/2 and the promenade is 1/50. This is the base bathymetry which is applicable to both configurations (i.e., bathymetry Q and W).

Tests with bathymetry Q is aimed to obtain flow and overtopping properties (i.e., h, u, V, q) at the end of the promenade for both 0 m and 20 m, and thus no vertical wall at the end of them. Tests with bathymetry W is aimed to obtain overtopping flow properties (i.e., h, u) in front of a vertical wall on the promenade.

Figure 1 shows the sketch of each bathymetry Q and W with different promenade width. The diamond points indicate the measurement points of flow and overtopping properties.

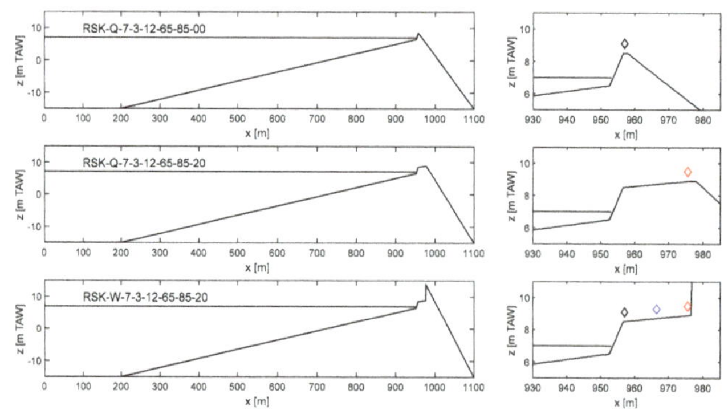

Figure 1. Bathymetries (type Q and W) and measurement points (black points: beginning, blue point: middle, red points: end of the promenade).

2.5. Post-Processing

In SWASH, two equidistant layers are used for this simulation and the flow parameter u to be used in this study is the averaged value of the two velocities of the two layers.

In order to calculate the individual overtopping volume, a water level criterion of 0.01 m is applied as a threshold. When the overtopping flow depth does not exceed this value, the overtopping is not counted as one overtopping event.

V, h and u are the time dependent parameters, however, in order to discuss overtopping hazard, maximum values during 1000 waves' test are used and they are expressed as V_{max}, h_{max} and u_{max}, respectively. The values are the maximum ones, so it is sensitive to the exceedance probability: when a lower number of waves are applied, the maximum value will be lower.

2.6. Physical Model

To validate the numerical model, we employed data from a physical model test campaign carried out in Belgium within the framework of the CREST project, see also the details in [23]. All the relevant data set to this study comprise average overtopping discharge and individual volume measured at the end of dike slope (i.e., only promenade width 0 m). Since the SWASH model to be used in this study is run in 2DV (i.e., flume like configuration) based on the second order wave generation, only cases with long crested waves and second order generation from the 3D wave basin physical model are used for the validation. The bottom configuration of the physical model is not exactly the same as one in numerical model (Figure 2), however the main features of the physical model test such as the water depth and the main bottom slope and the dike slope are the same as this numerical experiment. The numerical experiment has a wide range of different configurations (i.e., toe level × 2; dike crest level × 4, and promenade width × 2) and it includes the case of physical model (toe 6.5 m TAW, and dike crest 9.0 m TAW, promenade 0 m).

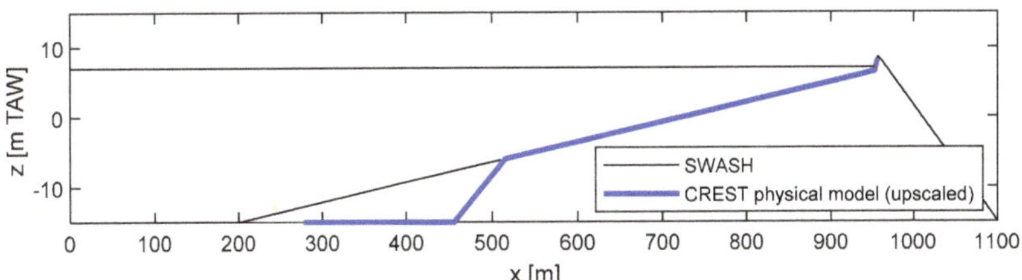

Figure 2. Comparison of the cross section of foreshore and dike profile between the SWASH model and the CREST physical model.

3. Results

3.1. Validation

The physical model test results of maximum individual overtopping volume V_{max} (data only limited to long crested and second order wave generation cases) are further processed and linked to the average overtopping discharge, see Figure 3. See [24] for further details of the data processing.

As can be seen in the figure the maximum individual overtopping volumes estimated by SWASH are in line with the physical model data. As shown in [25], SWASH can represent not only mean overtopping discharge over the dikes, but also wave run-up processes such as overtopping flow depth and velocity on the promenade, and wave force acting on a vertical wall on a dike. Therefore it can be concluded that the SWASH model is accurate enough. It is possible to explore further the wave overtopping characteristics on a dike with gentle and shallow foreshores based on the model.

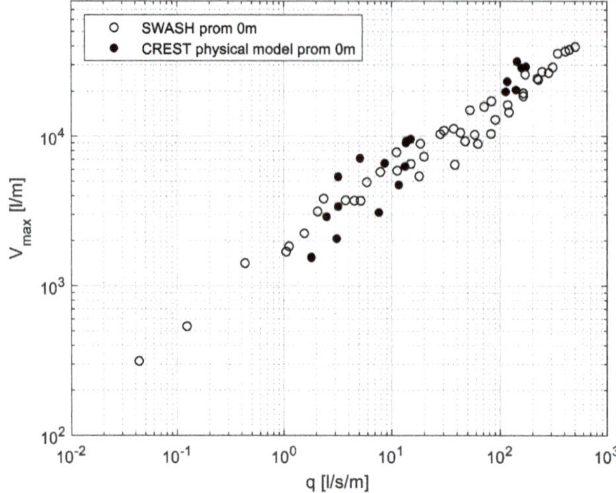

Figure 3. Comparison between SWASH and physical model on q-V_{max} (Average overtopping discharge-maximum individual overtopping volume) for the case of promenade width 0 m.

3.2. *Overtopping Flow Characteristics on a Promenade (without a Vertical Wall)*

In this section, overtopping flow properties on a promenade is investigated by SWASH. The used bathymetry is type Q, and thus there is no wall at the end of the promenade. This numerical experiment gives insights concerning how overtopping flow properties which are not disturbed by a vertical wall behave on the slightly sloped promenade.

3.2.1. q-V_{max} Relationship

The relationship between q and V_{max} (maximum individual volume) for the cases with promenade width 0 and 20 m is shown in Figure 4.

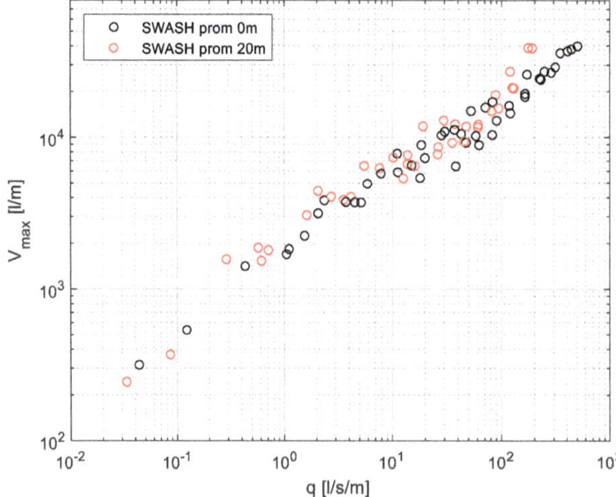

Figure 4. Comparison between promenade width 0 m and 20 m on q-V_{max} (Average overtopping discharge-maximum individual overtopping volume).

As shown in Figure 4, 1 l/s/m gives maximum individual volume V_{max} around 2000 l/m, and 10 l/s/m gives V_{max} around 6000 l/m for both promenade cases: there is no significant difference between the two promenade widths. From this result it can be concluded that V_{max} is determined by q in the gentle and shallow foreshore case and the promenade width does not make significant difference on q-V_{max} relationship for the wide range of the input hydraulic conditions and bathymetries. Even though Allsop et al. [5] indicated that the maximum individual overtopping volumes are more suitable hazard indicators, yet in this case V_{max} and q both give the same information. This might be due to the fact that the incident significant wave height in this shallow foreshore case is not significantly different at the toe of the dike (toe depth is 0.5 m) for different offshore wave conditions: wave height is limited by the shallow water depth.

3.2.2. q-h_{max} and q-u_{max} Relationships

Next, the relationship q-h_{max} (maximum overtopping flow depth) and q-u_{max} (maximum overtopping flow velocity) for the cases with promenade width 0 and 20 m are shown in Figure 5.

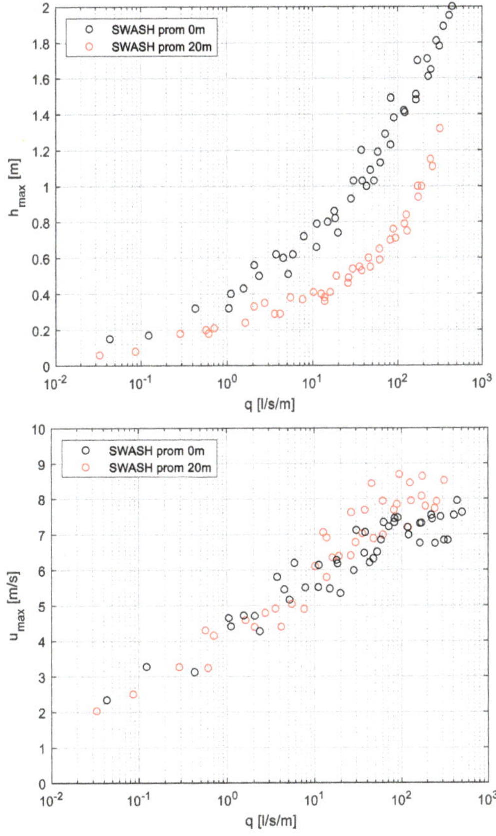

Figure 5. Comparison between promenade width 0 m and 20 m on q-h_{max} (**upper figure**) and q-u_{max} (**lower figure**).

As shown in the figure of q-h_{max}, the difference of h_{max} between promenade width 0 m and 20 m is significant: to up ~100 l/s/m the ratio is almost 2. Looking at the figure of q-u_{max}, the difference of the maximum velocity is not significant unless the highest overtopping discharges around ~100 l/s/m.

3.2.3. Time Evolution of Overtopping Flow Characteristics

It is interesting that on one hand the q-V_{max} gives very similar relationship between different promenade widths and on the other hand q-h_{max} shows a strong influence of the promenade width. In order to understand these differences, the time series of flow properties (time dependent overtopping flow depth h, velocity u and acceleration) under an overtopping event of similar V (both case around 1000 l/m, see Table 2) is visualized in Figure 6. Note that the V in this specific example is not V_{max} (maximum overtopping volume) in each case. In addition to u and h, the drag and inertia force acting on a person standing on the promenade is also calculated using the Morison equation since time evolution of the forces will be more relevant to the stability of a person standing on a promenade. In this case, two times of a cylinder with the diameter of 0.1 m are used to representing a person with two legs. Due to the nature of the equation, importance of u is higher than h (cfr. F is proportional to u^2 and h). As can be seen, the drag force is dominant and the inertia force is somewhat smaller in this case.

Table 2. Individual overtopping volumes for selected cases.

Case [-]	Promenade Width [m]	V [l/m]
RSK_7_5_12_69_00_00	0	1043
RSK_7_5_12_65_95_20	20	1109

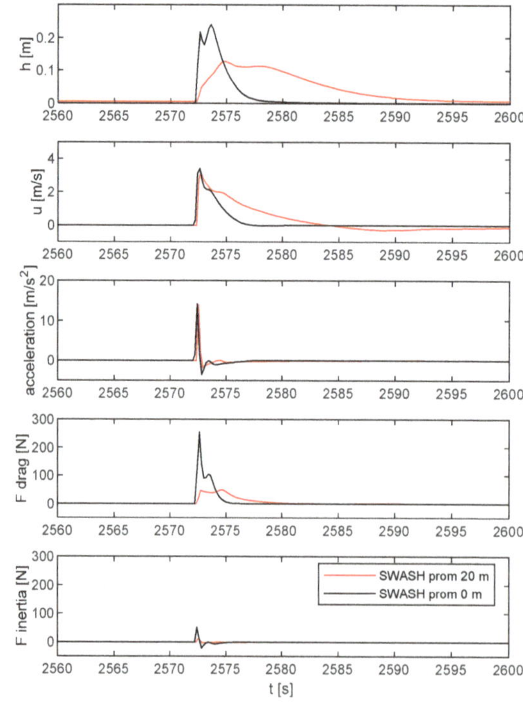

Figure 6. Time series of h, u, acceleration, F_drag and F_inertia for the case which V is around 1000 l/m (the red lines are slightly shifted to be able to compare with the black lines easily).

From Figure 6, it is obvious that the flow depth of the case with promenade width 0 m gives a higher peak while the flow duration is significantly different. The overtopping of the case with promenade width 20 m lasts about four times longer than one in promenade 0 m and this is how it gives the similar V. The overtopping flow depth of the overtopping waves is decreasing due to the gravity acting on the overtopped bore when it is propagating over the promenade. These relationships indicate that the overtopping flow depth and flow velocity will be more relevant to describe the overtopping hazard compared to the individual overtopping volume V, in the case of gentle and shallow foreshores.

Risk on pedestrians on the promenade is often evaluated by the overtopping flow depth and flow velocity [3–5]. Those are relevant parameters for the stability of a person exposed to the flows: the higher flow depth and flow velocity the lower the stability of a person. However, looking at Figure 6, one can see that the timing of the maximum layer thickness and layer velocity of the selected time window is different. This indicates that only the combination of the maximum values does not describe the hazard properly.

3.2.4. Overtopping Flow Characteristics and Stability

Stability is one of the key factors for the safety of the people. Endoh and Takahashi [3] discussed the human stability taking into account different human instability mode, slipping and tumbling. Sandoval and Bruce [4] revisited it taking into account the buoyancy and its position, and shows different criteria by ages and genders. The dashed line shown in Figure 7 is criterion for a tall adult. The criterion is expressed as a line by the combination of u and h. However, as explained earlier, in general u_{max} and h_{max} do not occur at the exactly same moment (there is a time-lag). If one wants to check stability properly, then it is advised to use a model which can describe the combination of u and h in a time series. The red line shows the time series of the u and h obtained at the end of the 20 m promenade from the SWASH model. Since it is based on 1000 waves, the line goes the same trajectory many times. The case shown in the figure corresponds to 16.2 l/s/m with V_{max} = 6491 l/m and the highest part the of the time dependent u-h line is located at the edge of the stability curve. In case the stability is evaluated by stand-alone h_{max} (horizontal red dotted line) in combination with stand-alone u_{max} (vertical red dotted line), then the hazard is overestimated as can be seen in the figure.

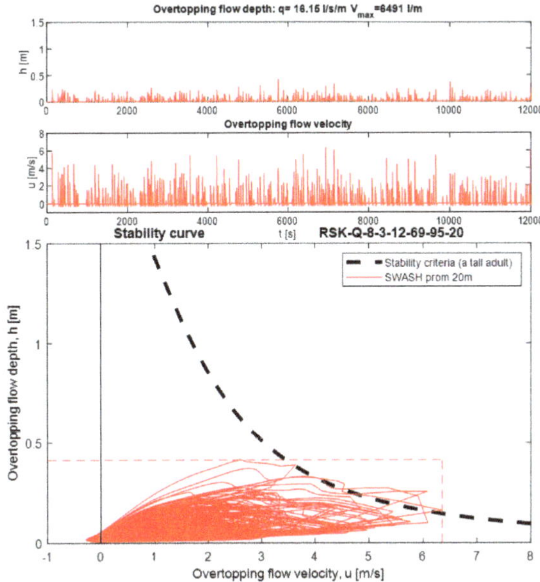

Figure 7. Time series of h and u (the first and second figures) and h-u relationship calculated in SWASH versus stability curve of a tall adult (the third figure): RSK_Q_8_3_12_69_95_20 00 (a case in which V is around 6500 l/s/m, and the promenade width of 0 m).

Figure 8 shows a case with very similar q and V_{max} (q = 15.0 l/s/m and V_{max} = 6535 l/m) but the promenade width is 0 m. In this case the h-u line exceeds the stability curve clearly and thus the risk is higher. This is the same observation as described in Section 3.2.3: the overtopping hazard is not always a function of the overtopping discharge nor maximum individual overtopping volume, but on the u and h (in the gentle and shallow foreshore case at least).

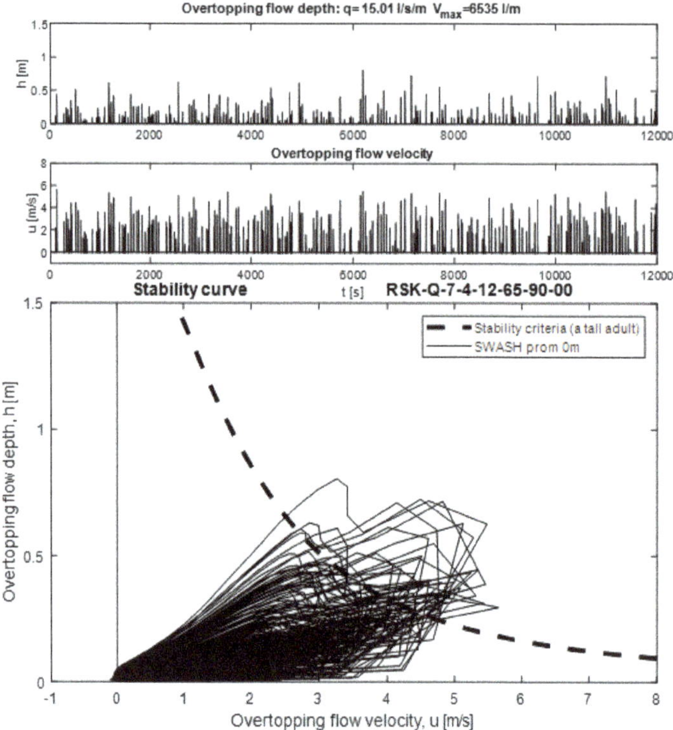

Figure 8. Time series of h and u (the first and second figures) and h-u relationship calculated in SWASH versus stability curve of a tall adult (the third figure): RSK_Q_7_4_12_65_90_00 (a case in which V is around 6500 l/s/m, and the promenade width of 0 m).

3.3. Overtopping Flow Properties in Front of a Vertical Wall

In case there is a vertical wall at the end of the promenade, waves are reflected at the wall and go back to the sea. Figure 9 shows the h-u time series of the same wave case in Figure 7 but with a vertical wall at the end of the promenade. Note that the h-u output point is just in front of the vertical wall. As can be seen in the figure, the water level (i.e., h) in front of the wall becomes very high due to the reflection. The height becomes more than two times of the one in the case without a vertical wall. Actually, the incident waves are not any more the shape of the wave but a bore, and thus the wave height can exceed two times of the incident wave height (cfr. standing wave). Especially at the end of the promenade, the duration of the bore becomes much longer (e.g., four times longer than one in the case of promenade width 0 m), as explained in Section 3.2.2. Thus the flow of the long bore locks up the water mass in front of the vertical wall and eventually the water level becomes much higher (the highest level h~1.3 m) than the incident bore height (the highest level h~0.5 m). During this process, the h-u line exceeds the stability curve significantly. It is an example of extra possible risk in the overtopping on the promenade: a structure can increase the hazard. When it reached the highest water level, the velocity becomes around zero, and then the reflected waves go back to the sea as if it is a dam break flow.

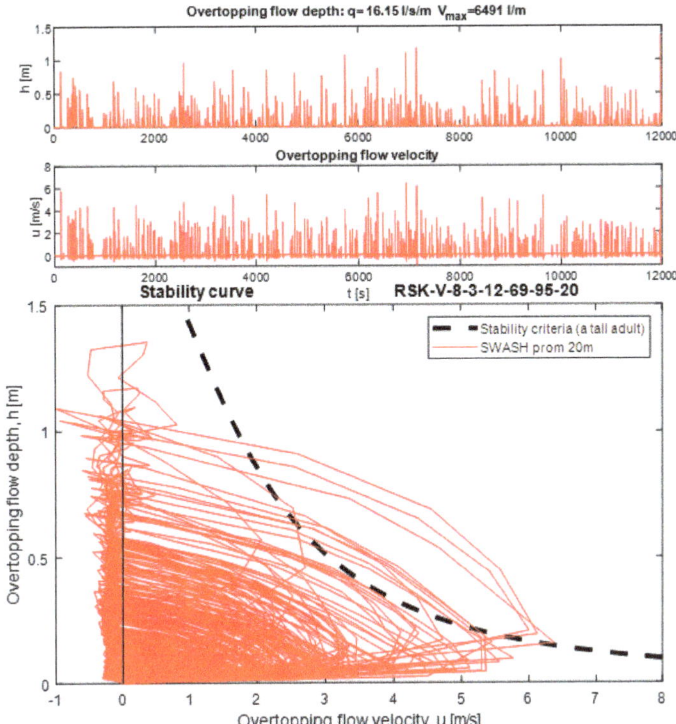

Figure 9. Time series of h and u (the first and second figures) and h-u relationship calculated in SWASH versus the stability curve of a tall adult (the third figure): RSK_W_8_3_12_69_95_20 (a vertical wall case corresponding to the case in which V is around 6500 l/s/m, and the promenade width of 20 m).

Figure 10 shows the two extra time series of h-u. The blue line shows h-u at the middle of the promenade and the black line shows h-u at the beginning of the promenade. These lines exceeded the stability curve for the incident bore since these are located more seaside. However extra attention is necessary at the h-u line for the return flow (i.e., line where u < 0 m/s). The negative velocity at the middle of the promenade exceeds 2 m/s and one at the beginning of the promenade exceeds 3 m/s in this case. Even though the return flow does not reach to the mirrored stability curve, but the values are not small: a person already fallen down due to the incident bore can be pulled into the sea by the return flow.

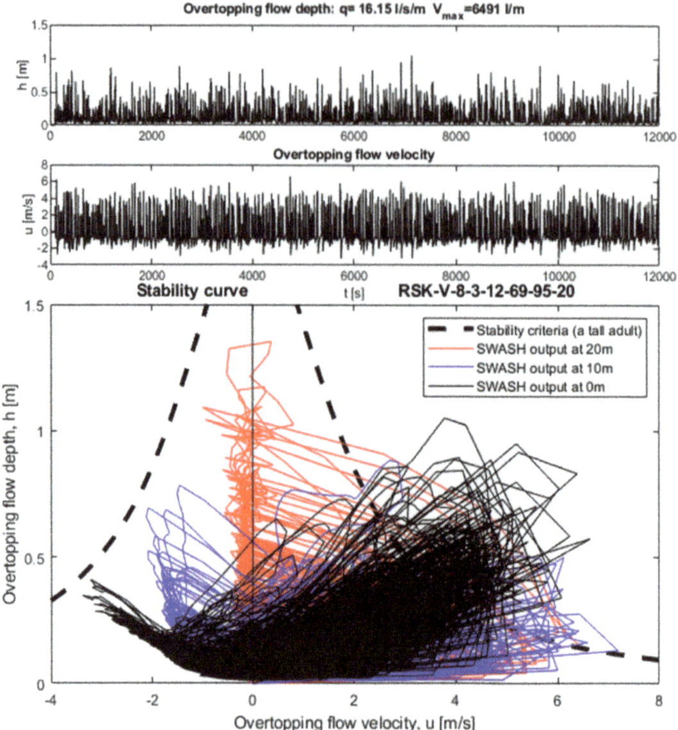

Figure 10. Time series of h and u (the first and second figures) and h-u relationship calculated in SWASH versus the stability curve of a tall adult (the third figure) at 3 output points: RSK_W_8_3_12_69_95_20 (a vertical wall case corresponding to the case in which V is around 6500 l/s/m, and the promenade width of 20 m).

4. Discussion

4.1. On Accuracy of the Model

In this study, the validated model is used to derive u and h. The accuracy of the model in terms of wave transformation and run-up on the dike is confirmed in [16,25]. Looking into [16], the overtopping estimation has a certain scatter especially when q is small. However, further validation using the CREST data confirmed the relationship between q and V_{max}, which shows an excellent match to the physical model test data. Note that the accuracy of h and u in time domain has not been confirmed yet in the present study.

The modelling is conducted in 2DV and thus the directional spreading effect is not included. The wave propagation and interaction process with the reflected waves in 3D is different from in 2DV as shown in [23]. The directional spreading effect is twofold – one is wave transformation from offshore to the toe, and the other one is from the toe to the end of the dike slope where overtopping is measured [6]. In this study mean overtopping discharge q and maximum individual volume V_{max} are compared for promenade 0 m and 20 m cases. However, the influence of the directional spreading is expected to be limited because the relation between q and V_{max} is similar. The green points depicted in Figure 11 show the q-V_{max} relationship from the CREST physical model where directional spreading is greater than 0 degree (i.e., 12, 16, 20, and 31.5 degrees). As can be seen, the green points are shown in the cloud of the black points (i.e., directional spreading is zero) while the majority of the green points are located in the lower part of the entire cloud. Strictly speaking directional spreading effect on q-V can be different for the case of 20 m promenade since the oblique wave can make the overtopping trajectory longer than one of perpendicular attack, namely the effective promenade width will be longer. When the width of the promenade is longer, the V can be slightly smaller.

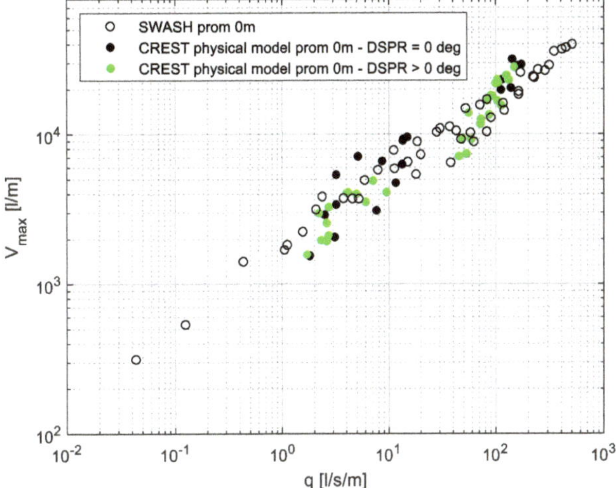

Figure 11. Comparison between DSPR = 0 and DSPR > 0 of physical models on q-V_{max} (Average overtopping discharge - maximum individual overtopping volume) for the case of promenade width 0 m.

4.2. On the Overtopping Parameters

Needless to say that q and V_{max} are still very important parameters to have the first idea to estimate how severe will be the overtopping event. However, time dependent value u and h are more relevant to understand the risk on the dike in details as shown in this study. These parameters have a direct link to the stability of a person. On top of these parameters, acceleration (a) and overtopping event duration (t_o) can be also parameters to give extra information to characterize the overtopping flow on a dike: acceleration in combination with h can give extra forcing of inertia and duration is also a useful parameter to understand how the overtopping is distributed in time.

4.3. On the Risk

Oppenheimer et al. [26] identified six sea level rise (SLR) scenarios. Some of them indicated to have structures in front of the properties to defend from the SLR. From the present study, it became clear that the time dependent h and u needs to be evaluated on top of average overtopping discharge and V_{max}, in order to understand the risk. Apparently the influence of the promenade is positive in a sense that it reduces not only q but also h, as also indicated in [6]. Eventually, the forcing on pedestrians, vehicles, and structures will be reduced due to the effect of the promenade. The effect can be strengthened if extra obstacles are placed on top of the promenade, for instance sea walls and vegetation. The key will be how to reduce q and V_{max} and also make overtopping event duration (t_o) longer, so that h-u line stays in a small range.

The forcing on the structure is governed by the hydrodynamics [27]. In this study, the façade is assumed to be broken. However, if the façade is strong enough the safety of the people inside the building is maintained. According to Streicher [28], the force acting on a vertical wall on the promenade can be the quasi-hydrostatic in the gentle and shallow foreshore case. Therefore, the force in such case can be estimated roughly if h_{max} is known. One of the effective evacuation strategies is the vertical evacuation, however in order to make sure this evacuation method is safe, first the stability of the building needs to be guaranteed. That is the reason why estimation of force acting on a building is necessary. Suzuki et al. [25] indicated that SWASH is also capable to estimate F on a vertical structure, but it is not explored in this study. In case detailed hydrodynamics needs to be obtained for complex structures, detailed hydrodynamic modelling (e.g., [29,30]) will be an alternative.

5. Conclusions

In this study, overtopping risk on a dike with gentle and shallow foreshores is investigated using SWASH, a NLSW equation solver. The model has been validated in different studies applied for shallow foreshores but it is further validated in terms of maximum individual volume based on a physical model. One of the benefits to use SWASH in this study is that the model can output h and u in time series while measurement of u on a dike in physical model test is often a challenge. Using SWASH, the risk on the dike can be evaluated in details, in function of time. On top, SWASH is relatively a light wave model, and thus it is possible to obtain overtopping flows in different wave conditions and bathymetries: in this study total ~200 cases are simulated.

It is often the case in practice that the coastal safety is evaluated by the average overtopping discharge and maximum individual volume V_{max}. However, it becomes clear from this study that overtopping risk is not only characterized by q and V: time dependent h and u are also useful and even better parameters to characterize risks on dikes more in details. For instance, two cases in the example of this study show different h, even though the two cases show very similar q and V_{max}. This was due to the influence of the promenade which made h smaller and the duration longer. It is noted that the combination of stand-alone h_{max} and u_{max} can lead an overestimation of the hazard and therefore time dependent h and u are better for the proper assessment.

In addition to the overtopping flow on plain dikes, the influence of a vertical wall at the end of the promenade is also evaluated in this study. The results show that the vertical wall can influence on the people's safety on the promenade in a negative way. The bore can create a higher flow depth compared to the case without a vertical wall and it becomes an extra risk o. Depending on the position on the promenade, a relatively high negative velocity was also observed in the simulation.

Further study of the characterization of overtopping waves, such as cases with deep water conditions, storm walls on the dike, and vegetation in front of and on the dike [31], will be useful since a proper assessment of wave overtopping is an essential key for designing coastal structures which provides safety for people in coastal areas. Numerical modelling is a strong tool to evaluate risks in different scenarios.

Author Contributions: Conceptualization, T.S.; methodology, T.S., C.A., and T.Y.; analysis, T.S.; investigation, T.S.; writing—original draft preparation, T.S.; writing—review and editing, C.A., T.Y, and T.V.; visualization, T.S.; supervision, T.V.; project administration, T.V. All authors have read and agreed to the published version of the manuscript.

Funding: This research was part of the CREST (Climate REsilient CoaST) project (http://www.crestproject.be/en), funded by the Flemish Agency for Innovation by Science and Technology, grant number 150028. Corrado Altomare acknowledges funding from the European Union's Horizon 2020 research and innovation programme under the Marie Sklodowska-Curie grant agreement No.: 792370.

Conflicts of Interest: The authors declare no conflict of interest. The funders had no role in the design of the study; in the collection, analyses, or interpretation of data; in the writing of the manuscript, or in the decision to publish the results.

References

1. Weisse, R.; von Storch, H.; Niemeyer, H.D.; Knaack, H. Changing North Sea storm surge climate: An increasing hazard? *Ocean Coast. Manag.* **2012**, *68*, 58–68. [CrossRef]
2. Neumann, B.; Vafeidis, A.T.; Zimmermann, J.; Nicholls, R.J. Future coastal population growth and exposure to sea-level rise and coastal flooding—A global assessment. *PLoS ONE* **2015**, *10*. [CrossRef] [PubMed]

3. Endoh, K.; Takahashi, S. Numerically Modeling Personnel Danger on a Promenade Breakwater Due to Overtopping Waves. In Proceedings of the 24th International Conference on Coastal Engineering, Kobe, Japan, 23–28 October 1994; pp. 1016–1029.
4. Sandoval, C.; Bruce, T. Wave overtopping hazard to pedestrians: Video evidence from real accidents. In *Coasts, Marine Structures and Breakwaters 2017*; Thomas Telford Ltd.: Liverpool, UK, 2018; pp. 501–512.
5. Allsop, N.W.H.; Bruce, T.; Pullen, T.; van der Meer, J. Direct Hazards from Wave Overtopping—The Forgotten Aspect of Coastal Flood Risk Assessment? In Proceedings of the 43rd Defra Flood and Coastal Management Conference, Manchester, UK, 1–3 July 2008; pp. 1–11.
6. Altomare, C.; Gironella, X.; Suzuki, T.; Viccione, G.; Saponieri, A. Overtopping metrics and coastal safety: A case of study from the catalan coast. *J. Mar. Sci. Eng.* **2020**, *8*, 556. [CrossRef]
7. Arrighi, C.; Oumeraci, H.; Castelli, F. Hydrodynamics of pedestrians' instability in floodwaters. *Hydrol. Earth Syst. Sci.* **2017**, *21*, 515–531. [CrossRef]
8. Arrighi, C.; Pregnolato, M.; Dawson, R.J.; Castelli, F. Preparedness against mobility disruption by floods. *Sci. Total Environ.* **2019**, *654*, 1010–1022. [CrossRef]
9. Schüttrumpf, H.; Oumeraci, H. Scale and Model Effects in Crest Level Design. In Proceedings of the 2nd Coastal Symposium, Höfn, Iceland, 5–8 June 2005; pp. 1–12.
10. Nørgaard, J.Q.H.; Lykke Andersen, T.; Burcharth, H.F.; Steendam, G.J. Analysis of overtopping flow on sea dikes in oblique and short-crested waves. *Coast. Eng.* **2013**, *76*, 43–54. [CrossRef]
11. Mares-Nasarre, P.; Argente, G.; Gómez-Martín, M.E.; Medina, J.R. Overtopping layer thickness and overtopping flow velocity on mound breakwaters. *Coast. Eng.* **2019**, *154*. [CrossRef]
12. van Bergeijk, V.M.; Warmink, J.J.; van Gent, M.R.; Hulscher, S.J. An analytical model of wave overtopping flow velocities on dike crests and landward slopes. *Coast. Eng.* **2019**, *149*, 28–38. [CrossRef]
13. Chen, X.; Jonkman, S.N.; Pasterkamp, S.; Suzuki, T.; Altomare, C. Vulnerability of buildings on coastal dikes due to wave overtopping. *Water (Switz.)* **2017**, *9*, 394. [CrossRef]
14. Van der Meer, J.W.; Allsop, N.W.; Bruce, T.; De Rouck, J.; Kortenhaus, A.; Pullen, T.; Schüttrumpf, H.; Troch, P.; Zanuttigh, B. *Manual on Wave Overtopping of Sea Defences and Related Structures. An Overtopping Manual Largely Based on European Research, but for Worldwide Application*, 2nd ed.; EurOtop: London, UK, 2018.
15. Altomare, C.; Suzuki, T.; Chen, X.; Verwaest, T.; Kortenhaus, A. Wave overtopping of sea dikes with very shallow foreshores. *Coast. Eng.* **2016**, *116*, 236–257. [CrossRef]
16. Suzuki, T.; Altomare, C.; Veale, W.; Verwaest, T.; Trouw, K.; Troch, P.; Zijlema, M. Efficient and robust wave overtopping estimation for impermeable coastal structures in shallow foreshores using SWASH. *Coast. Eng.* **2017**, *122*. [CrossRef]
17. Lashley, C.H.; Bricker, J.D.; Bricker, J.D.; Van Der Meer, J.; Van Der Meer, J.; Altomare, C.; Altomare, C.; Suzuki, T.; Suzuki, T. Relative Magnitude of Infragravity Waves at Coastal Dikes with Shallow Foreshores: A Prediction Tool. *J. Waterw. Port Coast. Ocean Eng.* **2020**, *146*, 1–17. [CrossRef]
18. Zijlema, M.; Stelling, G.; Smit, P. SWASH: An operational public domain code for simulating wave fields and rapidly varied flows in coastal waters. *Coast. Eng.* **2011**, *58*, 992–1012. [CrossRef]
19. Smit, P.; Zijlema, M.; Stelling, G. Depth-induced wave breaking in a non-hydrostatic, near-shore wave model. *Coast. Eng.* **2013**, *76*, 1–16. [CrossRef]
20. Suzuki, T.; De Roo, S.; Altomare, C.; Zhao, G.; Kolokythas, G.K.; Willems, M.; Verwaest, T.; Mostaert, F. *Toetsing Kustveiligheid 2015–Methodologie: Toetsingsmethodologie Voor Dijken en Duinen*; WL Rapporten; 10.0; Waterbouwkundig Laboratorium: Antwerpen, Belgium, 2016.
21. Rijnsdorp, D.P.; Smit, P.B.; Zijlema, M. Non-hydrostatic modelling of infragravity waves under laboratory conditions. *Coast. Eng.* **2014**, *85*, 30–42. [CrossRef]
22. Vasarmidis, P.; Stratigaki, V.; Suzuki, T.; Zijlema, M.; Troch, P. Internal wave generation in a non-hydrostatic wave model. *Water (Switz.)* **2019**, *11*, 986. [CrossRef]
23. Altomare, C.; Suzuki, T.; Verwaest, T. Influence of directional spreading on wave overtopping of sea dikes with gentle and shallow foreshores. *Coast. Eng.* **2020**, *157*, 103654. [CrossRef]
24. Tortora, S. Statistical characterisation of overtopping volumes for sea dikes in very shallow foreshore condition under short- and long-crested waves action. Master Thesis, University of L'Aquila, Rome, Italy, 2018.
25. Suzuki, T.; Altomare, C.; De Roo, S.; Vanneste, D.; Mostaert, F. *Manning's Roughness Coefficient in SWASH: Application to Overtopping Calculation*; FHR reports; Version 2; Flanders Hydraulics Research: Antwerp, Belgium, 2018.

26. Oppenheimer, M.; Glavovic, B.; Hinkel, J.; van de Wal, R.; Magnan, A.K.; Abd-Elgawad, A.; Cai, R.; Cifuentes-Jara, M.; Deconto, R.M.; Ghosh, T.; et al. Sea Level Rise and Implications for Low-Lying Islands, Coasts and Communities. In *IPCC Special Report on the Ocean and Cryosphere in a Changing Climate*; Pörtner, H.-O., Roberts, D.C., Masson-Delmotte, V., Zhai, P., Tignor, M., Poloczanska, E., Mintenbeck, K., Alegría, A., Nicolai, M., Okem, A., et al., Eds.; In press.
27. Chen, X.; Hofland, B.; Altomare, C.; Suzuki, T.; Uijttewaal, W. Forces on a vertical wall on a dike crest due to overtopping flow. *Coast. Eng.* **2015**, *95*, 94–104. [CrossRef]
28. Streicher, M.; Kortenhaus, A.; Marinov, K.; Hirt, M.; Hughes, S.; Hofland, B.; Scheres, B.; Schüttrumpf, H. Classification of bore patterns induced by storm waves overtopping a dike crest and their impact types on dike mounted vertical walls—A large-scale model study. *Coast. Eng. J.* **2019**, *61*, 321–339. [CrossRef]
29. Gruwez, V.; Altomare, C.; Suzuki, T.; Streicher, M.; Cappietti, L.; Kortenhaus, A.; Troch, P. Validation of RANS Modelling for Wave Interactions with Sea Dikes on Shallow Foreshores Using a Large-Scale Experimental Dataset. *J. Mar. Sci. Eng.* **2020**, *8*, 650. [CrossRef]
30. Altomare, C.; Crespo, A.J.C.; Domínguez, J.M.; Gómez-Gesteira, M.; Suzuki, T.; Verwaest, T. Applicability of Smoothed Particle Hydrodynamics for estimation of sea wave impact on coastal structures. *Coast. Eng.* **2015**, *96*, 1–12. [CrossRef]
31. Suzuki, T.; Hu, Z.; Kumada, K.; Phan, L.K.; Zijlema, M. Non-hydrostatic modeling of drag, inertia and porous effects in wave propagation over dense vegetation fields. *Coast. Eng.* **2019**, *149*, 49–64. [CrossRef]

Journal of
Marine Science and Engineering

Article

Interaction of a Solitary Wave with Vertical Fully/Partially Submerged Circular Cylinders with/without a Hollow Zone

Chih-Hua Chang [1,2]

[1] Department of Information Management, Ling-Tung University, Taichung 408, Taiwan; changbox@teamail.ltu.edu.tw
[2] Natural Science Division in General Education Center, Ling-Tung University, Taichung 408, Taiwan

Received: 22 November 2020; Accepted: 11 December 2020; Published: 15 December 2020

Abstract: In this article, a three-dimensional, fully nonlinear potential wave model is applied based on a curvilinear grid system. This model calculates the wave action on a fully/partially submerged vertical cylinder with or without a hollow zone. As basic verification, a solitary wave hitting a single fully or partially submerged circular cylinder is tested, and our numerical results agree with the experimental results obtained by others. The influence of cylinder immersion depth and size on the wave elevation change on the cylinder surface is considered. The model is also applied to investigate the wave energy of a solitary wave passing through a hollow circular cylinder to determine the effect of the size and draft on the wave oscillating in the hollow zone.

Keywords: solitary wave; fully nonlinear wave; three-dimensional wave; partially submerged cylinder; hollow circular cylinder

---

1. Introduction

Upright pillars are often used as the foundation of marine platforms. An abundance of studies have been conducted on the waves diffracted by vertical circular cylinders. An early focus was the analysis of linear waves diffracted by vertical cylinders in water. To address this topic, many researchers considered waves in water of infinite depth, including the early studies by Havelock [1], Ursell [2], and MacCamy and Fuchs [3], and the more recent work of Finnegan et al. [4]. Other researchers have focused on considering situations in water of finite depth for bottom-mounted, fixed-floating, or free-floating cylinders, including Miles and Gilbert [5], Garrett [6], Black et al. [7], Molin [8], Yeung [9], Sabuncu and Calisal [10], Williams and Demirbilek [11], Bhatta and Rahman [12], Bhatta and Rahman [13], Bhatta [14], Jiang et al. [15], Li and Liu [16], and Ghadimi et al. [17]. In addition, a twin-cylinder modeled as a wave energy converter device was analyzed by Xu et al. [18]. These studies have focused on the analysis of linear waves with circular cylinders. When considering nonlinear waves, it is relatively difficult to find analytical solutions. Taylor and Hung [19,20] investigated the second-order diffraction forces on a vertical cylinder for regular waves or bichromatic waves. A closed-form solution was derived by Kriebel [21] for the velocity potential resulting from the interaction of second-order Stokes waves with a large vertical circular cylinder. The phenomenon of nonlinear waves with a circular cylinder has been analyzed primarily by numerical (e.g., Kim [22]) or experimental (e.g., Yan [23]) methods.

When waves approach the shore, nonlinear and shallow-water wave characteristics will predominate. To protect coastal facilities, engineers must understand the influence of shallow-water waves. One frequently used nonlinear water wave model is the solitary-wave model. To simplify this physical problem, it is simulated as a long wave with a single hump. In this article, we apply a numerical model to calculate the interaction of a solitary wave with an immersed vertical cylinder that is or is not hollow.

Many studies have been conducted using the depth-averaged method to determine the interaction of a three-dimensional (3D) solitary wave with vertical cylinders. A shallow-water wave concept and a depth-averaged model have been used to simplify this 3D problem into two dimensions. For example, Wang et al. [24] combined a generalized Boussinesq (gB) model with horizontal curvilinear coordinate grids to study the scattering of a solitary wave as it meets a circular cylinder. Later, Yates and Wang [25] used gauges located around a cylinder to conduct an experiment to measure solitary wave elevations and observe their evolution. Some researchers have investigated structures using cylinder arrays. For example, in their study, Wang and Jiang [26] used a gB model for dual cylinders. Zhao et al. [27] built a mesh system with Cartesian unstructured grids and an O-shaped body-fitted grid, and solved the gB equation using the finite element method to explore the interaction between solitary waves and either two or four cylinders that make contact with the seabed. Neil et al. [28] established curvilinear coordinate grids to solve gB and Green–Naghdi (GN) equations for situations involving a solitary wave passing through three cylinders arranged one behind the other. However, all the above studies were restricted by their application of Boussineq-like models. Experimental work on the interaction of solitary waves with multiple cylinders is rare. One example is the work by Yuan and Huang [29], who measured the solitary wave forces on an array of vertical cylinders.

To theoretically analyze the problem of a solitary wave versus a vertical circular cylinder, Isaacson [30] used diffraction theory. In recent years, with the advances in computer technology, 3D, fully nonlinear, inviscid, or viscous wave models have gradually been developed. Many researchers have numerically investigated the solitary waves radiated by circular cylinders. Isaacson [31] and Isaacson and Cheung [32] applied the boundary integral method to investigate the interaction of a solitary wave with a vertical circular cylinder. Ohyama [33] used the boundary element method (BEM) to solve potential 3D wave equations for a solitary wave acting on a huge vertical cylinder. Using a similar method with curvilinear grids, Yang and Ertekin [34] analyzed fully submerged cylinders hit by nonlinear waves, including solitary and Stokes waves. Later, other researchers began to consider the fluid viscous effect and used Navier–Stokes equations to solve this problem. For example, Wang and Huang [35] employed the SIMPLER algorithm [36] to solve Navier–Stokes equations and used the Marker and Cell method to trace free-surface particles in a solitary wave diffracted by a vertical square cylinder. Mo [37] solved Navier–Stokes equations and combined the Cartesian grid with a quadrangular irregular mesh surrounding a single cylinder or three cylinders. Cao and Wan [38] performed a numerical simulation based on the OpenFOAM code to solve Reynolds-averaged Navier—Stokes (RANS) equations for the run-up of a solitary wave on a circular cylinder with different incident wave heights and circular radii. Almost simultaneously, Leschka and Oumeraci [39] and Leschka et al. [40] applied the OpenFOAM code to simulate a solitary wave traveling through three or more cylinders in a staggered arrangement. A particle-in-cell Navier–Stokes solver was developed by Chen et al. [41] to simulate a solitary wave travelling through a single cylinder or a group of eleven cylinders. However, most of these papers discuss the diffraction of solitary waves by bottom-mounted cylinders.

In recent years, the development of offshore marine resources has required platforms for floating bodies, so attention has been paid to the study of waves affected by fixed or free-floating structures. Kang et al. [42] extended the σ-coordinate method (developed by Lin, [43]) to solve a 3D Navier–Stokes model using the immersed boundary method to deal with solid boundaries and to solve the problem of nonlinear waves interacting with a fully or partially submerged circular cylinder. A BEM approach was applied by Zhou et al. [44] to solve 3D Euler equations for fixed/floating and fully/partially submerged circular cylinders. In their report of the model results, the authors did not discuss the effect of the immersion draft. Chen and Wang [45] developed a 3D fully nonlinear wave model using multi-block grids to solve the problem of a solitary wave interacting with a half-submerged circular cylinder. These authors also measured the wave elevations near the front and rear of the cylinder for comparison with their numerical solutions. There has been scant discussion of the case of a cylinder with a gap between its bottom and the seabed.

This article also presents an analysis of the water-level evolutions of a solitary wave passing through a hollow circular cylinder. A wave passing through a hollow structure causes the water level to oscillate in the hollow region, which is relevant to the research of wave energy development. The pattern of waves arriving near shore is often similar to that of shallow water wavers (such as solitary waves or Cnoidal waves). Although a solitary wave is nonperiodic, the scenario can still be simulated as multiple yet individual solitary waves (solitons) that are continuously interacting with coastal structures. Wave energy can be converted into electrical energy in many ways, with the oscillating water column (OWC) being one of the most common. For example, a dielectric elastomer generator (DEG) device developed by Scottish and Italian scientists (https://newatlas.com/dielectric-elastomer-generator-wave-power/58465/) basically consists of an anchored vertical circular cylinder in which a column of air is trapped. The head of the cylinder is sealed with a rubber membrane, and the base is open to the surrounding ocean. The efficient conversion of the power of the water column to drive a propeller in the air chamber to obtain electrical energy is a structural design problem. Basically, if vertical two-dimensional (2D) consideration is taken, the OWC mechanism can be simplified as a problem of the interaction between a wave and two vertical piercing plates. However, this article focuses on the 3D problem and does not review the 2D literature.

In 1970, Garrett [46] was likely the first to explore analytical solutions for small-amplitude waves incident on a hollow cylinder partially immersed in water of finite depth. Zhu and Mitchell [47] revisited Garrett's work using a different approach, which had the advantage of requiring less work in the analysis. The authors also noted that there are two key considerations to determining the maximum wave energy of the OWC: The properties of the OWC chamber and the location of the OWC in the ocean. Later, Simon [48] and Miles [49] used a variational technique to analyze the radiation of surface waves from a submerged cylindrical duct. Harun [50] solved the mild-slope equation for the OWC problem to obtain an analytical solution for a linear long wave diffracted around a hollow cylinder. Recently, a twin-cylinder wave energy device was considered by Xu et al. [18]. Mavrakos's research group conducted a number of experimental and numerical studies of the hydrodynamic problem of the OWC device where regular waves act on an arrangement of a single concentric cylinder or a set of compound concentric cylinders [51–53]. Another possible OWC device that consists of two concentric cylindrical shells placed in water has been studied by Shipway and Evans [54] and McIver and Newman [55]. A floating hollow cylinder placed above a solid bottom cylinder was investigated by Hassan and Bora [56,57]. The type, shape, size, draft depth, and position of the chamber (such as its distance from shore) are known to affect the elevation of the water column in the hollow cylinder. Previous experience indicates that even a simple hollow cylinder with different design factors can influence the wave diffraction around the cylinder and the oscillation in the column. Understanding the wave oscillation behavior in a column is the most important issue for controlling the OWC. In recent years, many 3D OWC studies have been conducted. For example, Simonetti et al. [58] used OpenForm to calculate 3D Navier–Stokes equations by large-eddy simulation to investigate the problem of wave and OWC interaction. Kamath et al. [59] used REEF3D software (an open-source computational fluid dynamics program) to calculate a RANS model to explore the interaction between periodic waves and an OWC and to compare the differences between the 2D and 3D results. Lee et al. [60] also used REEF3D to calculate the interaction of regular waves with a 3D OWC device. These authors noted that the efficiency of the current OWC system requires further study by scientists in many fields. Kim et al. [61] conducted experimental work on regular waves passing through two concentric cylinders. They measured the wave heaves within the ring region defined by the cylinders and analyzed the interaction between a solitary wave and an OWC device. In addition to understanding the oscillation effect, with respect to the design of an OWC device for disaster prevention, it is important to simulate the impact of an extreme wave (solitary wave) on OWC facilities to understand the status of the impacting wave.

To this end, this article considers the influence of the height of the solitary wave, the size of the hollow cylinder, and the draft on the wave diffraction and wave oscillation in hollow regions.

Overall, when a wave propagates to the continental shelf, the shallow-water characteristics of the wave are significant. It is surprising that only a few studies have investigated the interaction of a solitary wave with cylinders with a draft effect. Thus far, it seems that the analysis of the interaction between a solitary wave and OWC is rare. Consequently, in this paper, a 3D fully nonlinear wave model is applied to analyze the cases of a solitary wave hitting a solid or hollow circular cylinder with different drafts. The results obtained in this study fill this knowledge gap. From an engineering point of view, this article also has two main applications: (1) analysis of a long wave encountering the circular support cylinder (pile) of a marine platform and (2) simulation of the heave and drop of water levels when a long wave passes through an OWC device (a hollow cylinder).

2. Flow-Field Equations

A 3D free-surface flow region with a vertical cylinder is considered, as illustrated in Figure 1. The fluid is assumed to be inviscid, incompressible, and the motion is irrotational. The model formulations are based on a set of fully nonlinear potential flow equations; all variables are made dimensionless by introducing the referencing length scale H^*, the undisturbed water depth, the velocity scale, $\sqrt{gH^*}$ (here g is a constant due to gravity), and $\sqrt{H^*/g}$ as the time scale. In this model, the right-hand referencing frame has its x-axis pointing in the positive (right) direction, the y-axis is expanding laterally, and the z-axis is pointing up; the coordinate origin is located at the water level of the undisturbed fluid region. The non-dimensionalized initial-boundary-value problem can be formulated as the governing equation, initial condition, and associated boundary conditions. These equations can be referenced in many books on classical wave mechanics, such as Stokers [62]. These equations are listed as follows:

$$\phi_{xx} + \phi_{yy} + \phi_{zz} = 0, \text{ in the flow-field region} \tag{1}$$

$$\phi_z = \zeta_t + \phi_x \zeta_x + \phi_y \zeta_y, \text{ at } z = \zeta \tag{2}$$

$$\phi_t + ({\phi_x}^2 + {\phi_y}^2 + {\phi_z}^2)/2 + \zeta = 0, \text{ at } z = \zeta \tag{3}$$

$$\Omega_t \pm \sqrt{(1+\zeta)}\Omega_x = 0, \text{ at x-direction lateral boundaries} \tag{4}$$

$$\frac{\partial \Omega}{\partial \overline{n}} = 0, \text{ on the bottom, sidewalls, and cylinder surface} \tag{5}$$

where ϕ is the velocity potential function. Subscript characters with the coordinates denote the partial differentiation; $z = \zeta$ is the wave surface, Ω stands for either the ϕ or ζ, the $\pm$ sign in Equation (4) indicates the lateral boundary used for right (+) and left (−) outgoing conditions, and $\overline{n}$ in Equation (5) is the solid surface unit normal vector of fluid. According to Schember [63] and Wang [24,64], the right-moving solitary waveform (ζ) and the depth-averaged potential function $\overline{\phi}$ are expressed as:

$$\zeta = \left\{\operatorname{sech}^2 k(x - Ct - X_0) + A_0 \operatorname{sech}^4 k(x - Ct - X_0)\right\}/(1 + A_0) \tag{6}$$

$$\overline{\phi}(x) = \sqrt{4A_0/3}\tanh k(x - Ct - X_0) \tag{7}$$

where $k = \sqrt{3A_0/[4(1+A_0)]}$, wave celerity is $C = \sqrt{1+A_0}$, A_0 = the incident wave height, and X_0 = the wave's starting position. Wu [65] derived the relation between the potential function and its average value in the flow region as:

$$\phi(x,z) = \overline{\phi} - A_0\left(\frac{1}{3} + z + \frac{z^2}{2}\right)\overline{\phi}_{xx} + O(\varepsilon^5) \tag{8}$$

in which, ε = water depth/wavelength. Substituting Equation (7) into Equation (8) for $t = 0$, we get:

$$\phi(x,z) = \sqrt{4A_0/3}\tanh k(x - X_0)\left\{1 + A_0k^2(\frac{2}{3} + 2z + z^2)\mathrm{sech}^2 k(x - X_0)\right\} \quad (9)$$

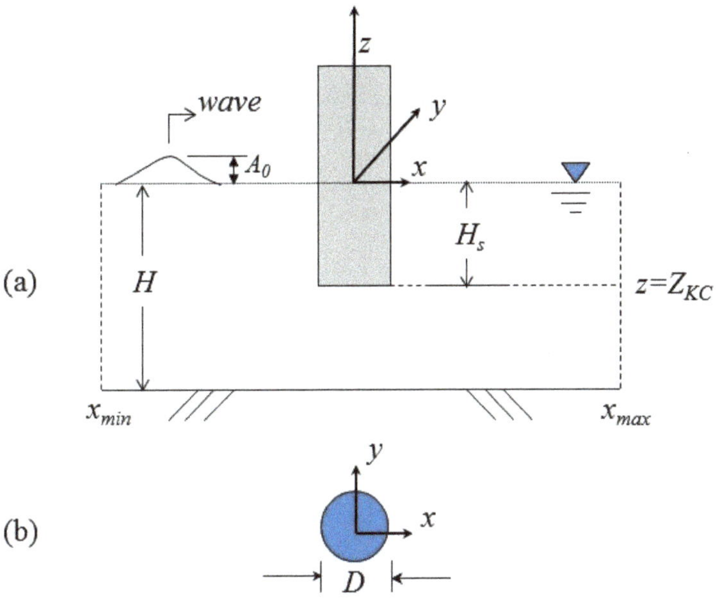

Figure 1. Schematic diagram of a wave passing through a vertical circular cylinder. (**a**) side view and (**b**) top view.

Therefore, Equation (6) for $t = 0$ and Equation (9) are specified as the initial wave profile and potential function in one x-z plane, respectively, which are equal in all the x-z planes and form the initial 3D conditions.

When calculating waves, there are many methods available for obtaining a 3D solution to the potential energy flow equation, such as the high-order spectral (HOS) method of Dommermuth and Yue [66], which has attracted much attention in recent years. Its advantage is that it can be used for the calculations of a large range of water types. As such, researchers have used the HOS method to calculate a large-scale wave field in combination with computer fluid dynamics to calculate details regarding areas of a wave that encounter a structure (Zhuang and Wan [67]). This article does not focus on innovations in calculation methods, and does not compare our proposed method with others. The numerical model of our work is based on the developmental work of Chang and Wang [68], wherein a transient boundary-fitted curvilinear grid system is used to conform to the moving free surface. The boundary-fitted curvilinear grid was first proposed by Thompson [69]. Here, the explicit–implicit hybrid finite difference scheme is adopted to solve the complete nonlinear free- surface boundary conditions, and discretely solve the internal flow field using the 15-point central difference method in each local computational element. For a detailed explanation of the grid generation process and the finite difference method applied in this numerical method, please refer to the Appendix A.

3. Validations

First, the structure of a single-bottomed mounted circular cylinder is considered, and the computation results are compared with those obtained in the experiment conducted by Mo [37]. In Mo's experimental setup, the water depth was 0.75 m, the solitary wave height was 0.3 m, and the cylindrical diameter was 1.22 m. In terms of the dimensionless scale of the water depth, the reference length is a water depth $H = 1$, an incident wave height $A_0 = 0.4$, and a cylindrical diameter $D = 1.63$. As the cylinder is bottom-mounted, $H = H_s = 1.0$. This case is typically adopted to test a 3D nonlinear water wave model. Mo [37] examined this case both experimentally and numerically, and provided abundant data for a solitary wave encountering a single cylinder or multiple piles. First, to verify our model, it is compared with that used by Mo. The calculation range is $(x_{max}, x_{min}) = (-30, 30)$ and $(y_{max}, y_{min}) = (-30, 30)$. This study used six wave-gauge positions (see Figure 2) around the cylinder (as shown in Figure 3). In Figure 3, t' is the time shift with respect to the peak time at Gauge 4. Although our model does not consider fluid viscosity, generally, it can efficiently identify the consistent trends at all six gauge positions.

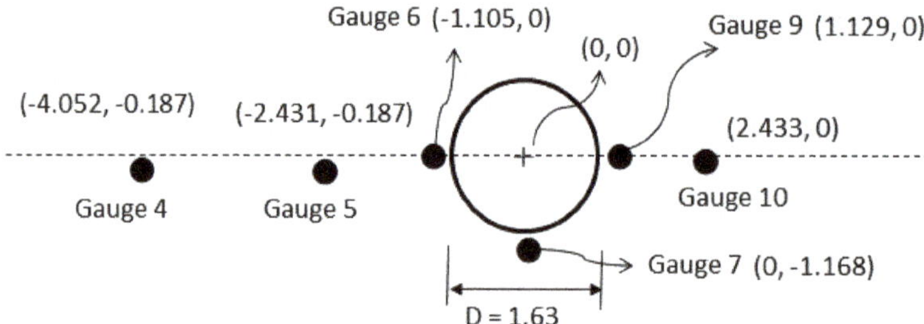

Figure 2. Plane view of wave-gauge locations for the experiment by Mo et al. (2010) in a dimensionless scale.

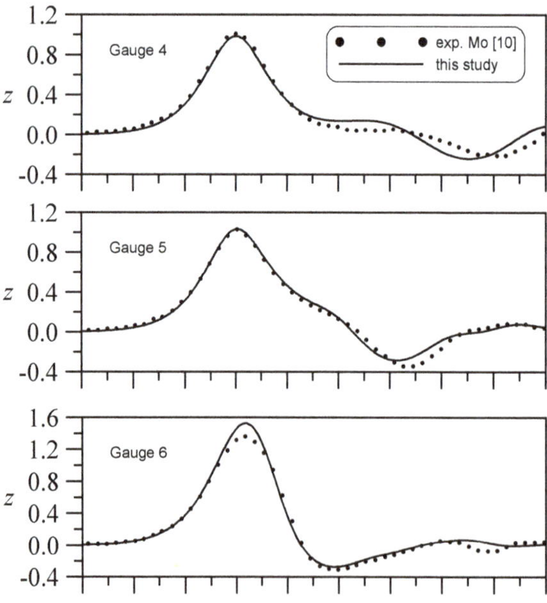

Figure 3. *Cont.*

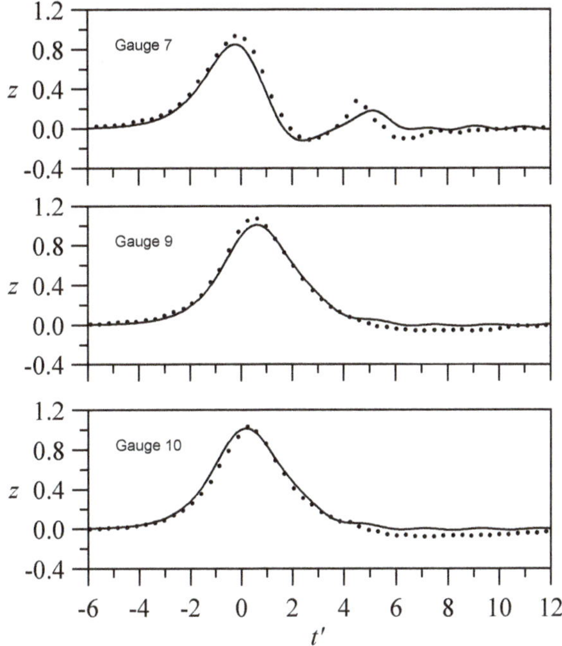

Figure 3. Comparisons of wave elevations at six wave-gauge positions.

Next, a single cylinder with a circular structure is considered with a gap between the cylinder bottom and channel bed. Chen and Wang [45] experimentally measured the wave elevations of a solitary wave passing through a cylinder that does not make complete contact with the bottom bed. The undisturbed water depth of their experiment was $H^* = 7.62$ cm, and according to its dimensionless channel experimental conditions, its length was (−40, 40), width was (−2.5, 2.5), wave gauge positions G1 and G2 were −1.411 and 1.411, respectively, and the cylinder diameter D was 1.5. The wave gauge measurements reported by the authors are for the case of $H_s = 0.5$. Compared with Chen and Wang's measured results shown in Figure 4, the results of the numerical simulation obtained in this study, which are consistent with their experimental results, are shown in Figure 4a,b with $A_0 = 0.19$ and Figure 4c,d with $A_0 = 0.31$. Because their channel width was $W = 5$ (only five times the still water depth), the cylinder with $D = 1.5$ was very close to the side wall of the channel. Figure 5 compares the difference in results obtained for $W = 5$ and 60, and shows that the wide channel $W = 60$ causes the scattering dispersive waves around the cylinder to be relatively small and weak.

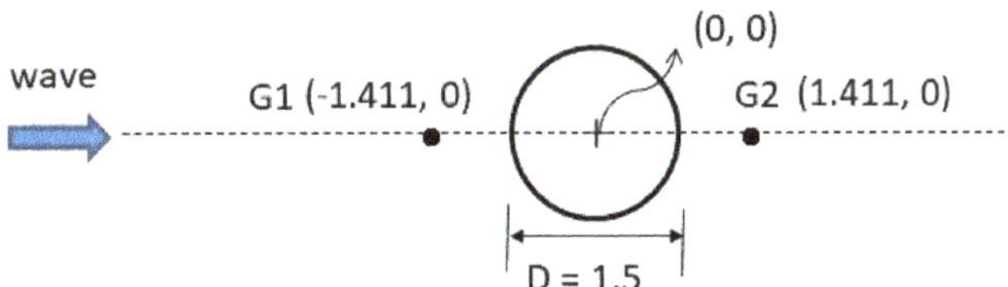

Figure 4. *Cont.*

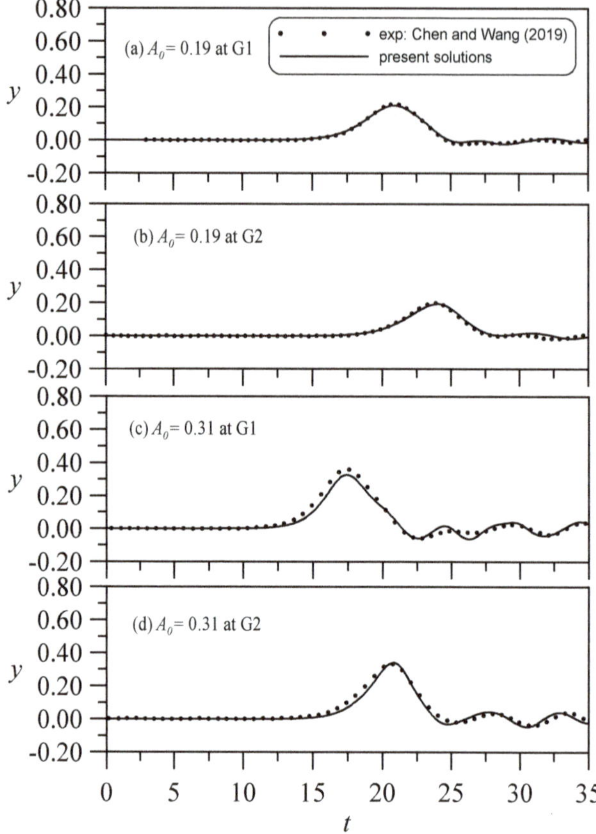

Figure 4. Comparisons of wave elevations at two gauges for a solitary wave passing through a non-touching seabed cylinder (H_s = 0.5).

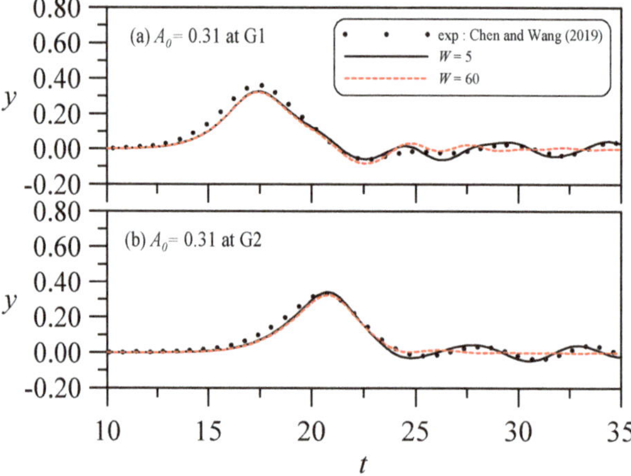

Figure 5. Comparisons of wave elevations at two gauges for a solitary wave passing through a non-touching seabed cylinder (H_s = 0.5). The results of W = 5 and 60 calculated and compared with the experimental condition with W = 5.

4. Results and Discussion

4.1. Solitary Wave Hitting a Circular Cylinder with Different Drafts and Sizes

Here, the influence of the gap between the bottom of the cylinder and the seabed on a solitary wave hitting the cylinder is analyzed. A wave with $A_0 = 0.3$ passes through a cylinder of $D = 4$, for which different immersion depths (H_s) are simulated. Figure 6 shows a plot of the wave motions for a circular cylinder that is semi-immersed at $H_s = 0.5$, with the left side of each figure showing the top view to enable observation of the evolutions of the crest lines, and the right side showing a 3D perspective. For example, in Figure 6 at $t = 0$, the distance of a solitary wave from the center of the cylinder 15 times the water depth. In this figure, it can be seen that when $t = 3$, the solitary wave approaches the cylinder but is not significantly affected by it. When $t = 12$, the wave touches the cylinder's surface, and runs up the front of the cylinder. Then, most of the wave passes through the cylinder to form a transmitted wave. The wave is also scattered by the cylinder, and the water elevation initially runs up the front of the cylinder and then drops and forms a system of diffraction waves around it. At $t = 18$, the wave passed through the cylinder, and the crest lines behind the cylinder are just slightly lower than those on both sides, such that the cylinder does not seriously eradicate the crest lines. After the wave has completely passed through the cylinder (such as at $t = 30$), the shape of the crest line that was damaged by the cylindrical structure returns to that of the original incident wave and continue to propagate forward. That is, the crest lines gradually become straight again. Eventually, a system of cylindrical diffraction waves radiates around the cylinder.

When a solitary wave passes through a cylinder, an impact runup is generated in front of the cylinder, and a water level squeeze and rise occurs at the cylinder's rear surface due to diffraction. The rising wave elevation has a great impact on the structure of the cylinder, so it is important to analyze the change in water levels at the front and rear of the cylinder for different submerged depths, H_s. Figure 7a shows the water level histories of the front cylinder at point A (−2, 0), and Figure 7b shows those of point B (2, 0) behind the cylinder. This figure shows the differences among $H_s = 1.0$ (complete immersion), 0.7, and 0.5. At an incident wave height of 0.3, although the H_s value changes, there is no obvious difference in the overall phenomenon. The maximum water level at point A can increase to about 0.5, which is about 1.7 times the height of the original wave. Then, the water level at point A drops below the still water level and gradually returns to the still water level. The evolution at point B gradually increases and then decreases, but all the recorded water levels at point B during this interaction are higher than the still water level.

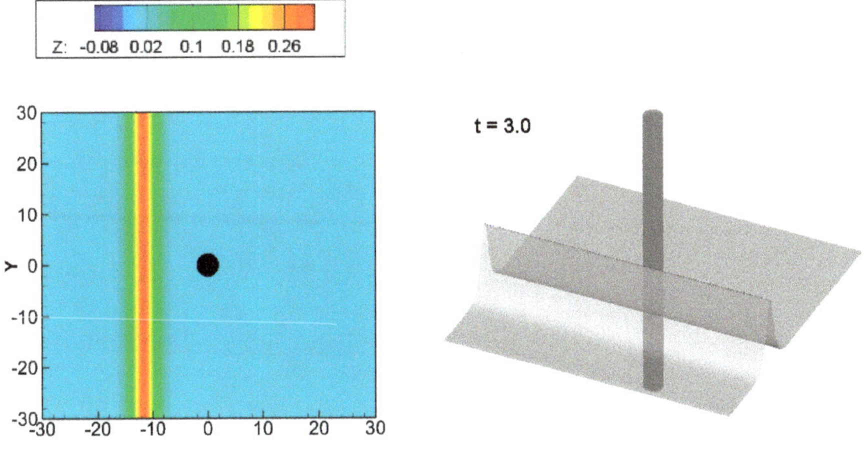

Figure 6. *Cont.*

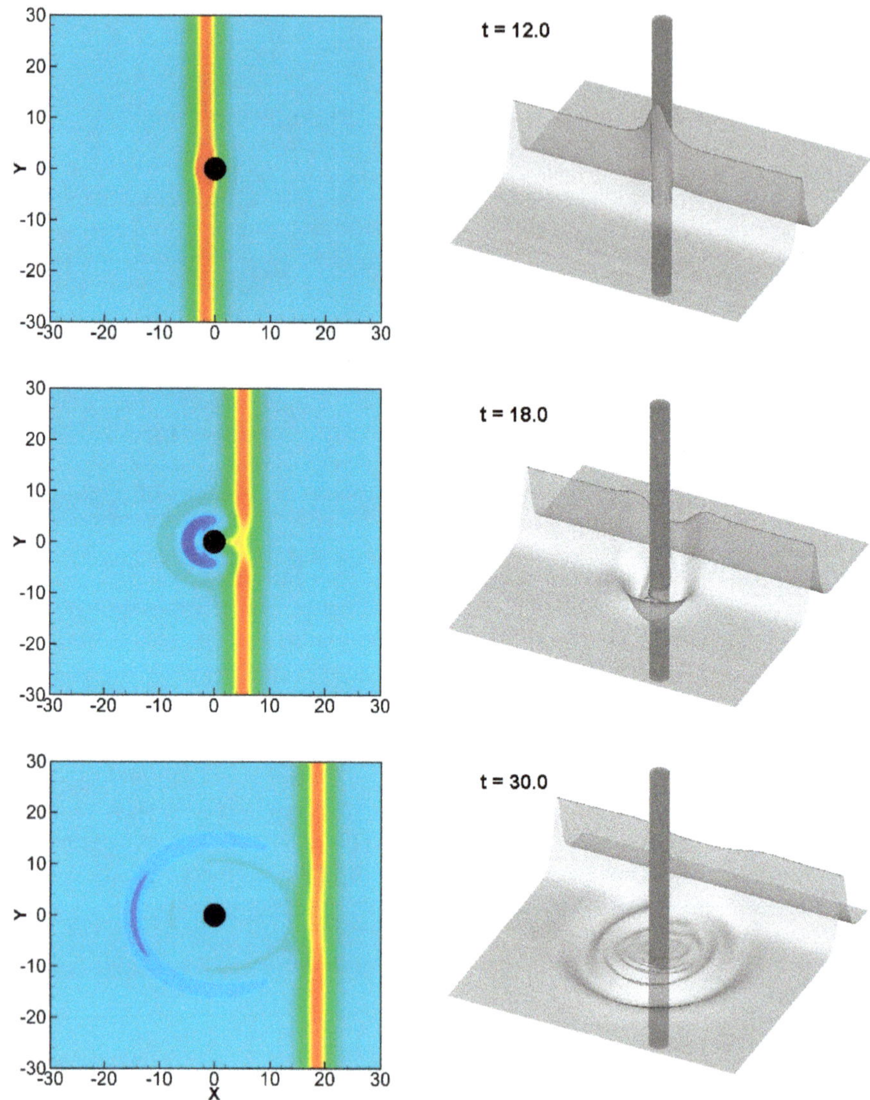

Figure 6. Wave patterns of a solitary wave passing through a partially immersed cylinder for $A_0 = 0.3$, $D = 4$, and $H_s = 0.5$. The left is a plane view, and the right is a 3D perspective.

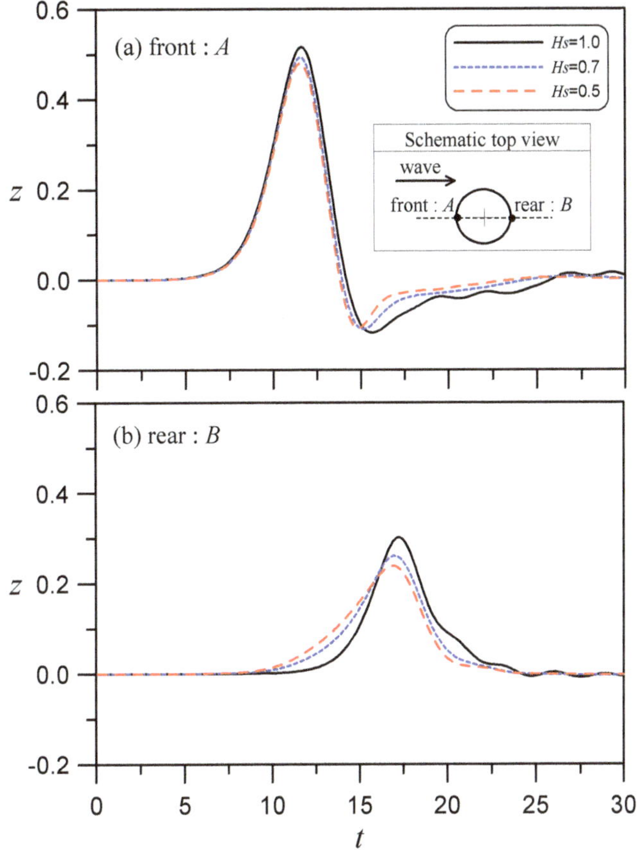

Figure 7. Front and rear wave elevations of a solitary wave passing through a vertically immersed circular cylinder in water with different H_s for $A_0 = 0.3$ and $D = 4$.

A careful examination of the effects of H_s for a deep immersion reveals that it produces a large runup and depression at point A and a large increase in the runup at point B. When $H_s = 1.0$, the maximum runup elevation of point B nearly reaches the height of the incident wave. However, if H_s is smaller, the runup of point B is slightly reduced, and the water level starts to rise earlier. This means that a larger gap (shallower H_s) allows more wave-induced flow to pass under the cylinder to form earlier wave surges at the rear of the cylinder. As such, the water level rise occurs earlier, but the volume of water squeezing and rising from the lateral sides of the cylinder after diffraction is weak, so the maximum runup at point B is low.

Figure 8 shows the vertical-profile differences for a solitary wave passing through a cylinder with immersions $H_s = 1.0$, 0.7, and 0.5. In this figure, the free-surface profiles at the symmetry plane ($y = 0$) are plotted at various moments. When $t = 9$, it can be seen that the wave touches the cylinder, so there is a rising runup at the front of the cylinder. Prior to this time, there is no obvious difference among the three immersions ($H_s = 1.0$, 0.7, and 0.5). However, when $t = 12$, the wave rises higher. For example, in the case of $H_s = 0.5$, there is a gap between the bottom of the cylinder and the seabed, so the current caused by the wave will pass through the gap. Therefore, the larger the gap (i.e., shallower H_s), the stronger the transmitted wave that initially emerges at the rear of the cylinder. However, the rear water level changes over time. By $t = 15$, a wave reflection and transmission mechanism appears. However, when $t = 18$ and $H_s = 1.0$, complete diffraction occurs (no current moves through the gap), so more diffraction waves will accumulate from the surrounding lateral areas of the cylinder, which will result in greater surge behind the cylinder, such that the transmitted wave behind the cylinder will become large. That is, at this time, the large gap creates a weaker runup at the rear of the cylinder. At $t = 30$, we can see that the wave bypasses the cylinder and is destroyed but still appears as a solitary wave, even though the wave height is slightly lower than the incident wave height (as denoted by the horizontal dashed line in Figure 8f, which represents the original wave height). A detailed comparison of the cases of $H_s = 0.5$ and $H_s = 1.0$ reveals that the transmission wave height at $H_s = 1.0$ is slightly high and is accompanied by some trailing waves, but the reflected waves produce more trailing waves for the case of $H_s = 0.5$. Generally, there is no significant difference between them.

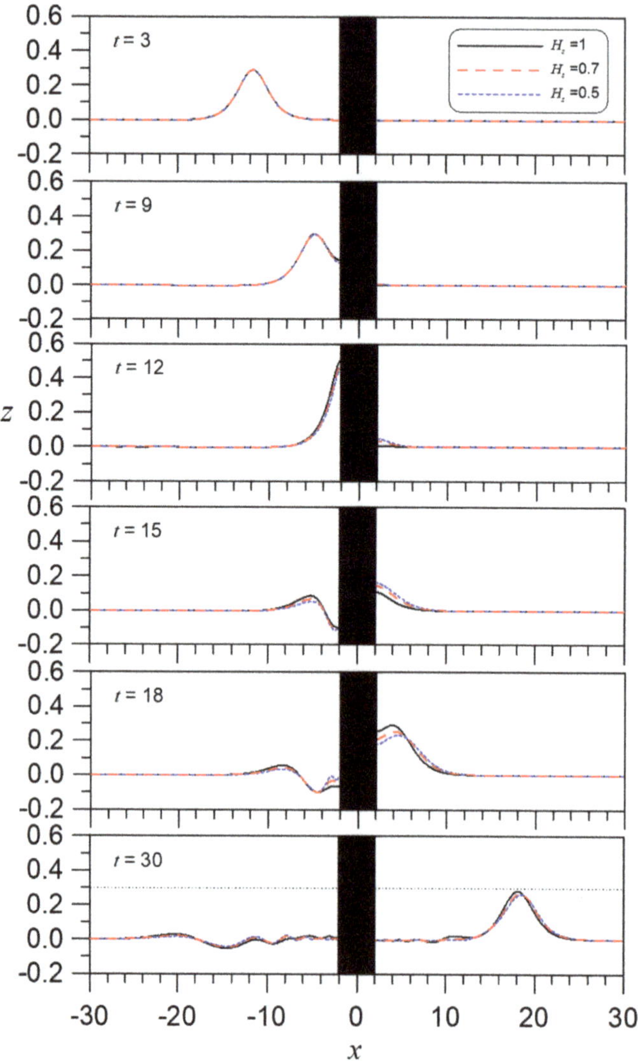

Figure 8. The solitary wave $A_0 = 0.3$ passes through an upright circular cylinder with $D = 4$ for $H_s = 1.0$, 0.7, and 0.5, comparing the wave profiles on $y = 0$.

Figure 9 shows a comparison of the impact of different cylinder sizes on the waves at the front and rear of the cylinder. We can see that the larger the cylinder, the larger the front runup and the smaller the rear runup. If D is fixed, as shown in the figure, the larger is H_s, the larger are both the front and rear runups. This echoes the results shown in Figure 7.

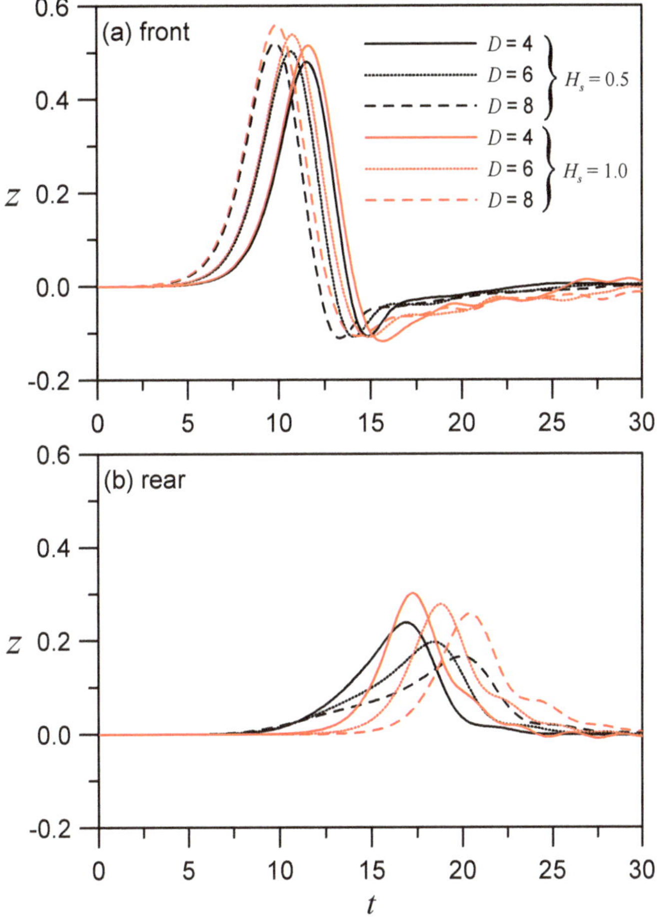

Figure 9. Front and rear wave elevations of a solitary wave passing through a vertically immersed circular cylinder in water for $A_0 = 0.3$; $H_s = 0.5$ with different D and H_s.

4.2. Solitary Wave Hits a Circular Cylinder with a Hollow Zone

A wave passing through a hollow cylinder will generate an oscillation effect in the hollow water column, which is helpful information for the design of an OWC device. This section discusses the problem of solitary waves passing through a hollow cylinder (Figure 10). In addition to the incident wave height (A_0), other possible influencing parameters are the still water depth H (H = 1 after normalization), the cylindrical immersion depth H_s, and the outer diameter r_1 and inner diameter r_2 of the concentric circular cylinder (cylinder thickness $dr = r_1 - r_2$).

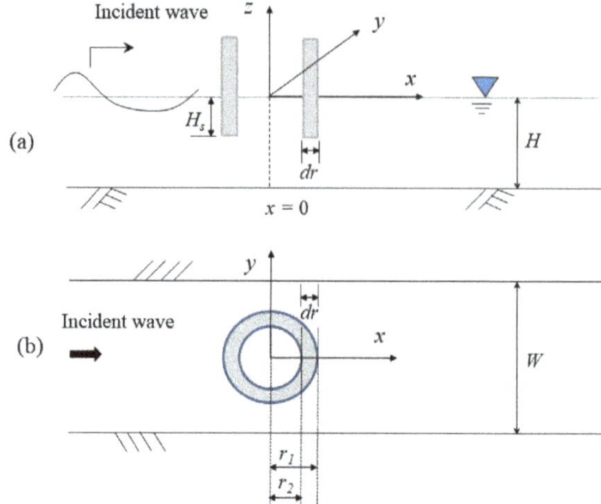

Figure 10. Schematic diagram of waves passing through a hollow cylinder: (**a**) side view on $y = 0$ and (**b**) top view.

For simplicity, this problem is discussed using $A_0 = 0.3$ as an example, mainly to analyze the influence of r_1, dr, H_s, and W on wave oscillation in the hollow circular cylinder. Figure 11 shows a hollow cylinder with $r_1 = 2$, $r_2 = 1$ ($dr = 1$), immersed at $H_s = 0.5$, with a channel width $W = 10$ to determine the characteristics of the change in the surrounding water level when a solitary wave passes through this cylinder. Figure 11a shows plots of the time history of the water level at three positions, a, b, and c, in the transverse direction. It is apparent that the wave runup and rundown at points a and c are similar to those obtained in the previous water-level analysis of a wave passing through a single solid cylinder at its front and rear positions, whereas Point b shows a change in the water-level oscillation in the hollow area, with its oscillation amplitude obviously larger than those of Points a and c. Figure 11b shows the water-level changes at Points e and f in the longitudinal direction. This figure shows that Point f in the hollow area has a significant oscillation amplitude similar to that of Point b, which indicates that the hollow area causes waves to generate an oscillation effect in the hollow column area.

To analyze the influence of dr (or r_2), Figure 12 shows the case of $r_1 = 3$, $H_s = 0.5$, and $W = 20$. That is, the outer diameter is held constant while the thickness (or inner diameter) is varied to observe how the water oscillates in the hollow column. A comparison of Figure 12a,b reveals that in Figure 12a, when $r_2 = 1$, the difference in the water levels at points a, b, and c of the hollow area is small. However, in Figure 12b, when $r_2 = 2$ (which is larger than that in Figure 12a), the water levels of the hollow area are significantly different at points a′, b′, and c′. This result indicates that when $r_2 = 1$ (Figure 12a), the water level of the hollow area oscillates more uniformly. Thus, if the hollow area is small (not greater than the still water depth), the hollow water column will fluctuate more uniformly. This phenomenon can be conceptualized and anticipated and can also be observed in the 2D water-level color contour map in Figure 13. Figures 13a–g and 13a′–g′ correspond to Figure 12a,b, respectively. The planar view shows the overall changes in the reflection, transmission, and diffraction of the wave as it encounters the hollow cylinder. This result is indicated by the color change in the hollow zone. The hollow area in Figure 13a–g is small ($r_2 = 1$) and always shows a single color in subsequent figures, which indicates that its water level is uniform in the hollow zone. In contrast, the color of the hollow area in Figure 13a′–g′ is not uniform.

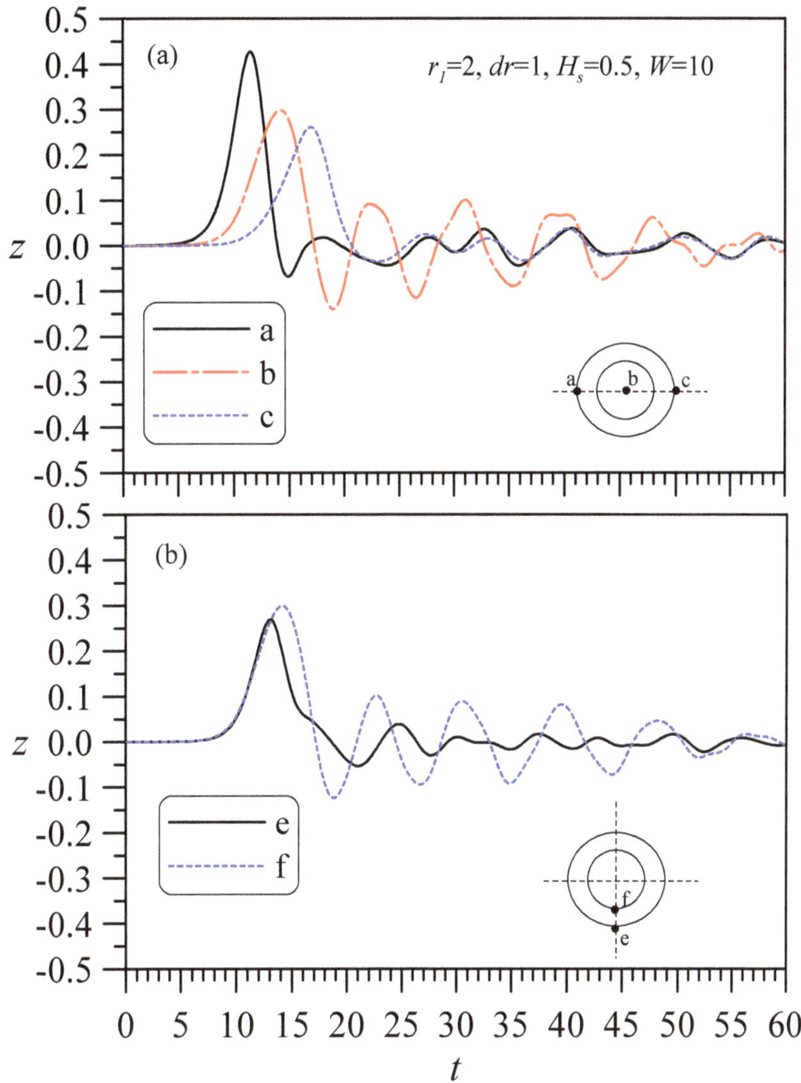

Figure 11. A solitary wave of height A_0 = 0.3 passes through a hollow circular cylinder generating the time histories of the water level at surrounding points.

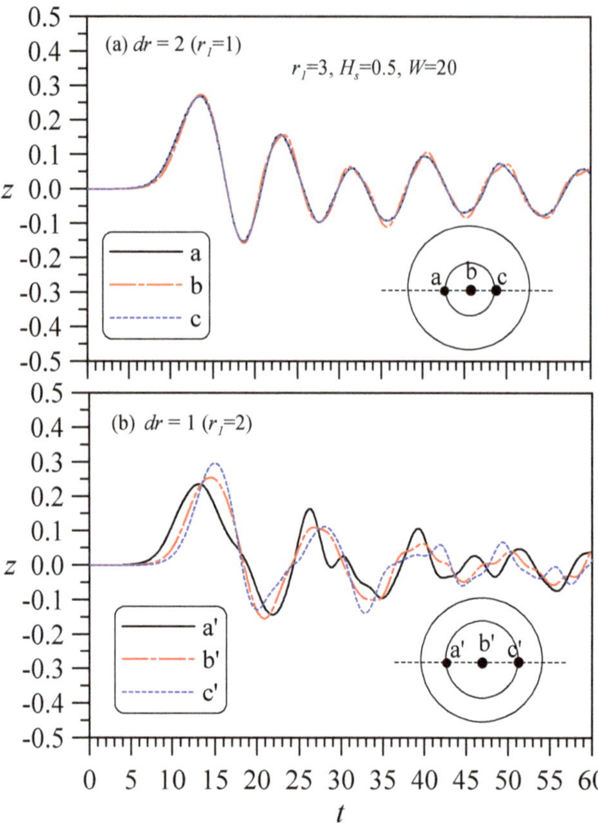

Figure 12. A solitary wave of height $A_0 = 0.3$ passes through the hollow circular cylinder recording wave elevations at points in the hollow area. A comparison of the wave elevations in the hollow area of the hollow cylinder with different inner diameters.

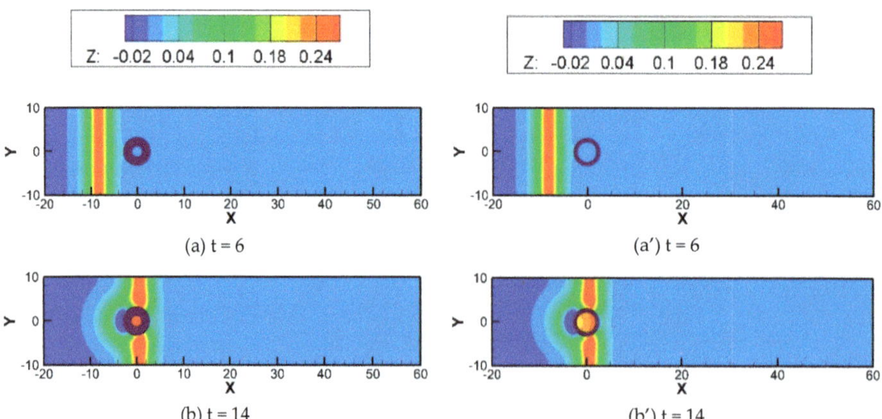

Figure 13. *Cont.*

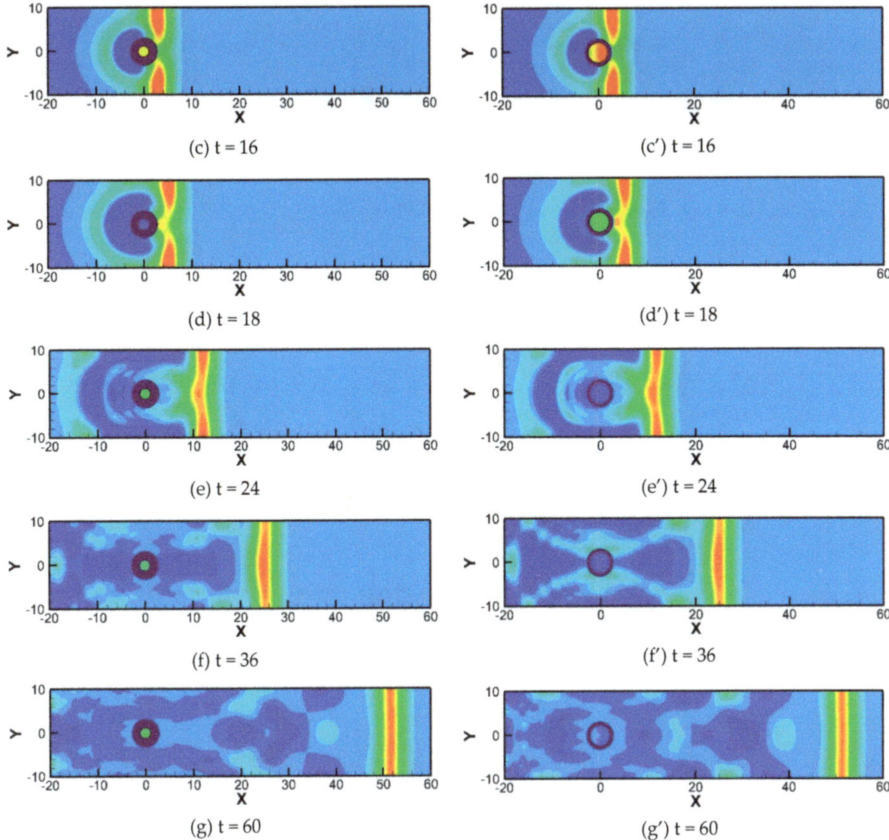

Figure 13. The wave-elevation contours at various times for a solitary wave A_0 = 0.3 propagating in the channel width W = 20 through the hollow cylinder with immersion depth H_s = 0.5. The cylinder's conditions are (**a–g**): r_1 = 3, r_2 = 1 and (**a′–g′**): r_1 = 3, r_2 = 2.

Next, the influence of the outer diameter, inner diameter, and cylinder thickness on the fluctuation is considered when W and one of the other factors are fixed.

4.2.1. Fixed Thickness, with Changes in the Outer and Inner Diameters

Figure 14 shows the water levels for a fixed channel width W = 15 and thickness dr = 1, with the outer diameters r_1 = 2, 3, and 4. Figure 8a–c show the effect of different H_s values on the rise and fall of the water column. These figures reveal that when dr is fixed and the outer diameter becomes larger, there will be a larger hollow area, which is not conducive to the formation of a uniform water level over the entire hollow area. A smaller r_1 value results in a better oscillation effect, and the larger is H_s (e.g., H_s = 0.7), the more significant the oscillation, i.e., a water column with a smaller r_1 and a larger H_s has a better oscillation effect.

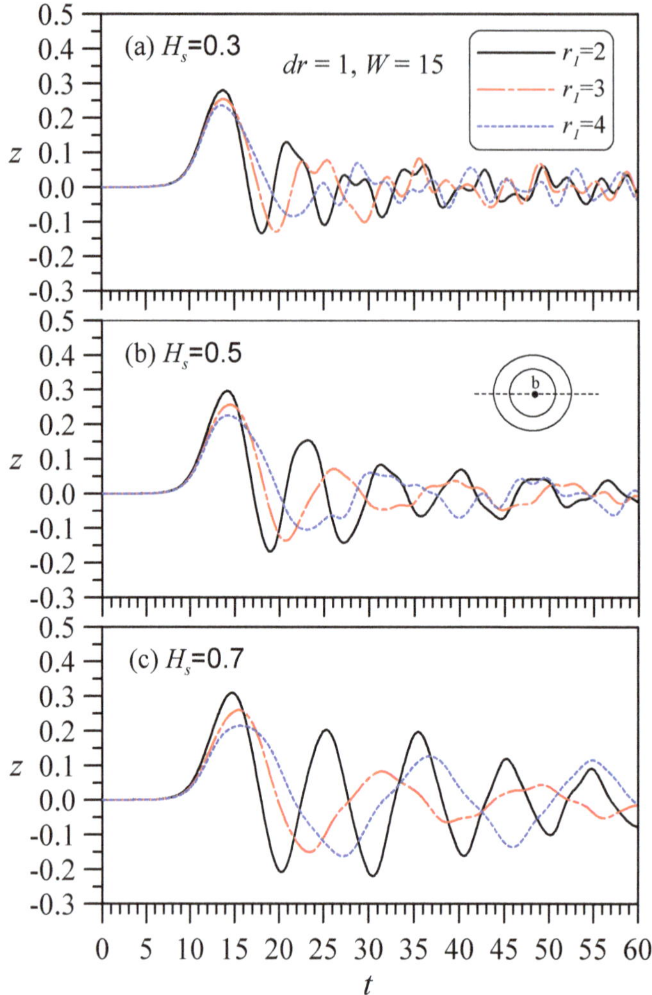

Figure 14. Influence of outer diameter (fixed wall thickness) is considered: A solitary wave of height $A_0 = 0.3$ passing through a hollow cylinder has the same cylindrical plate thickness ($dr = 1$) but different outer diameters (r_1). The water level of the hollow center point varies with time. The immersion depths (H_s) are not equal: (**a**) $H_s = 0.3$, (**b**) $H_s = 0.5$, and (**c**) $H_s = 0.7$.

4.2.2. Fixed Inner Diameter, with Changes in the Outer Diameter and Thickness

Another case for analyzing the influence of the outer diameter is obtained by fixing the inner diameter r_2 but changing the thickness. Similar to the analysis shown in Figure 14, Figure 15 shows the water levels for a fixed inner diameter $r_2 = 1$ and channel width $W = 15$. The inner diameter of the hollow area is fixed so the influence of the outer diameter r_1 can be observed. Figure 15a shows that when the cylinder is not deeply immersed (e.g., $H_s = 0.3$ in Figure 15a), a larger outer diameter helps to drive the amplitude to generate more regular motions. In Figure 15a, for the case of $H_s = 0.3$, the larger is r_1, the larger the amplitude (except for the main wave). If the immersion depth is deep, this phenomenon does not occur, but a smaller outer diameter brings about larger amplitudes. That is, if the immersion depth is not deep, the outer diameter can greatly improve the amplitude height and regularity, but a deeper immersion depth and smaller outer diameter can generate larger amplitudes.

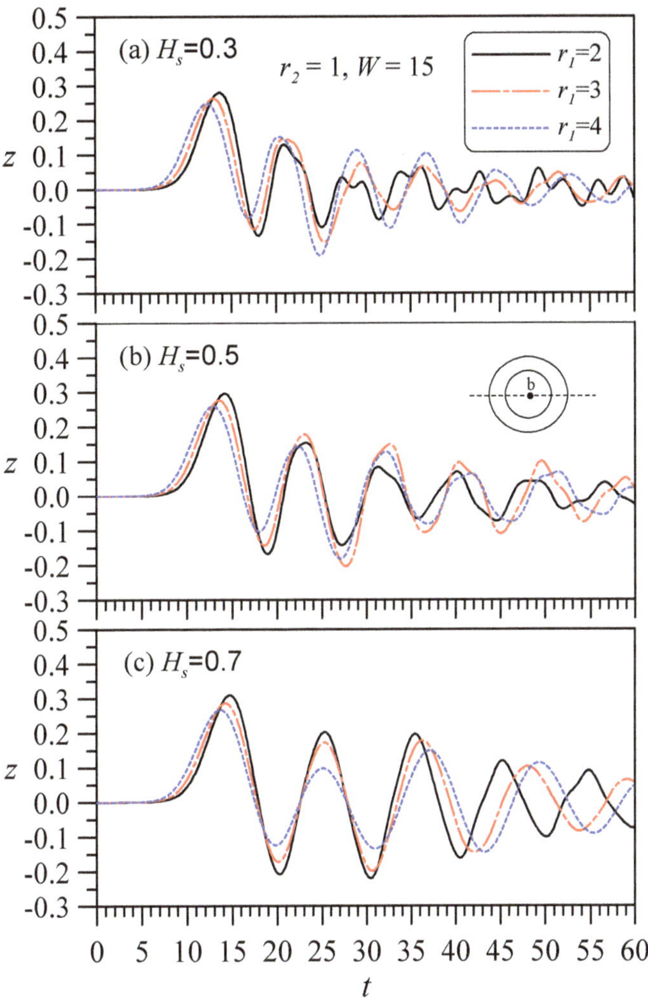

Figure 15. The influence of outer diameter r_1 (fixed r_2) is considered: A solitary wave height A_0 = 0.3 passes through a hollow cylinder with the same inner diameter (r_2 = 1) but different outer diameters (r_1) and the water level at the central point of the hollow is recorded over time. The immersion depths (H_s) are not equal: (**a**) H_s = 0.3, (**b**) H_s = 0.5, and (**c**) H_s = 0.7.

4.2.3. Fixed Outer Diameter, with Changes in the Inner Diameter and Thickness

For a fixed outer diameter r_1 = 4 and channel width W = 20, here, the effect of changing the thickness is analyzed. Figure 16 shows the case of a fixed outer diameter, which reveals that the greater the thickness (that is, the smaller the hollow area), the greater the oscillation, and if H_s is larger (e.g., Figure 16c), this phenomenon will be more obvious. However, the greater the thickness of the structure, the greater the manufacturing cost, which is not practical. Therefore, it is also recommended that the outer diameter not be too large.

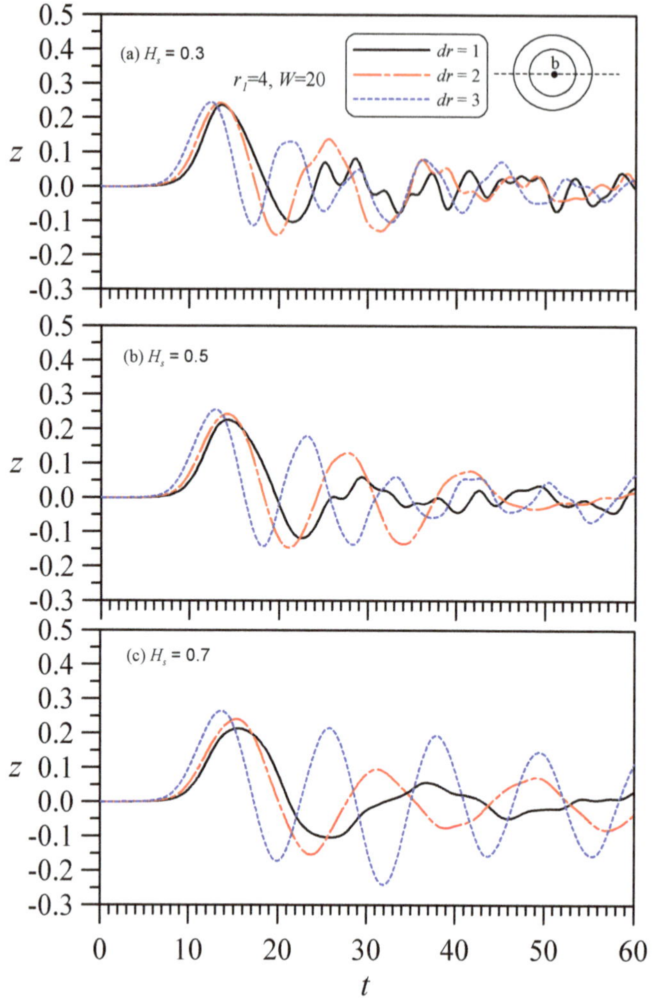

Figure 16. The influence of cylinder thickness (dr) is considered: A solitary wave of height $A_0 = 0.3$ passes through the hollow cylinder with the same outer diameter ($r_1 = 4$) but different cylinder plate thicknesses (dr). The water levels at the hollow center point are recorded over time. The immersion depths (H_s) are not equal: (**a**) $H_s = 0.3$, (**b**) $H_s = 0.5$, and (**c**) $H_s = 0.7$.

4.3. Influence of W on the Solitary Wave Hitting a Circular Cylinder with a Hollow Zone

Earlier in this article, the effect of W on a solitary wave against a solid cylinder was considered (Figure 5). The results indicated that a wide response channel smoothens the fluctuation in the diffraction of a solitary wave through a cylinder. In this subsection, the influence of W on a hollow cylinder is considered. If $r_1 = 2$ and $r_2 = 1$ ($dr = 1$) are fixed, the influence of W can be analyzed. Figure 17 shows that when $H_s = 0.3$ or 0.5, there is no obvious influence by W, but when $H_s = 0.7$, $W = 15$ produces larger oscillations than either $W = 10$ or 20. This result is somewhat confusing. To understand why $W = 15$ produces larger amplitudes, other W values were analyzed ($W = 10$–20) for the case of $H_s = 0.7$. Figure 18 shows that there are larger amplitudes when $W = 16$. Under the conditions of a fixed hollow cylinder size and immersion depth, there is an optimal W value that generates the maximum amplitude effect.

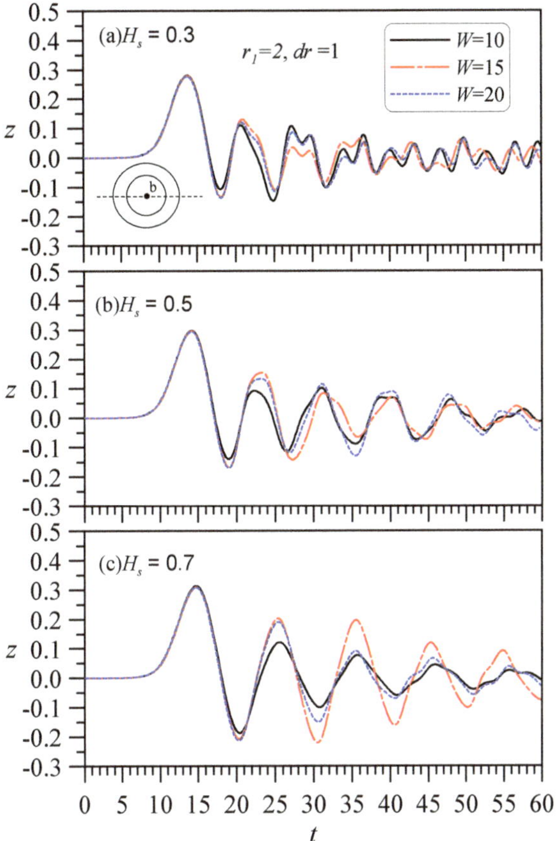

Figure 17. The influence of channel width W is considered: A solitary wave of height $A_0 = 0.3$ passes through a hollow cylinder with the same outer diameter ($r_1 = 2$) and inner diameter ($r_1 = 1$) but different channel widths (W). Observe the wave elevations at the center point varying with time. The immersion depths (H_s) are not equal: (**a**) $H_s = 0.3$, (**b**) $H_s = 0.5$, and (**c**) $H_s = 0.7$.

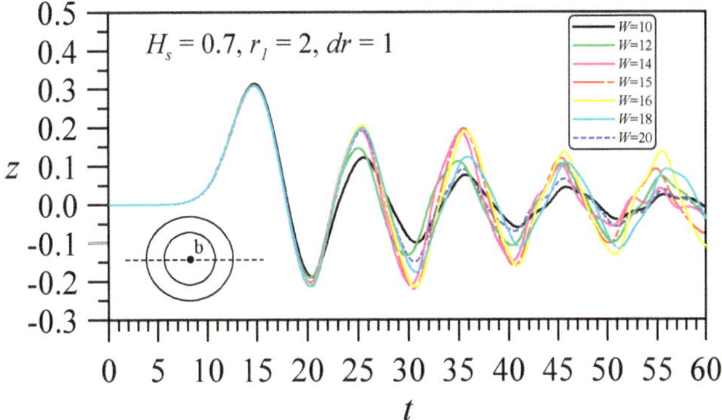

Figure 18. Comparisons of time histories of wave elevations at the central point for a solitary wave with $A_0 = 0.3$ passing through a hollow cylinder with the same outer diameter ($r_1 = 2$), inner diameter ($r_1 = 1$), and immersion depth $H_s = 0.7$ but different channel widths (W).

5. Conclusions

In this study, a 3D fully nonlinear potential wave model was applied to consider the interaction of a solitary wave with a vertical cylinder piercing the water at different depths with or without a hollow zone. A solitary wave with $A_0 = 0.3$ was used as the typical incident condition, and the model was first compared with the experimental data obtained by other researchers for verification, and was then applied to explore the wave–cylinder interactions. The conclusions are summarized below.

We first considered the interaction of a solitary wave with a single cylinder that has full contact with the seabed and obtained reasonable results. For a solitary wave $A_0 = 0.3$ passing through a cylinder of $D = 4$, the peak of the wave passing through the cylinder was determined to be slightly lower than the peaks on both sides that are undisturbed by the cylinder. When the wave has completely passed the cylinder, the waveform destroyed by the cylinder can nearly be restored. The observed runup elevations of the cylinder reveal that the maximum water level in front of the cylinder is about 0.5, which increases to approximately 1.7 times the original wave height (0.3) and then drops and gradually returns to the still water level. The runup evolutions behind the cylinder gradually increase and then decrease, but the recorded water elevations at the front and rear points remained higher than the still water level. This phenomenon changes when there is a gap between the bottom of the cylinder and the seabed. We analyzed the effect of the size of the gap between the bottom of a single cylinder and the seabed on the runup and rundown evolutions. Overall, the main effect of the immersion depth H_s is that when the H_s value is deep, greater runup and rundown occur in front of the cylinder, and the maximum water elevation behind the cylinder is larger. However, the water level behind the cylinder is always higher than the still water level. Regarding the effect of cylinder size D, the simulation results indicate that the larger the D value, the larger the front runup and the smaller the rear runup.

We also considered the oscillation effect of solitary waves passing through a hollow cylinder. The influencing parameters include A_0, r_1, r_2 (dr), H_s, and W, for which $A_0 = 0.3$; $r_1 = \sim 1–4$; $r_2 = \sim 1–3$; $H_s = 0.3$, 0.5, and 0.7; and $W = 10$, 15, and 20 were considered to analyze the oscillation in the hollow area under these conditions. The analysis showed that if r_2 is small (e.g., $r_2 = 1$), the hollow area obtains more uniform water level oscillation in the water column; the smaller the outer diameter (r_1), the better the oscillation effect; and the deeper the immersion depth (i.e., larger H_s), the more significant is this phenomenon. For long waves, when H_s is large and r_1 and r_2 are small ($r_1 > r_2$), large wave fluctuations can be produced in the hollow area. For the conditions considered in this study, when $r_1 = 2$, $r_2 = 1$, and $W = 16$, and the immersion depth is deeper, more uniform, and larger amplitude oscillation waves were observed in the hollow area. Therefore, the inner diameter should not exceed the depth of the water, and the outer diameter should not be too large (if the thickness can withstand the external forces, a less thick wall should be used). However, the effect of the W value was found to become more obvious when only H_s is larger. All results show that the larger is H_s, the larger the oscillating wave generated in the hollow area.

For nonlinear long waves, the wave energy is almost evenly distributed along the vertical depth, so a hollow cylinder can be more deeply immersed to block and squeeze more wave energy passing through the gap between the seabed and the cylinder bottom as it enters the hollow zone. For a deeper immersion, the vortex effect caused by the flow separation becomes relatively important. This is a problem that the model used in this study could not handle, and will be addressed by a more complete model in future work.

Author Contributions: The author (C.-H.C.) has reviewed and organized the literature, and has applied his numerical model to simulate cases, followed by plotting and presenting an analysis of the results. The author has read and agreed to the published version of the manuscript.

Funding: This research was funded by the Ministry of Science and Technology, Taiwan, Republic of China (Grant No. MOST 108-2221-E-275-002) and the APC for publishing this paper was also funded by the Ministry of Science and Technology, Taiwan, Republic of China (Grant No. MOST 108-2221-E-275-002).

Acknowledgments: This work was supported financially by the Ministry of Science and Technology, Taiwan, Republic of China (Serial No. MOST 108-2221-E-275-002).

Conflicts of Interest: The author declares no conflict of interest.

Appendix A

The following describes the grid generation, equation transformation, and numerical discretization methods.

Appendix A.1. Grid Generations

The numerical grids are established by curvilinear coordinates. An algebraic grid generation technique is used to generate the grids. The grid node numbers arranged along the ξ, η, and γ coordinates are i = 1~IM, j = 1~JM, and k = 1~KM. The range of the calculation area in the x, y, and z directions is assumed to be $x(1)$ to $x(IM)$, $y(1)$ to $y(JM)$ and $z(i, j, 1)$ to $z(i, j, KM)$. In this way, the grid distribution formula is as follows:

$$x = x(i) = x(1) + \Delta x(i-1),\ i = 1 \sim IM \tag{A1}$$

$$y = y(j) = y(1) + \Delta y(j-1),\ j = 1 \sim JM \tag{A2}$$

$$z = z(i,j,k) = z(i,j,1) + \{Z_{KC} - z(i,j,1)\}(k-1)/(KC-1),\ k = 1 \sim KC \tag{A3}$$

$$z = z(i,j,k) = Z_{KC} + \{z(i,j,KM) - Z_{KC}\}(k-KC)/(KM-KC),\ k = KC \sim KM \tag{A4}$$

here $\Delta x = \{x(IM) - x(1)\}/(IM-1)$, $\Delta y = \{y(JM) - y(1)\}/(JM-1)$, and Z_{KC} is the vertical position of the bottom of the cylinder (see Figure 1), and $z(i,j,KM)$ is equal to free-surface elevation, that is $\zeta_{i,j}$.

Appendix A.2. Equation Transformation

The calculation process needs to convert all the equations of the physical domain $(x, y, z; t)$ into the equations of the computational domain $(\xi, \eta, \gamma; \tau)$. After transformation, Equation (1) can be expressed as:

$$g^{11}\phi_{\xi\xi} + g^{22}\phi_{\eta\eta} + g^{33}\phi_{\gamma\gamma} + 2g^{12}\phi_{\xi\eta} + 2g^{13}\phi_{\xi\gamma} + 2g^{23}\phi_{\eta\gamma} + f^1\phi_\xi + f^2\phi_\eta + f^3\phi_\gamma = 0 \tag{A5}$$

where g^{ij} (i, j = 1, 2, 3) and f^i (i = 1, 2, 3) are grid geometric coefficients. Since x and y are uniform grids, and the gridlines in the z-direction are vertical and straight, the grid geometric coefficients in the formula can be simplified as:

$$g^{11} = 1/x_\xi^2,$$

$$g^{22} = 1/y_\eta^2,$$

$$g^{33} = [(z_\xi y_\eta)^2 + (y_\eta x_\xi)^2 + (x_\xi z_\eta)^2]/J^2,$$

$$g^{13} = -z_\xi/(x_\xi^2 z_\gamma),$$

$$g^{12} = 0,$$

$$g^{23} = -z_\eta/(y_\eta^2 z_\gamma),$$

$$f^1 = f^2 = 0,$$

and $f^3 = -\left(g^{11}z_{\xi\xi} + g^{33}z_{\gamma\gamma} + g^{22}z_{\eta\eta} + g^{13}z_{\xi\gamma} + g^{23}z_{\gamma\eta}\right)/z_\gamma$
where $J = x_\xi y_\eta z_\gamma$.

The free-surface boundary conditions of Equations (2) and (3) can be expressed as:

$$\zeta_\tau = w - u(\zeta_\xi/x_\xi) - v(\zeta_\eta/y_\eta) \tag{A6}$$

$$\phi_\tau - w\zeta_\tau + \frac{1}{2}(u^2 + v^2 + w^2) + \zeta = 0 \tag{A7}$$

In Equations (12) and (13), the water particle velocity (u, v, w) can be expressed as:

$$u = \phi_x = (\phi_\xi z_\gamma y_\eta - \phi_\gamma z_\xi y_\eta)/J \tag{A8}$$

$$v = \phi_y = (-\phi_\gamma x_\xi z_\eta + \phi_\eta x_\xi z_\gamma)/J \tag{A9}$$

$$w = \phi_z = \phi_\gamma y_\eta x_\xi/J \tag{A10}$$

The lateral conditions of Equation (4) can be derived for ϕ and ζ, respectively, as:

$$\phi_\tau - w\zeta_\tau \pm \left(\frac{\phi_\xi}{x_\xi} - \frac{\phi_\gamma\zeta_\xi}{x_\xi\zeta_\gamma}\right)\sqrt{(1+\zeta)} = 0 \tag{A11}$$

$$\zeta_\tau \pm \sqrt{(1+\zeta)}(\zeta_\xi/x_\xi) = 0 \tag{A12}$$

If the solid boundary grid is orthogonal (such as the seabed, bottom and wall of the cylinder, and side boundary in the y-direction), the boundary solution can be directly obtained from the Neumann condition of Equation (5).

Appendix A.3. Numerical Discretization

The transformed equations are discretized by the finite difference method. The details of the numerical discretization process can be found in work by Chang and Wang [68]. The vital processes are briefly described here; Equation (1) solves the points (i, j, k) in an element (Figure A1). Using the central difference method, the point solution can be arranged as:

$$\phi_{i,j,k} = \left\{ \begin{array}{ll} C_{22B}\phi_{i,j,k-1} & +C_{22T}\phi_{i,j,k+1} + C_{32C}\phi_{i+1,j,k} + C_{12C}\phi_{i-1,j,k} + C_{23C}\phi_{i,j+1,k} + C_{21C}\phi_{i,j-1,k} \\ & +C_{32T}\phi_{i+1,j,k+1} + C_{12T}\phi_{i-1,j,k+1} + C_{32B}\phi_{i+1,j,k-1} + C_{12B}\phi_{i-1,j,k-1} + C_{23T}\phi_{i,j+1,k+1} \\ & +C_{21T}\phi_{i,j-1,k+1} + C_{23B}\phi_{i,j+1,k-1} + C_{21B}\phi_{i,j-1,k-1} \end{array} \right\} \Big/ \left\{2\left(g^{11} + g^{22} + g^{33}\right)\right\} \tag{A13}$$

In which,

$$C_{32C} = C_{12C} = g^{11};\ C_{22T} = g^{33} + \frac{f^3}{2};\ C_{22B} = g^{33} - \frac{f^3}{2};$$

$$C_{23C} = C_{21C} = g^{22};\ C_{23T} = C_{21B} = \frac{g^{23}}{2};\ C_{21T} = C_{23B} = \frac{-g^{23}}{2};$$

$$C_{32T} = C_{12B} = \frac{g^{13}}{2};\ C_{12T} = C_{32B} = \frac{-g^{13}}{2}.$$

The subscription symbols T, C, and B in Equation (A13) indicate the grid coefficients at the upper, middle, and lower levels, respectively (as labeled in Figure A1).

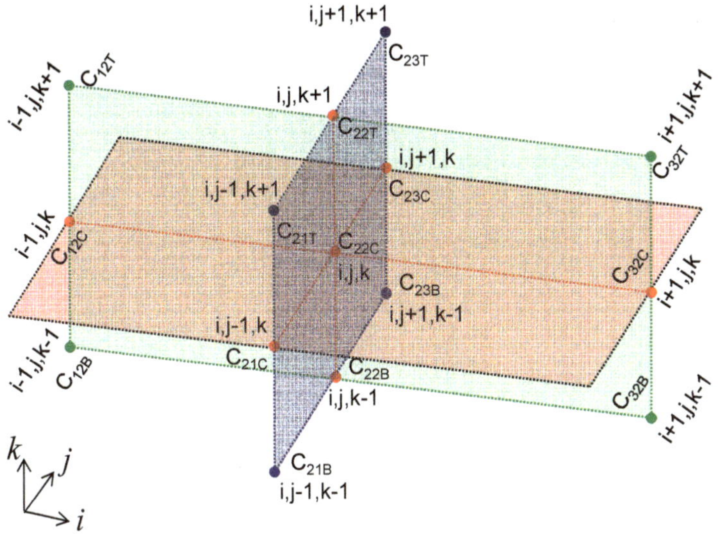

Figure A1. Schematic diagram of the numbered adjacent nodes in a grid element.

For free-surface boundary conditions, the first-order backward in the γ derivative and first-order central difference in the ξ and η derivatives are adopted. For x-lateral boundary conditions, the first forward and first backward methods are applied to the left and right truncated sections, respectively, in the computational domain. The finite difference in the time derivative can be solved explicitly and implicitly, and then the average value is used as the correct result. Based on these principles, the differential formula for the dummy variable Q is expressed by the operator as:

$$\delta_\xi Q = \frac{Q^{n+1}_{i+1,j,k} - Q^{n+1}_{i-1,j,k} + Q^{n}_{i+1,j,k} - Q^{n}_{i-1,j,k}}{4\Delta\xi} \tag{A14}$$

$$\delta_\eta Q = \frac{Q^{n+1}_{i,j+1,k} - Q^{n+1}_{i,j-1,k} + Q^{n}_{i,j+1,k} - Q^{n}_{i,j-1,k}}{4\Delta\eta} \tag{A15}$$

$$\delta_\gamma Q = \frac{Q^{n+1}_{i,j,k} - Q^{n+1}_{i,j,k-1} + Q^{n}_{i,j,k} - Q^{n}_{i,j,k-1}}{2\Delta\eta} \tag{A16}$$

$$\delta_\tau Q = \frac{Q^{n+1}_{i,j} - Q^{n}_{i,j}}{2\Delta\tau} \tag{A17}$$

$$\delta Q = \frac{Q^{n+1}_{i,j} + Q^{n}_{i,j}}{2} \tag{A18}$$

The superscript n in the formula represents the time step, $\Delta\xi = \Delta\eta = 1$, and $\Delta\tau$ is the calculation time interval. The free-surface boundary conditions are discretized using an explicit-implicit hybrid method. The numerical calculation process uses the free-surface kinematic boundary condition to solve ζ and the free-surface dynamic boundary condition to solve the boundary value ϕ. That is, Equations (A6) and (A7) can be expressed as:

$$\zeta^{n+1}_{i,j} = \zeta^{n}_{i,j} + \Delta\tau\left\{\delta w - \delta u\left(\frac{\delta_\xi \zeta}{\delta_\xi x}\right) - \delta v\left(\frac{\delta_\eta \zeta}{\delta_\eta y}\right)\right\} \tag{A19}$$

$$\phi_{i,j}^{n+1} = \phi_{i,j}^{n} + \Delta\tau\left\{\delta w \delta_\tau \zeta - \frac{1}{2}\delta\left(u^2 + v^2 + w^2\right) - \delta\zeta\right\} \tag{A20}$$

The opening boundary values can be derived from Equations (A11) and (A12) and then expressed as:

$$\begin{cases} \phi_{IM,j,k}^{n+1} = \phi_{IM,j,k}^{n} + \Delta\tau\left\{\delta w \delta_\tau \zeta - \sqrt{1 + \zeta_{IM,j}^{n+1}}\left(\frac{\delta_\xi \phi}{\delta_\xi x} - \frac{\delta_\gamma \phi \delta_\xi \zeta}{\delta_\xi x \delta_\gamma \zeta}\right)\right\} \\ \phi_{1,j,k}^{n+1} = \phi_{1,j,k}^{n} + \Delta\tau\left\{\delta w \delta_\tau \zeta + \sqrt{1 + \zeta_{1,j}^{n+1}}\left(\frac{\delta_\xi \phi}{\delta_\xi x} - \frac{\delta_\gamma \phi \delta_\xi \zeta}{\delta_\xi x \delta_\gamma \zeta}\right)\right\} \end{cases} \tag{A21}$$

$$\begin{cases} \zeta_{IM,j}^{n+1} = \zeta_{IM,j}^{n} - \Delta\tau\left\{\sqrt{1 + \zeta_{IM,j}^{n+1}}\left(\frac{\delta_\xi \zeta}{\delta_\xi x}\right)\right\} \\ \zeta_{1,j}^{n+1} = \zeta_{1,j}^{n} + \Delta\tau\left\{\sqrt{1 + \zeta_{1,j}^{n+1}}\left(\frac{\delta_\xi \zeta}{\delta_\xi x}\right)\right\} \end{cases} \tag{A22}$$

An over-relaxation iteration factor (SOR = 1.2~1.5) complements the solution process to accelerate iteration convergence. In this paper, the grid spacing $\Delta \approx 0.05 \sim 0.25$ is adopted, and the calculation time interval is $\Delta\tau = 0.05$~0.1. The numerical calculation process requires the convergence criteria $\left|\Omega^k - \Omega^{k-1}\right| \leq 10^{-6}$ at each time step, where Ω can be ϕ or ζ, and here the superscript k means the kth iteration of the loop.

References

1. Havelock, T.H. The pressure of water waves upon a fixed obstacle. *Proc. R. Soc. Lond.* **1940**, *175*, 409–421.
2. Ursell, F. On the heaving motion of a circular cylinder on the surface of a fluid. *Q. J. Mech. Appl. Math.* **1949**, *2*, 218–231. [CrossRef]
3. MacCamy, R.C.; Fuchs, R.A. *Wave Forces on a Pile: A Diffraction Theory*; Technical Memorandum 69; Beach Erosion Board Corps of Engineers: Washington, DC, USA, 1954.
4. Finnegan, W.; Meere, M.; Goggins, J. The wave excitation forces on a truncated vertical cylinder in water of infinite depth. *J. Fluids Struct.* **2013**, *40*, 201–213. [CrossRef]
5. Miles, J.W.; Gilbert, J.F. Scattering of gravity waves by a circular dock. *J. Fluid Mech.* **1968**, *34*, 783–793. [CrossRef]
6. Garrett, C.J.R. Wave forces on a circular dock. *J. Fluid Mech.* **1971**, *46*, 129–139. [CrossRef]
7. Black, J.L.; Mei, C.C.; Bray, C.G. Radiation and scattering of water waves by rigid bodies. *J. Fluid Mech.* **1971**, *46*, 151–164. [CrossRef]
8. Molin, B. Second-order diffraction loads upon three-dimensional bodies. *Appl. Ocean Res.* **1979**, *1*, 197–202. [CrossRef]
9. Yeung, R.W. Added mass and damping of a vertical cylinder in finite-depth waters. *Appl. Ocean Res.* **1981**, *3*, 119–133. [CrossRef]
10. Sabuncu, T.; Calisal, S. Hydrodynamic coefficients for a vertical circular cylinders at finite depth. *Ocean Eng.* **1981**, *8*, 25–63. [CrossRef]
11. Williams, A.N.; Demirbilek, Z. Hydrodynamic interactions in floating cylinder arrays–I. Wave scattering. *Ocean Eng.* **1988**, *15*, 549–583. [CrossRef]
12. Bhatta, D.D.; Rahman, M. Wave loadings on a vertical cylinder due to heave motion. *Int. J. Math. Math. Sci.* **1995**, *18*, 151–170. [CrossRef]
13. Bhatta, D.D.; Rahman, M. On scattering and radiation problem for a cylinder in water of finite depth. *Int. J. Eng. Sci.* **2003**, *41*, 931–967. [CrossRef]
14. Bhatta, D.D. Computations of hydrodynamic coefficients, displacement-amplitude ratios and forces for a floating cylinder due to wave diffraction and radiation. *Int. J. Non-Linear Mech.* **2011**, *46*, 1027–1041. [CrossRef]
15. Jiang, S.C.; Ning, D.Z. An analytical solution of wave diffraction problem on a submerged cylinder. *J. Eng. Mech.* **2014**, *140*, 225–232. [CrossRef]
16. Li, A.J.; Liu, Y. New analytical solutions to water wave diffraction by vertical truncated cylinders. *Int. J. Nav. Archit. Ocean Eng.* **2019**, *11*, 952–969. [CrossRef]

17. Ghadimi, P.; Bandari, H.P.; Rostami, A.B. Determination of the heave and pitch motions of a floating cylinder by analytical solution of its diffraction problem and examination of the effects of geometric parameters on its dynamics in regular waves. *Int. J. Appl. Math. Res.* **2012**, *1*, 611–633. [CrossRef]
18. Xu, D.; Stuhlmeier, R.; Stiassnie, M. Assessing the size of a twin-cylinder wave energy converter designed for real sea-states. *Ocean Eng.* **2018**, *147*, 243–255. [CrossRef]
19. Taylor, R.E.; Hung, S.M. Second order diffraction forces on a vertical cylinder in regular waves. *Appl. Ocean Res.* **1987**, *9*, 19–30. [CrossRef]
20. Taylor, R.E.; Hung, S.M. Semi-analytical formulation for second-order diffraction by a vertical cylinder in bichromatic waves. *J. Fluids Struct.* **1997**, *11*, 465–484. [CrossRef]
21. Kriebel, D.L. Nonlinear wave interaction with a vertical circular cylinder. Part I: Diffraction theory. *Ocean Eng.* **1990**, *17*, 345–377. [CrossRef]
22. Kim, J.W.; Kyoung, J.; Ertekin, R.C.; Bai, K.J. Wave diffraction of steep stokes waves by bottom-mounted vertical cylinders. In Proceedings of the ASME 22nd International Conference on Offshore Mechanics and Arctic Engineering, Cancun, Mexico, 8–13 June 2003.
23. Yan, K.; Shen, L.D.; Shang, J.W.; Ma, L.; Zou, Z.L. Experimental study on crescent waves diffracted by a circular cylinder. *China Ocean Eng.* **2018**, *32*, 624–632. [CrossRef]
24. Wang, K.H.; Wu, T.Y.; Yates, G.T. Three-dimensional scattering of solitary waves by vertical cylinder. *J. Waterw. Port Coast. Ocean Eng.* **1992**, *118*, 551–566. [CrossRef]
25. Yates, G.T.; Wang, K.H. Solitary wave scattering by a vertical cylinder: Experimental study. In Proceedings of the ISOPE-I-94-198, the Fourth International, Offshore and Polar Engineering Conference, Osaka, Japan, 10–15 April 1994.
26. Wang, K.H.; Jiang, L. Solitary wave interaction with an array of two vertical cylinders. *Appl. Ocean Res.* **1994**, *15*, 337–350. [CrossRef]
27. Zhao, M.; Cheng, L.; Teng, B. Numerical simulation of solitary wave scattering by a circular cylinder array. *Ocean Eng.* **2007**, *34*, 489–499. [CrossRef]
28. Neill, D.R.; Hayatdavoodi, M.; Ertekin, R.C. On solitary wave diffraction by multiple, in-line vertical cylinders. *Nonlinear Dyn.* **2018**, *91*, 975–994. [CrossRef]
29. Yuan, Z.; Huang, Z. Solitary wave forces on an array of closely-spaced circular cylinders. In *Asian and Pacific Coasts, 2009*; World Scientific: Singapore, 2009; pp. 136–142.
30. Isaacson, M.D.S.Q. Solitary wave diffraction around large cylinders. *Waterw. Port Coast. Ocean Eng.* **1983**, *109*, 121–127. [CrossRef]
31. Isaacson, M.D.S.Q. Nonlinear-wave effects on fixed and floating bodies. *J. Fluid Mech.* **1982**, *20*, 267–281. [CrossRef]
32. Isaacson, M.Q.; Cheung, K.F. Time-domain second-order wave diffraction in three dimensions. *J. Waterw. Port Coast. Ocean Eng.* **1992**, *118*, 496–517. [CrossRef]
33. Ohyama, T. A numerical method of solitary wave forces acting on a large vertical cylinder. *Coast. Eng. Proc.* **1990**, *1*, 840–852.
34. Yang, C.; Ertekin, R.C. Numerical simulation of nonlinear wave diffraction by a vertical cylinder. *J. Offshore Mech. Arctic Eng.* **1992**, *114*, 36–44. [CrossRef]
35. Wang, H.W.; Huang, C.J. Wave and flow fields induced by a solitary wave propagating over a submerged 3D breakwater. *J. Coast. Ocean Eng.* **2006**, *6*, 1–22. (In Chinese)
36. Patankar, S.V. *Numerical Heat Transfer and Fluid Flow*; Hemisphere: Washington, DC, USA, 1980.
37. Mo, W.H. Numerical Investigation of Solitary Wave Interaction with Group of Cylinders. Ph.D. Thesis, Graduate School of Civil and Environmental Engineering of Cornell University, Ithaca, NY, USA, August 2010.
38. Cao, H.J.; Wan, D.C. RANS-VOF solver for solitary wave run-up on a circular cylinder. *China Ocean Eng.* **2015**, *29*, 183–196. [CrossRef]
39. Leschka, S.; Oumeraci, H. Solitary waves and bores passing three cylinders—Effect of distance and arrangement. *Coast. Eng. Proc.* **2014**, *1*. [CrossRef]
40. Leschka, S.; Oumeraci, H.; Larsen, O. Hydrodynamic forces on a group of three emerged cylinders by solitary waves and bores: Effect of cylinder arrangements and distances. *J. Earthq. Tsunami* **2014**, *8*, 1440005. [CrossRef]
41. Chen, Q.; Zang, J.; Kelly, D.M.; Dimakopoulos, A.S. A 3-D numerical study of solitary wave interaction with vertical cylinders using a parallelized particle-in-cell solver. *J. Hydrodyn.* **2017**, *29*, 790–799. [CrossRef]

42. Kang, A.; Lin, P.; Lee, Y.J.; Zhu, B. Numerical simulation of wave interaction with vertical circular cylinders of different submergences using immersed boundary method. *Comput. Fluids* **2015**, *106*, 41–53. [CrossRef]
43. Lin, P. A multiple-layer σ-coordinate model for simulation of wave–structure interaction. *Comput. Fluids* **2006**, *35*, 147–167. [CrossRef]
44. Zhou, B.Z.; Wu, G.X.; Meng, Q.C. Interactions of fully nonlinear solitary wave with a freely floating vertical cylinder. *Eng. Anal. Bound. Elem.* **2016**, *69*, 119–131. [CrossRef]
45. Chen, Y.H.; Wang, K.H. Experiments and computations of solitary wave interaction with fixed, partially submerged, vertical cylinders. *J. Ocean Eng. Mar. Energy* **2019**, *5*, 189–204. [CrossRef]
46. Garrett, C.J.R. Bottomless harbours. *J. Fluid Mech.* **1970**, *43*, 433–449. [CrossRef]
47. Zhu, S.; Mitchell, L. Diffraction of ocean waves around a hollow cylindrical shell structure. *Wave Motion* **2009**, *46*, 78–88. [CrossRef]
48. Simon, M.J. Wave-energy extraction by a submerged cylindrical duct. *J. Fluid Mech.* **1981**, *104*, 159–187. [CrossRef]
49. Miles, J.W. On surface-wave radiation from a submerged cylindrical duct. *J. Fluid Mech.* **1982**, *122*, 339–346. [CrossRef]
50. Harun, F.N. Analytical solution for wave diffraction around cylinder. *Res. J. Recent Sci.* **2013**, *2*, 5–10.
51. Mavrakos, S.A. Wave loads on a stationary floating bottomless cylindrical body with finite wall thickness. *Appl. Ocean Res.* **1985**, *7*, 213–224. [CrossRef]
52. Mavrakos, S.A. Hydrodynamic coefficients for a thick walled bottomless cylindrical body floating in water of finite depth. *Ocean Eng.* **1988**, *15*, 213–229. [CrossRef]
53. Mavrakos, S.A. Hydrodynamic coefficients in heave of two concentric surface-piercing truncated circular cylinders. *Appl. Ocean Res.* **2004**, *26*, 84–97. [CrossRef]
54. Shipway, B.J.; Evans, D.V. Wave trapping by axisymmetric concentric cylinders. In Proceedings of the 21st International Conference on Offshore Mechanics and Artic Engineering (OMAE'2002), Oslo, Norway, 23–28 June 2002.
55. McIver, P.; Newman, J.N. Trapping structures in the three-dimensional water-wave problem. *J. Fluid Mech.* **2003**, *484*, 283–301. [CrossRef]
56. Hassan, M.; Bora, S.N. Exciting forces for a pair of coaxial hollow cylinder and bottom-mounted cylinder in water of finite depth. *Ocean Eng.* **2012**, *50*, 38–43. [CrossRef]
57. Hassan, M.; Bora, S.N. Hydrodynamic coefficients in surge for a radiating hollow cylinder placed above a coaxial cylinder at finite ocean depth. *J. Mar. Sci. Technol.* **2014**, *19*, 450–461. [CrossRef]
58. Simonetti, I.; Cappietti, I.; El Safti, H.; Oumeraci, H. 3D numerical modelling of oscillating water column wave energy conversion devices: Current knowledge and OpenFoam implementation. In Proceedings of the 1st International Conference on Renewable Energies Offshore—RENEW2014, Lisbon, Portugal, 24–26 November 2014.
59. Kamath, A.; Bihs, H.; Arntsen, Ø.A. Comparison of 2D and 3D simulations of an OWC device in different configurations. *Coast. Eng. Proc.* **2014**, *1*, 66. [CrossRef]
60. Lee, K.H.; Kim, T.G. Three-dimensional numerical simulation of airflow in oscillating water column device. *J. Coast. Res.* **2018**, *85*, 1346–1350. [CrossRef]
61. Kim, J.R.; Koh, H.J.; Ruy, W.S.; Cho, I.H. Experimental study on the hydrodynamic behaviors of two concentric cylinders. In Proceedings of the 2013 World Congress on Advances in Structural Engineering and Mechanics (ASEM13), Jeju, Korea, 8–12 September 2013.
62. Stokers, J.J. *Water Waves: The Mathematical Theory with Application*; John Wiley & Sons: New York, NY, USA, 1957; pp. 9–12. ISBN 978-0-471-57034-9.
63. Schember, H.R. A New Model for Three-Dimensional Nonlinear Dispersive Long Waves. Ph.D. Thesis, California Institute of Technology, Pasadena, CA, USA, 1982; p. 24.
64. Wang, K.H. Diffraction of solitary waves by breakwaters. *J. Waterw. Port Coast. Ocean Eng.* **1993**, *119*, 49–69. [CrossRef]
65. Wu, T.Y. Long waves in ocean and coastal waters. *J. Eng. Mech. Div. ASCE* **1981**, *107*, 501–522.
66. Dommermuth, D.G.; Yue, D.K.P. A high-order spectral method for the study of nonlinear gravity waves. *J. Fluid Mech.* **1987**, *184*, 267–288. [CrossRef]
67. Zhuang, Y.; Wan, D. Review: Recent development of high-order-spectral method combined with computational fluid dynamics method for wave-structure interactions. *J. Harbin Inst. Technol.* **2020**, *27*, 170–188.

68. Chang, C.H.; Wang, K.H. Fully nonlinear water waves generated by a submerged object moving in an unbounded fluid domain. *ASCE J. Eng. Mech.* **2011**, *137*, 101–112. [CrossRef]
69. Thompson, J.F.; Thames, F.C.; Mastin, C.W. TOMCAT—A code for numerical generation of boundary-fitted curvilinear coordinate systems on fields containing any number of arbitrary two-dimensional bodies. *J. Comput. Phys.* **1977**, *24*, 274–302. [CrossRef]

Publisher's Note: MDPI stays neutral with regard to jurisdictional claims in published maps and institutional affiliations.

Article

Numerical Analysis of Vertical Breakwater Stability under Extreme Waves

Meng-Syue Li [1], Cheng-Jung Hsu [2], Hung-Chu Hsu [3,*] and Li-Hung Tsai [4]

1 Marine Science and Information Research Center, National Academy of Marine Research, Kaohsiung 80661, Taiwan; d955040013@gmail.com
2 Department of Civil Engineering, National Cheng Kung University, Tainan 701, Taiwan; m995040011@student.nsysu.edu.tw
3 Department of Marine Environment and Engineering, National Sun Yat-sen University, Kaohsiung 80424, Taiwan
4 Harbor and Marine Technology Center, Institute of Transportation, Ministry of Transportation and Communications, Taichung 435058, Taiwan; ali@mail.ihmt.gov.tw
* Correspondence: hchsu@mail.nsysu.edu.tw; Tel.: +886-7-525-5172

Received: 6 October 2020; Accepted: 1 December 2020; Published: 3 December 2020

Abstract: The purpose of this study is to perform a numerical simulation of caisson breakwater stability concerning the effect of wave overtopping under extreme waves. A numerical model, which solves two-dimensional Reynolds-averaged Navier–Stokes equations with the $k-\varepsilon$ turbulence closure and uses the volume of fluid method for surface capturing, is validated with the laboratory observations. The numerical model is shown to accurately predict the measured free-surface profiles and the wave pressures around a caisson breakwater. Considering the dynamic loading on caisson breakwaters during overtopping waves, not only landward force and lift force but also the seaward force are calculated. Model results suggest that the forces induced by the wave overtopping on the back side of vertical breakwater and the phase lag of surface elevations have to be considered for calculating the breakwater stability. The numerical results also show that the failure of sliding is more dangerous than the failure of overturning in the vertical breakwater. Under extreme waves with more than 100 year return period, the caisson breakwater is sliding unstable, whereas it is safe in overturning stability. The influence of wave overtopping on the stability analysis is dominated by the force on the rear side of the caisson and the phase difference on the two ends of caisson. For the case of extreme conditions, if the impulse force happens at the moment of the minimum of load in the rear side, the safety factor might decrease significantly and the failure of sliding might cause breakwater damage. This paper demonstrates the potential stability failure of coastal structures under extreme sea states and provides adapted formulations of safety factors in dynamic form to involve the influence of overtopping waves.

Keywords: wave pressure; caisson breakwater; stability; wave overtopping; RANS model

---

1. Introduction

It is important to provide guidelines of structure stability for designing coastal protection structures or harbor breakwaters. Many previous studies on wave-structure interaction have discussed the seaward and lifting forces including those induced by the effects of wave overtopping and breaking from the physical or numerical models. Among these studies, the most important historical failures of the vertical structure have been documented [1]. Large-scale hydraulic experiments have been performed over the past several decades, leading to different empirical formulations that allow for the calculation of loads induced by wave breaking on vertical breakwaters [2–5]. Some semi-empirical equations for impulsive loads on vertical breakwaters are based on experimental data and, in some cases, with prototype-measured data [6]. Recently, the most commonly semi-empirical formulas for calculating pressures on a vertical structure wave have been developed [7–9]. When the wave obliquely enters the elongated structure, the maximum wave force decreases [10,11]. Lately, research results have reported that an increase in wave pressure caused by the diffracted wave can be one of the reasons for destruction of the conventional caisson breakwater [12].

In the theoretical analysis, there are some idealized analytical models which have been presented to study the forces exerted on the vertical structures [13,14]. However, these analytical approaches are complicated and inconvenient due to the complex geometries. In order to overcome the limitations, the numerical models have been developed since they are more flexible and efficient. The nonlinear shallow water or Boussinesq equations are developed to simulate the wave–structure interaction in coastal processes, especially for porous structures [15–22]. Besides, a boundary element method was used to solve the unsteady Forchheimer equations and described the wave transmission and reflection by a multilayer breakwater with arbitrary shapes [23]. Recently, the development of more efficient Navier–Stokes solvers have allowed a more sophisticated modelling of the wave-structure interaction problem [24–27]. For instance, the Reynolds-averaged Navier–Stokes (RANS) approach employing the volume of fluid (VOF) method for the modelling of complex (turbulent) flows has become popular in the past two decades [28–30]. A robust model was presented to investigate the functionality of rubble mound breakwaters with special attention focused on wave overtopping processes [31]. Then a numerical analysis considering wave loads corresponding to a low-mound and a rubble-mound breakwater with both regular and irregular incident wave conditions was carried out [32]. The results showed that this RANS equation model had a high potential to become a complementary tool to analyze the hydraulic response of caisson structures.

However, the existing research described above only considered the forces on the seaward and bottom of the caisson breakwater. The harbor-side loads induced by wave overtopping on a caisson breakwater have been investigated [33] and concluded that the backward loads should be considered as failure mode when designing the caisson breakwater. Very little information has been reported on wave forces acting on the backward side of caisson breakwater, which is an important factor in calculating the sliding and overturning stabilities for the engineering design. In this study, we shall extend the numerical model based on the Reynolds-averaged Navier–Stokes (RANS) equations to simulate the extreme wave overtopping induced loads on the front, bottom and rear sides of caisson breakwater. The numerical results are validated with the experimental data of surface elevations and wave pressures acting on the caisson breakwater [34]. The numerical model is applied to actual caisson breakwaters under various extreme wave conditions and the results are compared to the empirical formulas to discuss the stability of sliding and overturning [35]. Finally, some conclusions are made.

2. The Numerical Model

In this paper we used a numerical model (COBRAS model) which is a depth and time resolving 2DV numerical model solving the Reynolds-averaged Navier–Stokes (RANS) equations and the $k - \varepsilon$ turbulence closure model to check experimental data. The RANS equations for the ensemble-averaged velocity, $< u_i >$, and the ensemble-averaged pressure, $<p>$, are well-known and can be expressed as

$$\frac{\partial \langle u_i \rangle}{\partial x_i} = 0 \tag{1}$$

$$\frac{\partial \langle u_i \rangle}{\partial t} + \langle u_j \rangle \frac{\partial \langle u_i \rangle}{\partial x_j} = -\frac{1}{\rho}\frac{\partial \langle p \rangle}{\partial x_i} + g_i + \frac{1}{\rho}\frac{\partial \tau_{ij}}{\partial x_j} - \frac{\partial \langle u'_i u'_j \rangle}{\partial x_j} \tag{2}$$

in which i, j = 1, 2 for two-dimensional flow and τ_{ij} is the viscous stress. ρ, p and t denote the water density, pressure and time. The Reynolds stress $-\rho\langle u'_i u'_j \rangle$ is calculated with a nonlinear eddy viscosity relationship [36]. This nonlinear closure relationship for Reynolds stress mimics the more complete Reynolds stress closure without solving the transport equations of Reynolds stress. A demonstratation showed that standard linear eddy viscosity closure tends to over predict the diffusion of turbulence under breaking waves [29], and the nonlinear relationship suggested above predicts better turbulence statistics when compared with laboratory data of breaking wave over sloping beaches [37]. The governing equations are solved by a finite difference scheme. A two-step projection method is adopted to calculate the RANS equation. The evolution of free surface is calculated by the VOF method. The functionality of rubble mound breakwaters with special attention focused on wave overtopping processes uses the COBRAS model, and the numerical result comparison with experimental data has good agreement [38]. The mean overtopping discharge at the root of the South Breakwater of Póvoa de Varzim Harbour (Portugal) modelling by COBRAS model has good agreement with physical model tests [38].

A small-scale (1:36) physical model was carried out in a wave flume (100 m × 1.5 m × 2 m) to study the wave forces on a composite vertical breakwater [34], which was composed of an impermeable vertical caisson and permeable rubble foundation as shown in Figure 1. The specifications of the breakwater were as follows: the water depth was 0.526 m, the freeboard height was 0.153 m, the mound height was 0.138 m and the caisson weight was 780.66 kg. There were four wave gauges marked g1–g4 to measure the surface elevations and nine pressure transducers to measure wave pressures where five pressure transducers were along the front face of the caisson (U1–U5) and four pressure transducers were set on the bottom of the caisson (V1–V4). For the permeable rubble foundation, the median diameter was 0.7 cm and the porosity was 0.49. The incident wave height and period was experimental measured at wave gauge g1. Wave gauges (g2–g4) were used to calculate the reflection of the breakwater. A numerical flume was implemented with the same characteristics in the experimental setup. The computational grid system was discretized with non-uniform meshes in the x and z directions (Δx = 0.1 m, Δz = 0.05 m). Thus, the total number of cell meshes was 300 × 34. Additionally, the total simulation time was 64 s and the Courant number was 0.3 for all cases. The corresponding time step was automatically adjusted during calculations to satisfy stability constrains by both advection and diffusion processes, in which the maximum time step was adequately chosen as 10^{-2} s compared with experimental measurements. The aforementioned computational conditions were used throughout this study.

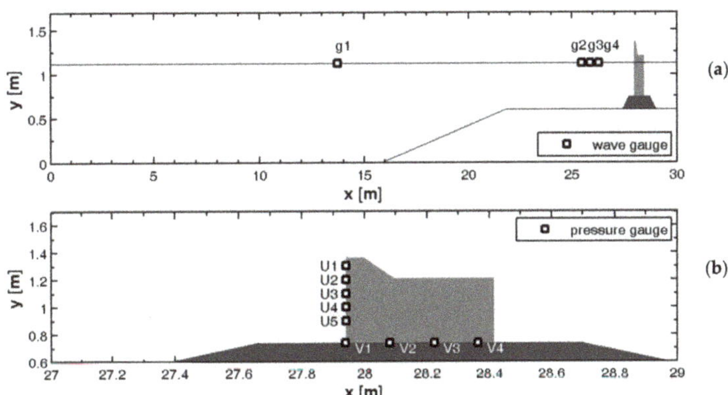

Figure 1. Experimental setup for regular wave [34]. (**a**) is the side view; (**b**) is the detailed view of the caisson breakwater.

The comparison of wave surface elevations for the four wave gauges between the numerical results and experimental data for wave overtopping is shown in Figure 2. The agreement between numerical results and experimental data was good except the wave gauge g2, where the vicinity of a node of the partial standing wave effected the interaction of the incident wave and reflection wave. Figures 3 and 4 show the comparison of wave dynamic pressure time series between the numerical simulations and experimental data for the regular overtopping wave condition (H = 0.25 m and T = 2 s). The pressure transducers U1~U2 were above the still water label. In Figure 3, the pressure transducers U3~U4 are exposed to air under the wave trough in that wave condition. The comparison includes five pressure transducers along the vertical front face of the caisson (U1–U5) and four pressure transducers underneath the caisson (V1–V4). The solid line represents the numerical simulations while the dashed line represents the laboratory measurements. From Figures 2–4, good qualitative comparisons between numerical simulation and experimental data are investigated. It indicates that the numerical model could simulate the waves propagating through the porous medium. In general, the COBRAS numerical model has a good prediction for the dynamic wave pressure time series at the vertical face of the caisson than those beneath the caisson. To quantify the comparisons, a validation method proposed by Wilmott was used as Equation (3) [39]

$$\varphi = 1 - \frac{\Sigma \left| X_{num} - X_{\exp} \right|^2}{\Sigma \left(\left| X_{num} - \overline{X_{num}} \right|^2 + \left| X_{\exp} - \overline{X_{\exp}} \right|^2 \right)} \quad (3)$$

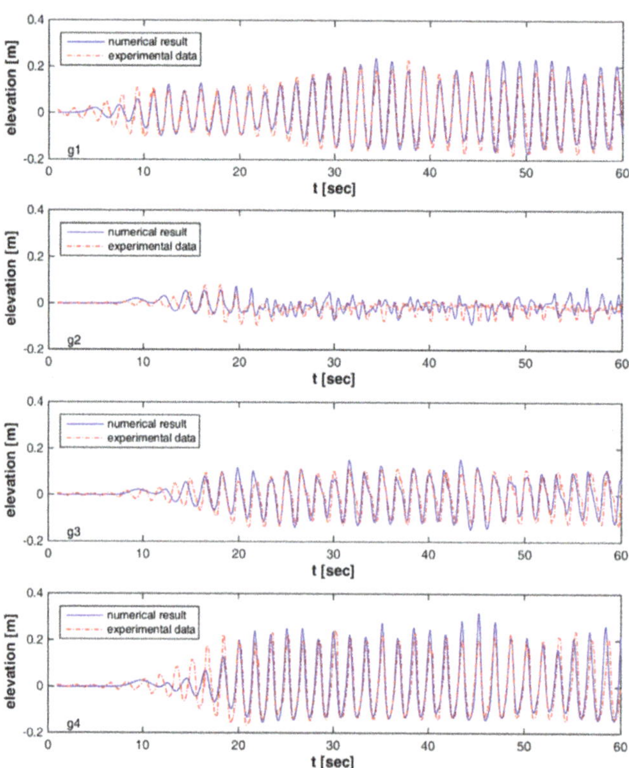

Figure 2. Comparisons of the surface elevation in regular wave condition (H = 0.19 m, T = 1.67 s) between numerical results (dashed line) and experimental data (solid line).

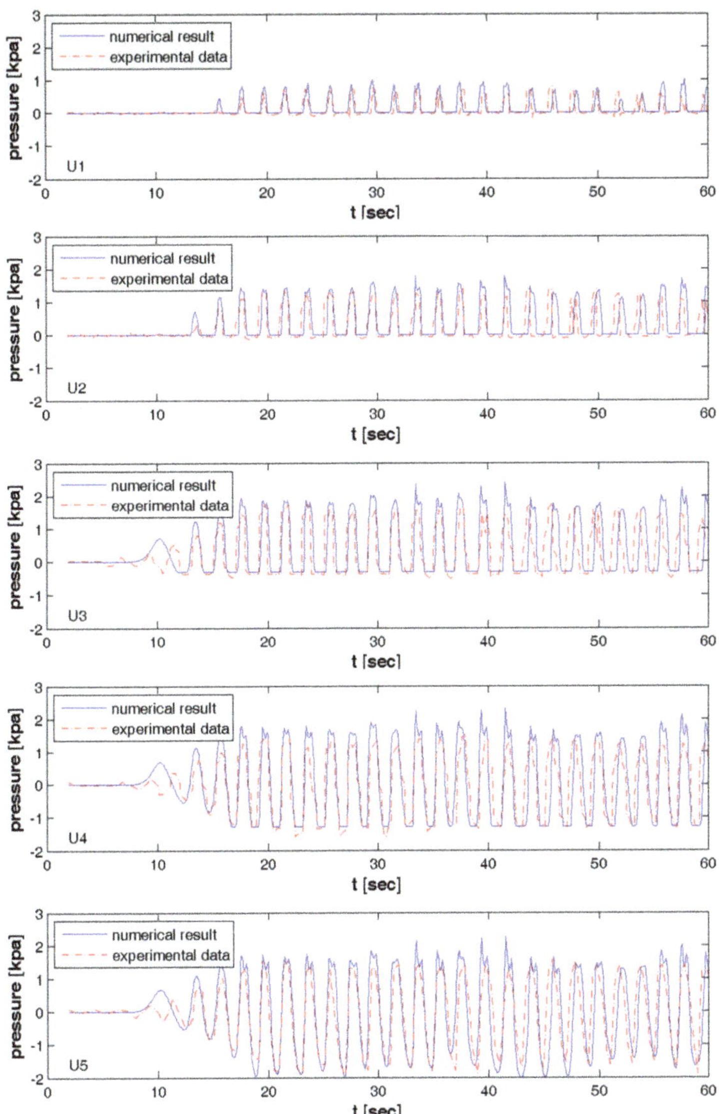

Figure 3. Comparisons of the dynamic wave pressure along the front face of the caisson breakwater in the typhoon wave condition (H = 0.25 m, T = 2 s) between numerical results (dashed line) and experimental data (solid line).

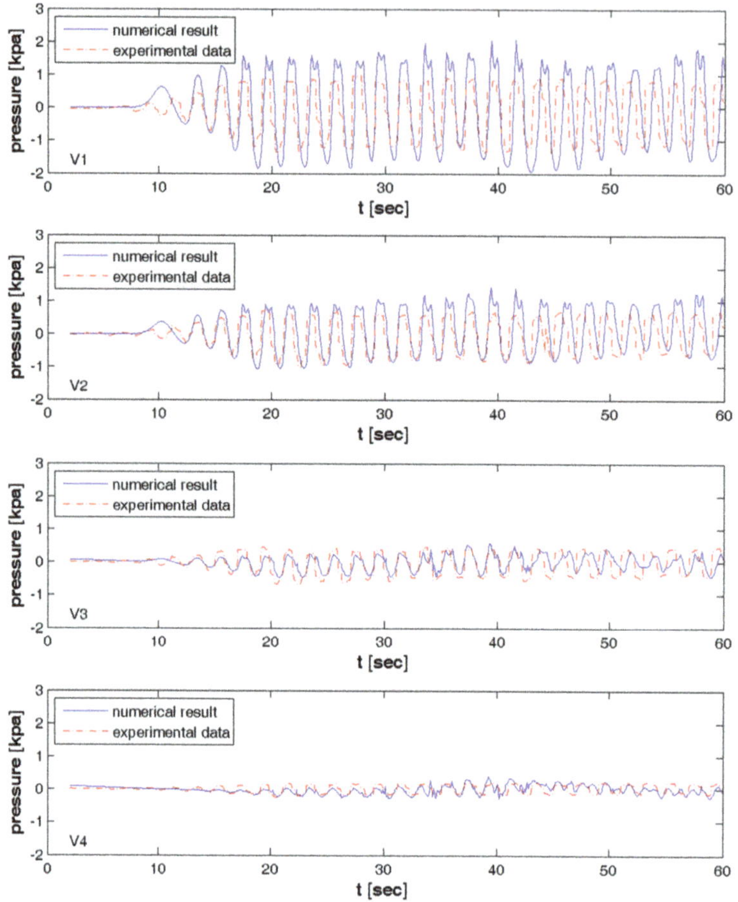

Figure 4. Comparisons of the dynamic wave pressure on the bottom of the caisson breakwater in the typhoon wave condition (H = 0.25 m, T = 2 s) between numerical results (dashed line) and experimental data (solid line).

In Equation (3), X_{num} is the numerical result, X_{exp} is the experimental data and the "−" is an averaging operator. The index of validation (φ), ranging from 0 to 1, represents the degree of agreement between numerical predictions and experimental data, and the unit is dimensionless. The validated results are shown as Table 1. The value of validated index, ranging from 0.71 to 0.96, is acceptable. However, it has to be noted that a smaller index is found for the wave dynamic pressure beneath the caisson breakwater. In general, the average value of all validated indexes is 0.8, which indicates that numerical predictions from the COBRAS model and experimental data are consistent. Consequently, the comparisons results provide confidence in the subsequent application of the model to the prototype scale, and which is described in the next section. In this study, we focus on the load on the rear sides of caisson breakwater induced by wave overtopping, so the overtopping discharge won't be concerned.

Table 1. The index of model validation.

Case	φ	Case	φ	Case	φ
g1	0.960	U1	0.723	V1	0.714
g2	0.638	U2	0.777	V2	0.741
g3	0.873	U3	0.772	V3	0.739
g4	0.920	U4	0.888	V4	0.708
-	-	U5	0.895	-	-

3. Stability Analysis of Prototype Vertical Breakwater

3.1. Prototype Wave Load Analysis

The validated numerical model was used to evaluate the total wave forces on the caisson breakwater in Taiwan at a prototype scale. There are two parts of the caisson breakwater: the porous rubble foundation and vertical caisson with parapet [40]. The vertical caisson is installed on a rubble foundation, which has a thin filter layer between caisson and foundation and is placed on a foreshore slope of 1 on 100. The design wave condition for the breakwater of Taichung harbor is 50 year return period wave condition. The specifications of the breakwater is as follows: the water depth is 22 m, the freeboard height is 11 m and the mound height is 6 m. Figure 5 shows a vertical breakwater cross-section at Taichung harbor in Taiwan. Tests were simulated using regular waves for various values of wave conditions as shown in Table 2. In order to understand the force contribution clearly, the regular wave was applied. Besides the condition of different return period typhoon waves, extreme waves due to climate change were also used [41]. According to [41], the wave height will increase up to 65% and the corresponding wave period might increase to 25% in 2039 around Taiwan. The numerical domain is 1125 m long and 80 m height. The mesh resolution at the structure is 0.2 m in the x-direction and 0.5 m in the y-direction. The rubble mounted layers beneath the caisson are defined using the porous flow parameter in Table 3.

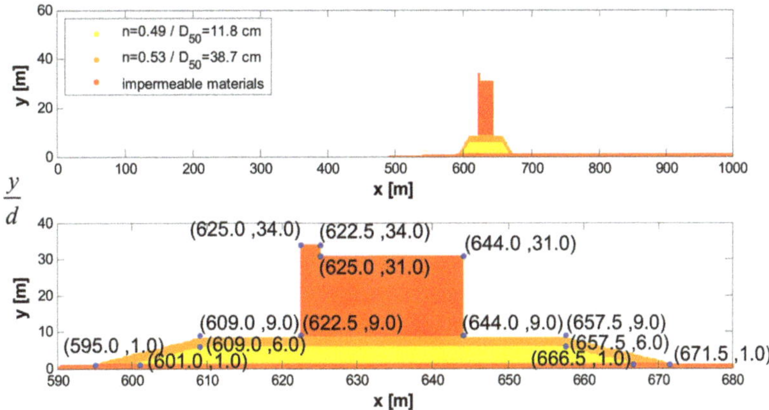

Figure 5. The wave-structure interaction model setup of Taichung harbor setup [41].

Table 2. The wave conditions (including return period of typhoon wave and extreme condition).

Case	H (m)	T (s)	Return Period (Year)
1	7.77	11.30	50
2	8.46	11.70	100
3	9.24	12.20	200
4	9.45	12.30	250
5	12.87	14.13	50 *
6	13.70	14.63	100 *
7	14.47	15.25	200 *
8	14.63	15.38	250 *

mean water depth = 28.86 m at x = 0 m. * represents the climate change effected extreme wave condition [41].

Table 3. The porous flow parameter for rubble mounted layers.

Layer	Porosity	The Median Diameter (m)
1	0.49	0.118
2	0.53	0.387
3	impermeable materials	

In the past, the existing researches only considered the forces on the seaward and bottom of the caisson breakwater. Very little information has been reported on overtopping wave forces acting on the backward side of caisson breakwater, which is an important factor in calculating the sliding and overturning stabilities for the engineering design. In order to bring more insight on the wave forces induced by the overtopping waves, the numerical model is extended for simulating the wave pressure acting on the front, bottom and back sides of the caisson breakwater. A free body diagram shown in Figure 6 was used to analyze the overtopping wave induced forces and moments acting on the caisson breakwater. In Figure 6, W denotes the weight of caisson, f_1 denotes the horizontal force acting on the front face of the caisson, f_2 denotes the horizontal force acting on the rear face of the caisson and f_3 denotes the uplift force acting on the bottom of the caisson. Buoyancy force is included in still water. The fixed reference point of the moment for the landward overturning is denoted by the point s_1, while that for seaward overturning is denoted by the point s_2. The moment of f_1 is m_1, and the moments of f_2 is m_2. The moment of f_3 at fixed point s_1 is denoted by m_3, and the moment of f_3 at fixed point s_2 is denoted by m_4. The moment of W at fixed point s_1 is m_{w1}, and the moment of W at fixed point s_2 is m_{w2}.

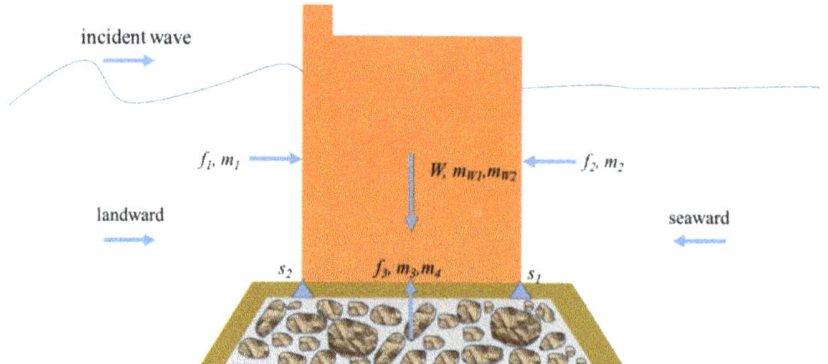

Figure 6. Free body diagram of load analysis.

The relative forces and moments acting on a vertical structure can be determined by integrating the numerically calculated pressure over a portion of vertical structure. Figure 7 presents the time histories of horizontal forces on the seaward-side (upper panel), harbor-side (middle panel) and uplift force for the studied case 1. In case 1, wave overtopping happens, indicating that all cases in Table 2 are necessary to investigate the influence of wave overtopping (IWO) on the structure stability. The force calculated included the behavior of flow transmitting through the rubble mound. In case 1, the overtopping wave happened after $t = 75$ s, so the transmitting flow disturbed a fluctuation of f_2 before overtopping. The calculated moments induced by the three forces are shown in Figure 7. From Figure 7, the oscillation of the rear force f_2 and the induced moment m_2 occurs after wave overtopping. Moreover, the peak values occasionally appear at some instantaneous time which might be caused by the jet of the overtopping wave. In addition, the differences between the water level on the front face and that on the rear side of caisson breakwater will influence the pressure distribution underneath the caisson. Therefore, the corresponding upward force f_3 and the induced moments (m_3 and m_4) will vary with the surface elevation difference for wave overtopping. The negative maximum of f_3 is larger than the positive maximum of f_3, which is the same as the experimental results in [34].

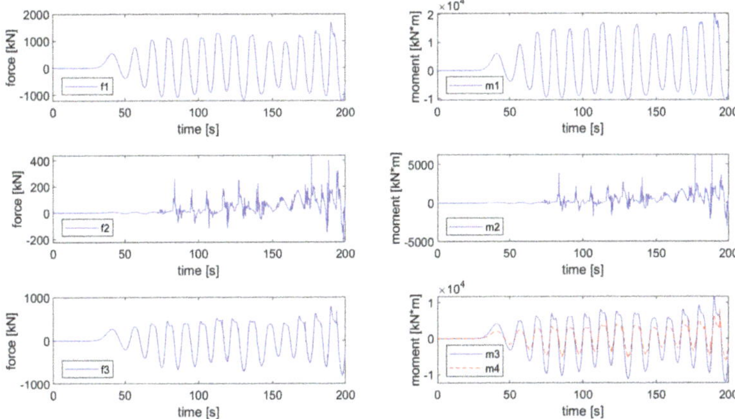

Figure 7. The time histories of wave-induced forces and their moment acting on caisson in case 1.

3.2. Stability Analysis

It is important to safely design a vertical breakwater with respect to the stability criteria of sliding and overturning. The movements of a caisson breakwater are influenced by the total active forces exerted on the seaward, harbor side and underneath the vertical structure. The minimum safety factors for sliding and overturning are suggested by [35] for engineering design and these values must not be less than 1.2 in practice. These safety factors for sliding and overturning are reviewed as the following:

Using an empirical formula to calculate the safety factors for sliding and overturning, the force acting on the rear side of the caisson was usually taken to be static for simplicity's sake. Few literatures studied the horizontal force acting on the rear side along the caisson for overturning wave. Therefore, the total force determination requires a good understanding of the overtopping wave behavior in the front and rear sides of the caisson as well as beneath the caisson. Here, the sliding safety factor and overturning safety factor without the force on the backward side can be presented as Equations (4) and (5).

$$f_s = \mu \frac{W - f_3}{f_p} \tag{4}$$

$$f_O = \frac{m_W - m_3}{m_p} \tag{5}$$

where μ is the coefficient of friction between the caisson and the foundation, W denotes the weight of caisson, f_p denotes the maximum horizontal force along the front side of caisson, m_p denotes the moment of f_p at the point s_1, and m_w denotes the moment of W at the point s_1.

Due to the rear side force being considered, the safety factor against seaward sliding and overturning should be accounted. Consequently, the corresponding safety factor against landward sliding and seaward are presented as Equations (6)–(9), where superscript "+" and "−" represent landward and seaward, respectively. The subscript "s" and "o" mean the sliding and overturning.

$$f_s^+ = \mu \frac{W - f_3}{f_1 - f_2} \tag{6}$$

$$f_s^- = \mu \frac{W - f_3}{f_2 - f_1} \tag{7}$$

$$f_o^+ = \frac{m_{W1} - m_3}{m_1 - m_2} \tag{8}$$

$$f_o^- = \frac{m_{W1} - m_3}{m_2 - m_1} \tag{9}$$

For example, the time histories of the safety factors against sliding including sliding safety factor (f_s), landward sliding safety factor ($f_s{}^+$) and seaward sliding safety factor ($f_s{}^-$) in case 1 are shown as Figure 8, and that against overturning including overturning safety factor f_0, landward overturning safety factor f_o^+ and seaward overturning safety factor f_o^- in case 1 are shown as Figure 9. The water wave in the rear side caused the sliding safety factor and overturning safety factor to oscillate, which means Equations (6)–(9) provide a time series variation value as shown in Figures 8 and 9. Overall, the caisson breakwater in Taichung harbor has sufficient stability for sliding and overturning on the IWO in the 50 year return period wave condition.

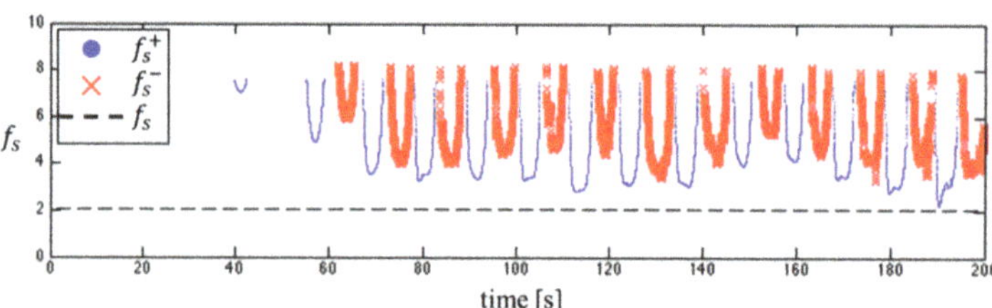

Figure 8. The time histories of the safety factors against sliding among $f_s{}^+$(·), $f_s{}^-$ (×) and f_s (—) for case 1.

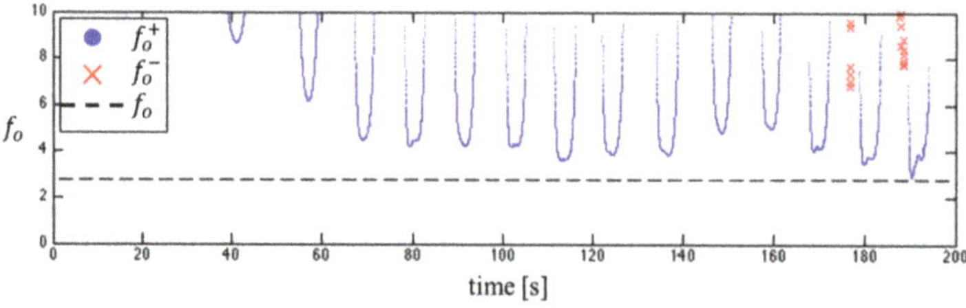

Figure 9. The time histories of the safety factors against overturning among $f_o{}^+$ (·), $f_o{}^-$ (×) and f_o (—) for case 1.

4. Results and Discussions

To investigate the structure stability on the IWO, the sliding and overturning safety factors considering the water level fluctuates on the rear side of caisson were calculated under 50 to 250 year return period typhoon wave conditions as well as extreme wave conditions. The represented safety factors for sliding on the IWO determined by the minimum of landward sliding safety factor and seaward sliding safety factor was denoted by $f_s^{\pm}$. In the same way, the overtopping safety factors determined by the landward overturning safety factor and seaward overturning safety factor, was denoted by $f_o^{\pm}$. The comparisons of the safety factors against sliding are shown as Figure 10. Generally, the safety factors following severe wave conditions get lower, and cases 6–8 are less than 1.2, indicating the risk of sliding failure might occur in extreme wave conditions. Although $f_s^{\pm}$ is closed to f_s, $f_s^{\pm}$ varies due to the IWO.

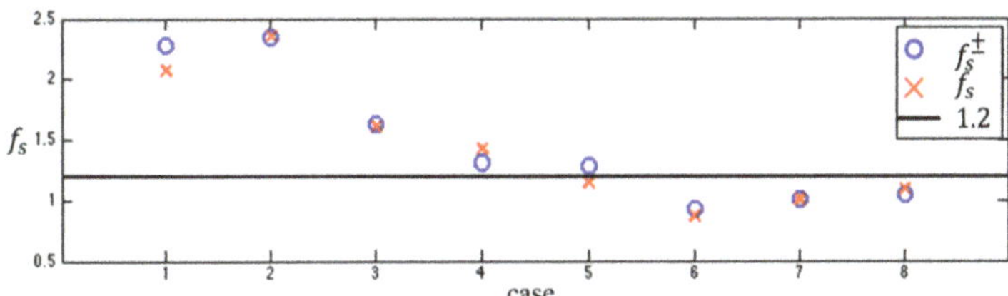

Figure 10. Comparisons of the safety factors against sliding among the minimum of $f_s^{\pm}$ (o), f_s (×) and safety criteria 1.2(—).

The safety factors against overturning are shown as Figure 11. All of the cases are greater than 1.2. Therefore, sliding failure is more dangerous than the overturning failure in Taichung harbor breakwater. Comparing between Figures 10 and 11, the tendency of safety factors between sliding and overturning are similar. This study shows the impact of overtopping waves in respect of the phase difference between the front and rear side. Not only wave condition but also the configuration of caisson such as its width and crown height might also affect the safety factor of breakwater. The impact on the fluctuation of the safety factor in case 2 is smaller than in case 1.

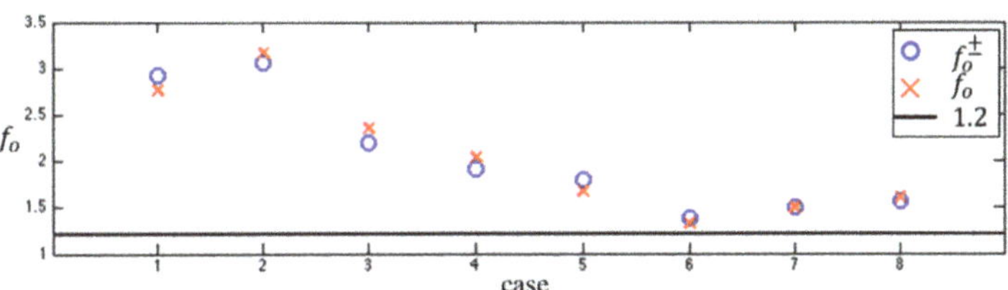

Figure 11. Comparisons of the safety factors against overturning among the minimum of f_o ± (o), f_o (×) and safety criteria 1.2(—).

To focus on the IWO on the stability analysis of vertical breakwaters, the load analysis of case 4, an example of $f_o^{\pm} < f_o$, is shown as Figure 12. When time is 189.5 s, the minimum safety factor against sliding on the IWO $f_s^{\pm} = 1.32$ is greater than the safety factor against sliding in static $f_s = 1.43$. At that time, the force f_1 in Figure 12b is 4395 kN and the force f_2 in Figure 12c in dynamic is 1840 kN, while that in static is 2084 kN, and the force f_3 in Figure 12d is 5614 kN. Because the force on the rear side of caisson considers the IWO is less than the force in still water, that caused the corresponding safety factor against sliding to be lower. Therefore, the impulse force happens at the moment of the minimum of load in the rear side, the safety factor drops significantly and the failure of sliding might cause breakwater damage. Furthermore, overtopping safety factors in case 4, an example of $f_o^{\pm} < f_o$, are shown at Figure 13. At 189.6 s, the minimum safety factor against overturning on the IWO $f_o^{\pm} = 1.921$ is less than the safety factor against overturning in static $f_o = 2.048$. At the time point, the moment m_1 in Figure 13b is 42,230 kNm, the moment m_2 in Figure 13c in dynamic is 12,035 kNm; while that in static is 14,317 kNm and the moment m_3 in Figure 13d is 69,279 kNm. Since the moment m_2 on the IWO is less than that in static, the corresponding safety factor against overturning is lower.

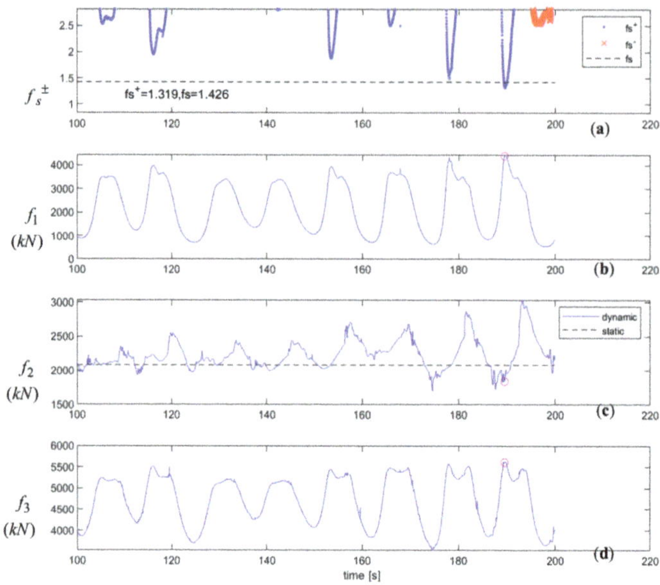

Figure 12. Comparisons of time histories of the safety factors against sliding and their loads for case 4. (**a**) The safety factors against sliding among $f_s^+(\cdot), f_s^-$ (×) and f_s (—); (**b**) The forces f_1 (—); (**c**) The forces f_2 in static (—) and in dynamic (—); (**d**) The forces f_3 (—).

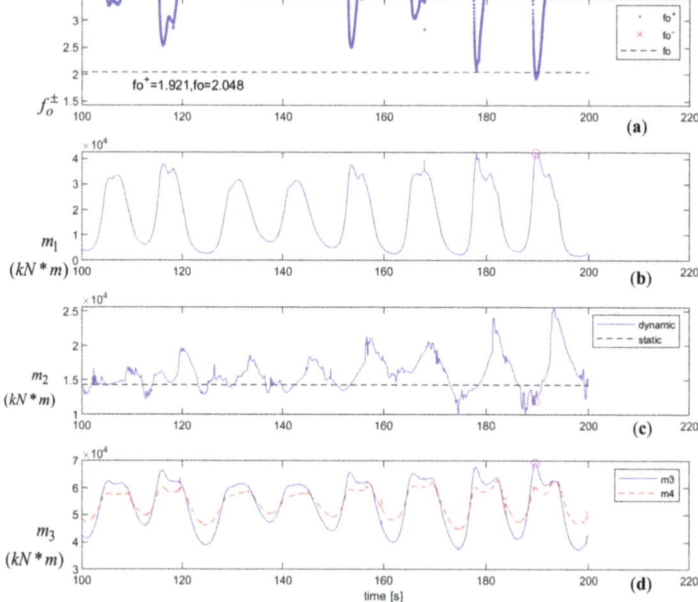

Figure 13. Comparisons of time histories of the safety factors against overturning and their moments for case 4. (**a**) The safety factors against overturning among f_o^+ $(\cdot), f_o^-$ (×) and f_o (—); (**b**) The moment m_1 (—); (**c**) The moment m_2 in static (—) and in dynamic (—); (**d**) The moment m_3 (—) and the moment m_4 (—).

In summary, the risk of sliding is greater than that of overturning in Taichung harbor breakwater. Under the extreme wave more than 100 year return period, the breakwater is unstable due to sliding, but safe from overturning. The IWO on the stability analysis is dominated by the force on the rear side of the caisson and the phase difference on the two ends of caisson.

5. Conclusions

To evaluate the stability of caisson breakwater concerning the effect of wave overtopping, a 2D VOF-type RANS model using a $k-\varepsilon$ turbulence closure (COBRAS) has been demonstrated to be suitable for simulating the complex hydrodynamics induced by overtopping waves. Experimental wave profiles and dynamic wave pressures on the seaward, backward and the bottom of the caisson breakwater can be suitably simulated by the present model. The horizontal forces on the front and rear sides and the uplift force as well as the moments on the caisson could also be analyzed. According to the numerical simulations, the risk of sliding is greater than that of overturning in the case of caisson breakwater in Taichung harbor. Under the extreme wave more than 100 year return period, the caisson breakwater is unstable due to sliding, but that is safe for overturning. Therefore, it is important to calculate the force on the rear side of the caisson and the phase difference of surface elevation on the two sides of the caisson as we analyze the stability analysis for wave overtopping for engineering design.

Author Contributions: Conceptualization, H.-C.H.; methodology, H.-C.H.; software, M.-S.L. and C.-J.H.; validation, M.-S.L. and C.-J.H.; formal analysis, H.-C.H. and L.-H.T.; writing—Original draft preparation, H.-C.H., M.-S.L. and C.-J.H.; writing—Review and editing, H.-C.H. and L.-H.T.; funding acquisition, H.-C.H. and L.-H.T. All authors have read and agreed to the published version of the manuscript.

Funding: This research was funded by the Harbor and Marine Technology Center, Ministry of Transportation and Communications, grant number MOTC-IOT-104-H2DB004a and the Ministry of Science and Technology, Taiwan, R.O.C., grant number 107-2221-E-110-077-MY3 and MOST 107-2221-E-006-087.

Conflicts of Interest: The authors declare no conflict of interest.

References

1. Oumeraci, H. Review and analysis of vertical breakwater failures—Lessons learned. *Coast. Eng.* **1994**, 22, 3–29.
2. Allsop, N.W.; Hawkes, P.I.; Jackson, F.A.; Franco, L. *Wave Run-Up on Steep Slopes-Model Tests under Random Waves*; Report No. SR2; Hydraulics Research Station: Wallingford, UK, 1985.
3. Bullock, G.; Obhrai, C.; Peregrine, D.; Bredmose, H. Violent breaking wave impacts. Part 1: Results from large-scale regular wave tests on vertical and sloping walls. *Coast. Eng.* **2007**, 54, 602–617. [CrossRef]
4. Cuomo, G.; Allsop, W.; Bruce, T.; Pearson, J. Breaking wave loads at vertical seawalls and breakwaters. *Coast. Eng.* **2010**, 57, 424–439. [CrossRef]
5. Oumeraci, H.; Kortenhaus, A.; Allsop, N.W.H.; De Groot, M.B.; Crouch, R.S.; Vrijling, J.K.; Voortman, H.G. *Probabilistic Design Tools for Vertical Breakwater*; Balkema: Rotterdam, The Netherlands, 2001.
6. Minikin, R.C.R. *Winds, Waves, and Maritime Structures: Studies in Harbour Making and in the Protection of Coasts*; Griffin: London, UK, 1963.
7. Goda, Y. New wave pressure formulae for composite breakwater. In *Coastal Engineering 1974, Proceedings of the 14th Conference on Coastal Engineering, Copenhagen, Denmark, 24–28 June 1974*; ASCE: New York, NY, USA, 1974; pp. 1702–1720.
8. Ewing, J.A. *Random Seas and Design of Maritime Structures*, 3rd ed.; World Scientific: Singapore, 1985.
9. Takahashi, S.; Tanimoto, K.; Simosako, K. A proposal of impulsive pressure coefficient for the design of composite breakwaters. In *In HYDRO-PORT '94: Proceedings of the International Conference on Hydro-Technical Engineering for Port and Harbor Construction: Yokosuka, Japan, 19–21 October 1994*; Coastal Development Institute of Technology: Yokosuka, Japan, 1994; Volume 1, pp. 489–504.
10. Takahashi, S.; Shimosako, K. *Reduction of Wave Force on a Long Caisson of Vertical Breakwater and Its Stability*; Technical Notes No. 685 of Port and Harbour Research Institute; Port and Harbour Research Institute: Nagase, Yokosuka, Japan, 1990; pp. 1–20. (In Japanese)

11. Burcharth, H.F.; Liu, Z. Force Reduction of Short-Crested Non-Breaking Waves on Caissons. In *Final Report of Mast III Project PROVERBS*; (PRObabilistic Design Tools for VERtical BreakwaterS); Technical University of Braunschweig: Braunschweig, Germany, 1999; Volume II, Chapter 4.3; pp. 1–26.
12. Mares-Nasarre, P.; Van Gent, M.R. Oblique wave attack on rubble mound breakwater crest walls of finite length. *Water* **2020**, *12*, 353. [CrossRef]
13. Sollitt, C.K.; Cross, R.H. Wave transmission through permeable breakwaters. In Proceedings of the 13th International Conference on Coastal Engineering, Vancouver, BC, Canada, 10–14 July 1972; ASCE: New York, NY, USA, 1972; pp. 1827–1846.
14. Vidal, C.; Losada, M.A.; Medina, R.; Rubio, J. Solitary wave transmission through porous breakwaters. In *Coastal Engineering 1988, Proceedings of 21th International Conference on Coastal Engineering, Costa del Sol-Malaga, Spain, 20–25 June 1988*; ASCE: New York, NY, USA, 1988; pp. 1073–1083.
15. Liu, P.L.-F.; Wen, J. Nonlinear diffusive surface waves in porous media. *J. Fluid Mech.* **1997**, *347*, 119–139. [CrossRef]
16. Kobayashi, N.; Wurjanto, A. Wave transmission over submerged breakwaters. *J. Waterw. Port Coast. Ocean Eng.* **1989**, *115*, 662–680. [CrossRef]
17. Kobayashi, N.; Wurjanto, A. Wave overtopping on coastal structures. *J. Waterw. Port Coast. Ocean Eng.* **1989**, *115*, 235–251. [CrossRef]
18. Kobayashi, N.; Wurjanto, A. Numerical model for wave on rough permeable slopes. *J. Coast. Res.* **1990**, *7*, 149–166.
19. Mingham, C.G.; Causon, D.M. High-resolution finite-volume method for shallow water flows. *J. Hydraul. Eng.* **1998**, *124*, 605–614. [CrossRef]
20. Hu, K.; Mingham, C.; Causon, D. Numerical simulation of wave overtopping of coastal structures using the non-linear shallow water equations. *Coast. Eng.* **2000**, *41*, 433–465. [CrossRef]
21. Hubbard, M.E.; Dodd, N. A 2D numerical model of wave run-up and overtopping. *Coast. Eng.* **2002**, *47*, 1–26. [CrossRef]
22. Stansby, P.K.; Feng, T. Surf zone wave overtopping a trapezoidal structure: 1-D modelling and PIV comparison. *Coast. Eng.* **2004**, *51*, 483–500. [CrossRef]
23. Sulisz, W. Wave reflection and transmission at permeable breakwaters of arbitrary aross-section. *Coast. Eng.* **1985**, *9*, 371–386. [CrossRef]
24. Shigematsu, T.; Liu, P.L.F.; Oda, K. Numerical modeling of the initial stages of dam-break waves. *J. Hydraul. Res.* **2004**, *42*, 183–195. [CrossRef]
25. Özgökmen, T.M.; Iliescu, T.; Fischer, P.F.; Srinivasan, A.; Duan, J. Large eddy simulation of stratified mixing in two-dimensional dam-break problem in a rectangular enclosed domain. *Ocean Model.* **2007**, *16*, 106–140. [CrossRef]
26. Khayyer, A.; Gotoh, H. On particle-based simulation of a dam break over a wet bed. *J. Hydraul. Res.* **2010**, *48*, 238–249. [CrossRef]
27. Shakibaeinia, A.; Jin, Y.-C. A mesh-free particle model for simulation of mobile-bed dam break. *Adv. Water Resour.* **2011**, *34*, 794–807. [CrossRef]
28. Lin, P.Z.; Liu, P.L.-F. A numerical study of breaking waves in the surf zone. *J. Fluid Mech.* **1998**, *359*, 239–264. [CrossRef]
29. Lin, P.; Liu, P.L.-F. Turbulence transport, vorticity dynamics, and solute mixing under plunging breaking waves in surf zone. *J. Geophys. Res. Space Phys.* **1998**, *103*, 15677–15694. [CrossRef]
30. Lin, P.; Xu, W. Newflume: A numerical water flume for two-dimensional turbulent free surface flows. *J. Hydraul. Res.* **2006**, *44*, 79–93. [CrossRef]
31. Losada, I.J.; Lara, J.L.; Guanche, R.; Gonzalez-Ondina, J.M. Numerical analysis of wave overtopping of rubble mound breakwaters. *Coast. Eng.* **2008**, *55*, 47–62. [CrossRef]
32. Guanche, R.; Losada, I.J.; Lara, J.L. Numerical analysis of wave loads for coastal structure stability. *Coast. Eng.* **2009**, *56*, 543–558. [CrossRef]
33. Walkden, M.; Wood, D.; Bruce, T.; Peregrine, D. Impulsive seaward loads induced by wave overtopping on caisson breakwaters. *Coast. Eng.* **2001**, *42*, 257–276. [CrossRef]
34. Chiu, Y.F.; Lee, J.Y.; Chang, S.C.; Lin, Y.J.; Chen, C.H. An experiment study of wave forces on vertical breakwater. *J. Mar. Sci. Technol.* **2007**, *15*, 158–170.
35. Goda, Y.; Takahashi, S. *Advanced Design of Maritime Structures in the 21st Century*; Port and Harbour Research Institute: Yokosuka, Japan, 2001.
36. Shih, T.-H.; Zhu, J.; Lumley, J.L. Calculation of wall-bounded complex flows and free shear flows. *Int. J. Numer. Methods Fluids* **1996**, *23*, 1133–1144. [CrossRef]

37. Ting, F.C.; Kirby, J.T. Observation of undertow and turbulence in a laboratory surf zone. *Coast. Eng.* **1994**, *24*, 51–80. [CrossRef]
38. Neves, M.D.G.; Reis, M.T.; Losada, I.J.; Hu, K. Wave overtopping of Póvoa de varzim breakwater: Physical and numerical simulations. *J. Waterw. Port Coast. Ocean Eng.* **2008**, *134*, 226–236. [CrossRef]
39. Willmott, C.J. On the validation of models. *Phys. Geogr.* **1981**, *2*, 184–194. [CrossRef]
40. Cheng, K.G.; Chien, C.C.; Su, C.H.; Tseng, H.M.; Chen, G.Y.; Chen, Y.C. *Taichung Harbour Extension Design Phase II—Primary*; Institute of Transportation; MOTC: Taipei City, Taiwan, 1999. (In Chinese)
41. Hsu, T.W. *A Study of Adaptation Capacity of Coastal Disasters Due to Climate Change in Order to Strengthen Northwest and Northeast Areas of Taiwan (2/2)*; Water Resources Agency; MOEA: Taipei City, Taiwan, 2013. (In Chinese)

Journal of
Marine Science and Engineering

Article

Validation of RANS Modelling for Wave Interactions with Sea Dikes on Shallow Foreshores Using a Large-Scale Experimental Dataset

Vincent Gruwez [1,*], Corrado Altomare [1,2], Tomohiro Suzuki [3,4], Maximilian Streicher [1], Lorenzo Cappietti [5], Andreas Kortenhaus [1] and Peter Troch [1]

1 Department of Civil Engineering, Ghent University, 9000 Ghent, Belgium; corrado.altomare@upc.edu (C.A.); maximilian.streicher@ugent.be (M.S.); andreas.kortenhaus@ugent.be (A.K.); peter.troch@ugent.be (P.T.)
2 Maritime Engineering Laboratory, Department of Civil and Environmental Engineering, Universitat Politecnica de Catalunya—BarcelonaTech (UPC), 08034 Barcelona, Spain
3 Flanders Hydraulics Research, 2140 Antwerp, Belgium; tomohiro.suzuki@mow.vlaanderen.be
4 Faculty of Civil Engineering and Geosciences, Delft University of Technology, 2628 CN Delft, The Netherlands
5 Department of Civil and Environmental Engineering, University of Florence, 50139 Florence, Italy; lorenzo.cappietti@unifi.it
* Correspondence: vincent.gruwez@ugent.be

Received: 17 July 2020; Accepted: 18 August 2020; Published: 24 August 2020

Abstract: In this paper, a Reynolds-averaged Navier–Stokes (RANS) equations solver, interFoam of OpenFOAM®, is validated for wave interactions with a dike, including a promenade and vertical wall, on a shallow foreshore. Such a coastal defence system is comprised of both an impermeable dike and a beach in front of it, forming the shallow foreshore depth at the dike toe. This case necessitates the simulation of several processes simultaneously: wave propagation, wave breaking over the beach slope, and wave interactions with the sea dike, consisting of wave overtopping, bore interactions on the promenade, and bore impacts on the dike-mounted vertical wall at the end of the promenade (storm wall or building). The validation is done using rare large-scale experimental data. Model performance and pattern statistics are employed to quantify the ability of the numerical model to reproduce the experimental data. In the evaluation method, a repeated test is used to estimate the experimental uncertainty. The solver interFoam is shown to generally have a very good model performance rating. A detailed analysis of the complex processes preceding the impacts on the vertical wall proves that a correct reproduction of the horizontal impact force and pressures is highly dependent on the accuracy of reproducing the bore interactions.

Keywords: validation; wave modelling; shallow foreshore; dike-mounted vertical wall; wave impact loads; OpenFOAM

1. Introduction

Low-elevation coastal zones often have mildly to steeply-sloping sandy beaches as part of their coastal defence system. For countries in north-western Europe, coastal urban areas typically have high-rise buildings close to the coastline. These buildings are usually fronted by a low-crested, steep-sloped, and impermeable sea dike with a relatively short promenade, where the long (nourished) beach in front of the dike acts as a mildly sloping shallow foreshore. This type of coastal defence system therefore combines hard and soft coastal protection against flooding. Such hybrid approaches are regarded by the Intergovernmental Panel on Climate Change (IPCC) with high agreement as a promising way forward in terms of response to sea level rise [1]. Along the cross-section of this hybrid beach-dike coastal defence system, storm waves undergo many transformation processes before they finally hit the buildings on top of the dike. Along the shallow waters of the mildly sloping foreshore in front of the dike, sea/swell or short waves (hereafter SW, $O(10^1$ s)) shoal and eventually break, transferring energy to both their super- and subharmonics (or long waves: hereafter LW, $O(10^2$ s)) by nonlinear wave-wave interactions. Further pre-overtopping hydrodynamic processes along the mildly sloping foreshore include wave dissipation by breaking (turbulent bore formation) and bottom friction, reflection against the foreshore and dike, and wave run-up on the dike slope. Finally, waves overtop the dike crest, and post-overtopping processes include bore propagation on the promenade, bore impact on a wall or building, and reflection back towards the sea interacting with incoming bores on the promenade.

For the (structural) design of storm walls or buildings on such coastal dikes, the wave impact force expected for specific design conditions needs to be estimated. Semi-empirical formulas, mostly based on physical model tests, are commonly used in practice to assess wave forces and pressures on coastal defences, at least in a preliminary design phase. However, semi-empirical formulas are usually restricted within very specific ranges of application, currently limiting force prediction to dikes with deep foreshore depths [2,3]. Such formulas do exist for dikes with very/extremely shallow foreshore depths as well [4,5], but their application is also strictly limited. For the final design, therefore, often detailed experimental campaigns are required [6]. Alternatively, during the last decade, numerical modelling of these combined processes has become feasible [3,7–11]. Numerical modelling is also able to provide a detailed and accurate assessment of a specific case. Moreover, numerical models can provide information on physical quantities that are difficult to measure in a scaled model or in prototype (e.g., detailed velocity fields, pressure distributions, etc.).

To study fully two-dimensional vertical (2DV) complex fluid flows, computational fluid dynamics (CFD) techniques are typically applied. Relatively new mesh-free Lagrangian numerical methods, such as Smoothed Particle Hydrodynamics (SPH) [12] and the particle finite element method [13], have been recently validated and applied to several coastal engineering problems [9,14–17], showing much promise. However, different from Eulerian grid-based methods, multi-phase air-fluid SPH models are still quite scarce and have a high computational cost [18]. The more traditional Eulerian numerical methods are already more consolidated. For example, volume-of-fluid methods (VOF) based on the Reynolds-averaged Navier–Stokes equations (RANS) have been widely employed during the last decades. Using RANS models, processes such as wave transformation [8,19,20], wave overtopping [7,21,22], and wave impact on coastal structures [3,23–26] have been modelled and validated, but never before at the same time (to the knowledge of the authors). They are computationally very expensive to apply, but have shown their value particularly for wave-structure interaction phenomena involving complex geometries. In addition, two-phase water-air RANS models allow taking the effects of air entrapment on the wave impact processes into account [27,28].

Validation of numerical models is crucial before they can be reliably applied. Even though plenty of works have been published on numerical modelling and validation of individual processes previously listed, there is still a lack of literature about RANS model validation for wave impacts on sea dikes and dike-mounted walls in presence of a very shallow foreshore. The main goal of this paper is to validate a two-phase (water-air) RANS model for this specific case. Such a modelling approach is deemed necessary to fully resolve the 2DV complex fluid flows of overtopped waves and bore interactions on top of the promenade. The RANS solver (interFoam) for two incompressible fluids within the open source CFD toolbox OpenFOAM® is chosen because of its increasing popularity for application to wave-structure interactions. Validation of this numerical model is done by reproducing large-scale experiments of overtopped wave impacts on coastal dikes with a very shallow foreshore from the WAve LOads on WAlls (WALOWA) project [29]. The large-scale nature of these experiments reduces the scale effects significantly compared to small-scale experiments, which can be particularly important to the wave impacts on the dike-mounted vertical wall, especially in case of plunging breaking bore patterns and impulsive impacts [30].

The paper is structured as follows. First, the methods used in the paper are explained in Section 2, starting with the experimental model setup and a description of the tests used for the validation. This is followed by a description of the applied RANS model and the numerical model setup. Finally, the statistical model performance methods applied in this study are discussed. Next, in Section 3 the results of the qualitative and quantitative numerical model validation are provided, including a comparison of model snapshots at key time instants during impacts on the vertical wall. This is finally followed by Section 4 with a discussion on these results and the conclusions in Section 5.

2. Methods

2.1. Large-Scale Laboratory Experiments

The laboratory experiments (Froude length scale 1/4.3) were done during the research project WALOWA in the Deltares Delta Flume, which is 291 m long, 9.5 m deep, and 5 m wide. This wave flume is equipped with a piston-type wave maker capable of up to second-order wave generation (in the frequency range 0.02 Hz–1.50 Hz) and includes active reflection compensation (ARC), which is an active wave absorption (AWA) system to minimise reflections against the wave paddle. For a detailed description of the model setup, reference is made to Streicher et al. [29]. The WALOWA dataset is open access and is described by Kortenhaus et al. [31].

The model geometry consisted of a moveable sandy foreshore with a transition slope of 1:10 and a slope of 1:35 up to the toe of the dike (Figure 1). The smooth impermeable concrete dike had a front slope of 1:2, a promenade width of 2.35 m with an inclination of 1:100 in order to help drain the water in case of wave overtopping, and finally a 1.60 m high wall. The wall height was designed to be high enough to prevent wave overtopping during testing, but small amounts of overtopped water could still be returned via a recirculation drainage pipe behind the wall.

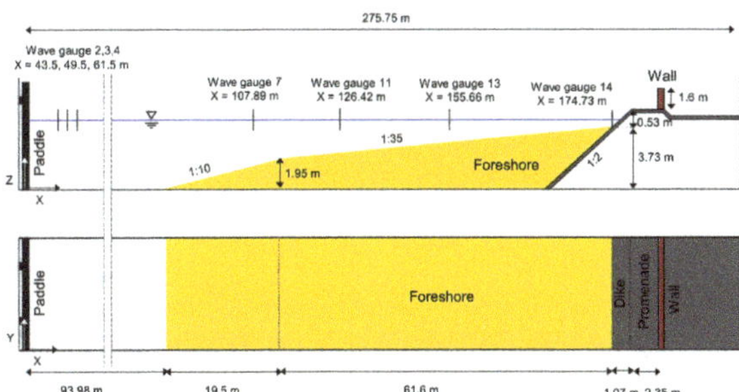

Figure 1. Overview of the geometrical parameters of the wave flume and WALOWA model set-up, with indicated wave gauge locations. Reprinted with permission from [29].

The WALOWA dataset includes both bichromatic and irregular wave tests. For validation of the numerical model, the bichromatic wave test Bi_02_6 (EXP) and its repetition Bi_02_6_R (REXP) were selected (Table 1). The bichromatic wave tests have the advantage to be relatively short in time, while still considering the effects of wave dispersion and bound LWs, and are therefore more representative of irregular waves than monochromatic waves. In this way, even numerical models with a high computational demand are able to simulate the tests in a reasonable amount of computational time. This specific bichromatic wave test was chosen because it is the only test that was conducted shortly after a foreshore profile measurement and at the same time immediately followed by its repetition and another foreshore profile measurement [32,33]. Since these bichromatic wave tests are relatively short in duration and only limited changes ($O(10^{-2}$ m)) were noted between the profile measurements before and after [32], a fixed bed is a reasonable assumption for the numerical modelling. In addition, the repeated test makes validation of the numerical model possible relative to the experimental uncertainty.

Table 1. Hydraulic parameters for the WALOWA bichromatic wave test (EXP) and its repetition (REXP): h_o is the offshore water depth, h_t the water depth at the dike toe, $H_{m0,o}$ the incident offshore significant wave height, R_c the dike crest freeboard, f_i the SW component frequency, a_i the SW component amplitude, and β_m (= a_2/a_1) the modulation factor.

TestID [-]	Duration [s]	h_o [m]	h_t [m]	$h_t/H_{m0,o}$ [-]	R_c [m]	f_1 [Hz]	a_1 [m]	f_2 [Hz]	a_2 [m]	β_m [-]
Bi_02_6 (EXP) & Bi_02_6_R (REXP)	209	4.14	0.43	0.33	0.117	0.19	0.45	0.155	0.428	0.951

During these tests, three bichromatic wave groups were generated with first order wave control over 125 s, including 10 s of tapering at the beginning and end of the wave generation. Plunging breakers occurred on the 1:10 transition slope (i.e., deep water Iribarren number $\xi_0 = \tan\alpha/(H/L_0)^{1/2}$ with α the foreshore slope angle, H the wave height, and L_0 the deep water wave length [34]: $0.5 < \xi_0 \approx 0.7 < 3.3$) and spilling breakers on the 1:35 foreshore slope ($\xi_0 \approx 0.2 < 0.5$). Considering this was a test of a dike with a very shallow foreshore depth (Table 1: $0.3 < h_t/H_{m0,o} < 1.0$ [35]), the wave energy at the toe of the dike was dominated by LW energy.

The measurement setup consisted of instruments to measure the water surface elevation along the flume and on the promenade, the velocity of the overtopped flow on the promenade, and the impact pressure and force on the vertical wall (Figure 2). All measurements were sampled at 1000 Hz frequency and were synchronized in time.

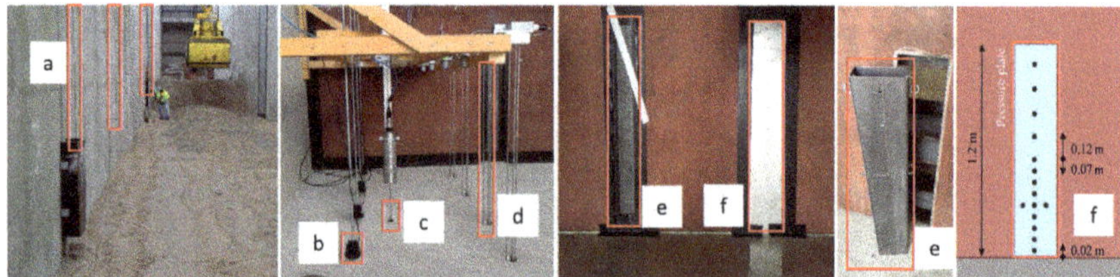

Figure 2. (**a**) WGs deployed along the flume side wall to measure η; (**b**) PWs; (**c**) ECM to measure U_x; (**d**) WLDMs installed on the promenade to measure η; (**e**) Hollow steel profile attached to two LCs and (**f**) aluminium plate equipped with pressure sensors (PS) to measure F_x and p.

The water surface elevation η (with the vertical origin at $z = h_o$) was measured with resistance-type wave gauges (WG) deployed at seven different locations along the Delta Flume side wall (Figures 1 and 2a). WG02–WG04 were installed over the flat bottom part of the flume close to the wave paddle. These wave gauges were positioned to allow a reflection analysis following the method of Mansard and Funke [36]. WG07 was installed along the transition slope; WG11 and WG13 along the foreshore slope. WG14 was installed close (~0.35 m) to the dike toe. The data of WG11 are not considered further in the present analysis because of faulty data. Furthermore, to remove unwanted noise in the η signals measured by the other WG's from the wave paddle up to the dike toe, a low-pass 3rd order Butterworth filter with a cut-off frequency of 1.50 Hz was applied. This frequency is well above the frequencies of the super-harmonics of the primary waves and frequency components due to triad interactions between the primary components and the difference frequency, which gain energy in the shoaling and surf zone [37].

Flow layer level measurements η on the promenade were obtained by four resistance-type water level distance meters (WLDM01–WLDM04, Figure 2d). Flow velocity measurements on the promenade were obtained by four paddle wheels (PW01–PW04, Figure 2b), measuring the horizontal flow velocity U_x in one direction (i.e., towards the wall) 0.026 m above the promenade. Additionally, a bidirectional electromagnetic current meter (ECM, Figure 2c) was installed at the same cross-shore location as WLDM02 and PW02 to obtain directional information of the incoming or reflected flow. The ECM disc was positioned 0.03 m above the promenade and sampled the horizontal velocity at 16 Hz. Further detailed information on the sensor setup on the promenade and the post-processing of the η and U_x data measured on top of the promenade was provided by Cappietti et al. [38]. During return flow, positive U_x values were possibly incorrectly measured by the PWs, indicated by the ECM that measured negative U_x values during return flow (compared to the measurements of the co-located PW02). This will be further discussed in the comparison with the numerical model result (Section 3.1). However, no such co-located measurements are available for other paddle wheels than PW02, so no correction of the PW measurements during return flows was attempted.

The overtopped wave impacts on the wall were measured by horizontal force F_x and pressure p measurement systems integrated into the wall. The horizontal impact force was measured by two compression-type load cells (LC) connecting the same hollow steel profile to the very stiff supporting structure (Figure 2e). Impact pressures were measured by 15 pressure sensors (PS). The first 13 PSs were spaced vertically over a metal plate flush mounted in the middle section of the steel wall, with PS14 and PS15 placed horizontally next to PS05 or the fifth PS from the bottom (Figure 2f). The initial post-processing of the F_x and p signals, including baseline correction and filtering, is discussed by Streicher [39]. Additional filtering was applied to remove the high frequency oscillations caused by stochastic processes during dynamic or impulsive impacts, so that the signal can be reproduced by a deterministic numerical model [40]. To achieve this, an additional 3rd order Butterworth low-pass filter with a cut-off frequency of 6.22 Hz was necessary. This corresponds to a cut-off frequency of 3.0 Hz at a prototype scale, which is still well above the natural frequency of about 1.0 Hz for typical buildings found along, e.g., the Belgian coast [41]. Furthermore, local spatial variability over the width of the flume of the resultant F_x (i.e., derived from the LCs and pressure integrated) and p (i.e., PS05, PS14, and PS15) time series was found to be low (not shown). This spatial variability over the width of the experimental flume was therefore further neglected in the quantitative numerical model validation: for F_x, the LC-derived signal was used and for p, the PS05 signal was used.

2.2. Numerical Model

2.2.1. Model Description

In this work, OpenFOAM v6 [42] was applied and validated, or more specifically interFoam, a solver of the RANS equations, where the advection and sharpness of the water–air interface are handled by an algebraic VOF method [43] based on the multidimensional universal limiter with explicit solution (MULES) [44–46]. InterFoam with MULES has already been successfully applied for wave propagation [45], wave breaking [20,47–50], wave run-up [20,50], wave overtopping [51,52], and bore impact on a vertical wall [26].

Several open source contributions of boundary conditions for wave generation and absorption exist for interFoam, of which the main developments are IHFOAM [53], olaFlow [54], and waves2Foam [55]. In the present study, olaFlow was chosen, which was found to be the most computational efficient [53,56,57] and feature complete package at the time of the simulations presented in this paper.

The turbulence is modelled by the k-ω SST turbulence closure model [58], which has been shown to be one of the most proficient in modelling wave breaking [47]. Two-equation turbulence closure models are known to cause over-predicted turbulence levels beneath computed surface waves, leading to unphysical wave decay for wave propagation over constant water depth and long distance [49,59,60]. Turbulence modelling was therefore stabilized in nearly potential flow regions by Larsen and Fuhrman [49], with their default parameter values [61]. Hereafter, the OpenFOAM numerical model as presented here is simply referred to as OF.

2.2.2. Computational Domain and Mesh

Wave breaking is an inherently three-dimensional (3D) process due to the formation of 3D vortices extending obliquely downward in the inner surf zone [62]. Even so, many examples exist where the wave kinematics during wave breaking could be approximated well by vertical two-dimensional (2DV) RANS modelling [8,19,47–50,63,64]. To reduce the computational time as much as possible, OF is therefore applied in a 2DV configuration (i.e., cross-shore section of the wave flume).

The OF model domain (Figure 3) starts at the wave paddle zero position ($x = 0.00$ m) and ends on top of the vertical wall ($x = 178.80$ m). The bottom boundary is at its lowest point ($z = 0.00$ m) along the flume bottom between the wave paddle and the foreshore toe, and extends up to z = 7.20 m, well above the maximum measured surface elevations along the flume. The bottom is further defined by the measured foreshore and dike geometry as described in Section 2.1. The vertical wall is included up to its height of 1.60 m including the top which was given a slight inclination towards the model boundary to allow overtopped water (limited to mainly spray in this case) to exit the model domain.

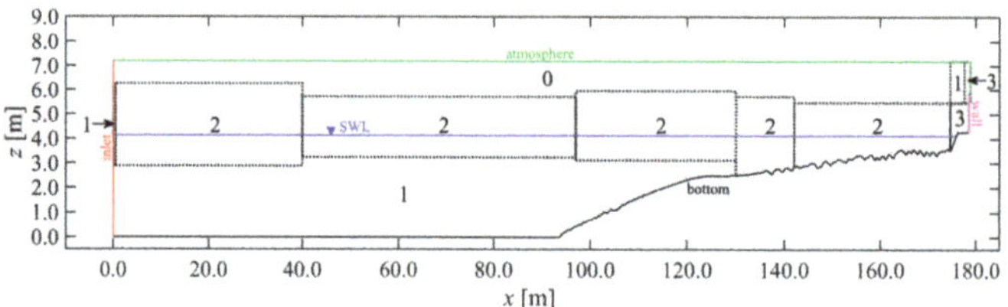

Figure 3. Definition of the OF 2DV computational domain, with coloured indication of the model boundary types. The still water level (SWL) is indicated in blue (z = 4.14 m). The number in each of the mesh subdomains of the model domain (demarcated by black dotted lines) is the refinement level β applied in each subdomain (for $\beta = 0$, 1, 2, and 3: $\Delta x = \Delta z = 0.18$ m, 0.09 m, 0.045 m, and 0.0225 m). Note: the axes are in a distorted scale.

The computational domain is discretised into a structured grid. To optimise the computational time, a variable grid resolution is applied, where a higher resolution is defined only where it is necessary. This is mostly the areas of the model domain where the water-air interface is expected to pass [46,56]. The expected location of the free surface along the flume during the entire test was estimated first by a fast preliminary one-layer depth-averaged SWASH calculation (not shown: see [65] for the SWASH model setup description). The minimum and maximum η along the flume and over the complete test duration were obtained from the SWASH model result to define areas in which mesh refinement should be done. These locations are delineated by the dotted lines in Figure 3, defining several areas around the still water level (*SWL*). In front of the wave paddle, the refinement area is slightly higher to accommodate the stabilisation of the newly generated waves, after which the refinement zone can decrease in height when the waves have fully developed. Then, the refinement area is increased in height again to allow room for wave shoaling and incipient wave breaking on the foreshore. The upper limit can subsequently be lowered again due to wave breaking, but the lower limit is extended to include the bottom boundary. This is done to properly resolve the entrained air pockets that have been shown to travel towards the bottom during the breaking process in the inner surf zone [66]. The height of the refinement zone on the dike was defined based on the maximum measured water level in the experiment by the WLDM's on the promenade and extended to the upper model boundary along the vertical wall to resolve the run-up and splashing against the vertical wall.

In terms of the grid cell size in these refinement zones, about 20 cells are typically recommended over the wave height H of a regular wave (i.e., $H/\Delta z = 20$, with Δz being the vertical cell size) [46,57]. Applied to the wave heights of the primary wave components of the bichromatic wave in Table 1, a minimal vertical cell size of $\Delta z = 0.045$ m to 0.043 m is obtained. Smaller wave heights in the bichromatic wave group are less resolved with this choice, but this is deemed acceptable because of their relatively low steepness. A value of $\Delta z = 0.045$ m was chosen, because the water depth at the wave paddle h_o is divisible by it (i.e., $h_o/\Delta z = 4.14/0.045 = 92$), meaning that the *SWL* can lie perfectly along cell boundaries, or in other words, α-values between 0 and 1 are thereby minimised at the start of the simulation, which simplifies the initialisation of the SWL and is beneficial for an effectively still SWL at the start of the simulation.

The mesh maintains an aspect ratio $\Delta x/\Delta z$ of 1 (with Δx being the horizontal cell size) throughout the entire computational domain, which has been shown necessary for accuracy [46,55,66] and numerical stability in this study. One exception is a higher aspect ratio along the bottom and wall, where layers were locally added to the mesh to resolve the boundary layer. Six layers were added over the vertical cell size along those boundaries, with a growth rate of 1.2, leading to a maximum aspect ratio of 18.

Outside the refinement zones, in the air and water phases, the mesh can be coarser [46,57]. The structured mesh was given a base grid resolution of 0.18 m. This base resolution is multiplied by a refinement ratio r, here defined as:

$$r = \frac{1}{2^{\beta}} \tag{1}$$

in which β signifies the refinement level. Each refinement level effectively refines every cell into four new cells. The applied refinement levels are provided for each mesh subdomain in Figure 3. For the air in the model domain, the base resolution was assumed ($\beta = 0$), except for a small area over the dike ($\beta = 1$). In the water phase, refinement level 1 was assumed ($\Delta x = \Delta z = 0.09$ m) and was further refined in the zone of the surface elevation up to the dike toe (level 2 or $\Delta x = \Delta z = 0.045$ m). Close to the inlet boundary, however, a lower refinement level was necessary for numerical stability ($\beta = 1$) over a very short distance (0 m $< x <$ 0.50 m) where locally high water velocities (i.e., low Courant numbers and low time steps) at the interface can occur due to the wave generation. On the dike up to the wall, the mesh was refined even more (level 3 or $\Delta x = \Delta z = 0.0225$ m) to resolve thin layer flows, the complex flows of bore interactions, and impacts on the vertical wall. In addition, a refinement level 3 was necessary to resolve the experimental pressure sensor locations along the vertical wall.

The mesh was generated by applying the *cartesian2DMesh* algorithm of cfMesh [67], which resulted in a mesh with 318,381 cells, for the refinement levels indicated in Figure 3.

The adaptive time stepping is controlled by a predefined maximum Courant number $maxCo$ ($Co = \Delta t\ |U|/\Delta X$, where Δt is the time step, $|U|$ is the magnitude of the velocity through that cell, and ΔX is the cell size in the direction of the velocity [68]) and a maximum Courant number in the interface cells $maxAlphaCo$. Generally $maxCo = maxAlphaCo$ is chosen, as well as in this paper. Larsen et al. [45] have shown that a relatively low $maxCo$ (~0.05) is necessary to obtain a stable wave profile over more than five wave periods' propagation duration. Here, however, a $maxCo$ of 0.25 is used to balance the accuracy and computational costs. Since the primary waves of the bichromatic wave group only propagate over about three wave lengths up to the mean breaking point location (x_b = ~120 m), this is considered an acceptable assumption. Both the refinement level in the refinement zones around the surface elevation zones (β_{sez}) and the $maxCo$ were verified in a convergence analysis (Appendix A).

2.2.3. Boundary Conditions

Since the model domain represents a 2DV simulation, no solution is necessary in the y-direction, and the lateral boundaries of numerical wave flume were assigned an "empty" boundary condition. Non-empty boundary conditions were defined for the remaining boundaries in the xz-plane (Figure 3).

The bichromatic waves from Table 1 were generated at the inlet by applying a Dirichlet-type boundary condition: the experimental wave paddle velocity was imposed. The paddle displacement time series is used by olaFlow to calculate the wave paddle velocity by a first-order forward derivative [69]. Since the reflection in the numerical wave flume is expected to behave close to, but not exactly the same as in the experiment, the theoretical paddle displacement without ARC was selected and the AWA by olaFlow was activated instead. In addition to the paddle displacement, the surface elevation at the wave paddle is provided, which allows olaFlow to trigger the AWA with fewer assumptions [69]. The AWA implementation in olaFlow is most effective for shallow water waves. The primary components of the bichromatic wave group are intermediate waves for the water depth at the wave paddle, but their reflection is expected to be low, since most of their wave energy dissipates over the foreshore in the surf zone. However, reflected free long (infragravity) waves are expected to be non-negligible (Section 3.2). They are shallow water waves and are by definition absorbed well by the AWA system in olaFlow, preventing their re-reflection and therefore replicating the behaviour of the ARC in the experiment.

Both the bottom and wall boundaries are fixed boundaries, including the sandy foreshore (Section 2.1), along which the velocity vector field U has a Dirichlet-type boundary condition (U = (0, 0, 0) m/s), while the pressure p and α are given a Neumann boundary condition. Along the foreshore, dike and wall, no-slip boundary conditions are assumed and a continuous scalable wall function based on Spalding's law [70] is implemented. The six boundary layers that were previously added in the mesh along these no-slip fixed boundaries make sure that the scalable wall function criterion for the dimensionless wall distance z^+ (i.e., $1 < z^+ < 300$) is complied with. For the remaining boundary conditions, initial conditions, and solver settings, the same settings were chosen as those reported by Devolder et al. [48].

The OF simulations were run in parallel on a 24-core Intel Xeon E5-2680 @ 2500 MHz computer with 128 GB of RAM. The scotch decomposition algorithm was used to divide the mesh into equal amounts of cells for each processor, while minimising the number of processor boundaries [42]. The cells along the inlet patch were forced onto the same processor, which benefits the computational efficiency. In this setup, the simulation required a CPU time of about 85 h.

2.2.4. Data Sampling and Processing

The same data were sampled in OF at the same cross-shore locations as in the experiment (Section 2.1). Applying the same sampling frequency of 1000 Hz in OF, however, would increase the calculation time to unpractical levels because it affects the time stepping. Instead, a sampling frequency of 80 Hz was maintained throughout, which is a compromise between the temporal resolution of the output data and the calculation time.

To obtain η in OF, α was recorded at a fixed interval over a vertical line at each wave gauge location. In post-processing, η was then obtained by vertical integration of α, thereby excluding air inclusions produced in the surf zone, but taking into account all water volumes (i.e., even air-borne water, e.g., in case of plunging waves, spray). This corresponds best to how η in the experiment was measured: resistive wave gauges give a response proportional to the wire wet length [71], thereby similarly excluding air pockets. However, it is acknowledged that some uncertainty remains regarding how resistive type wave gauges measure the free surface in the presence of air–water mixtures along the gauge. This could lead to discrepancies in the numerical–experimental model comparisons in the surf zone and on top of the promenade [72].

The resulting numerical time series were filtered in the same way as the experimental data (Section 2.1) and were synchronised to the experimental time reference. The synchronisation was done based on the η time series at the three most offshore located wave gauges (i.e., WG02–03–04) by means of a cross-correlation. The obtained numerical–experimental time lags for each of these WG locations were subsequently averaged and rounded to the nearest multiple of the time series time step. This time lag was then used to synchronise all numerical time series to the experimental time reference. This makes sure that numerical errors (such as phase lag), which are important for model validation, were retained.

Furthermore, to investigate the model performance for the SW and LW components separately, the η time series were separated into η_{SW} and η_{LW} by applying a 3rd-order Butterworth high- and low-pass filter, respectively. A separation frequency of 0.09 Hz was employed, which is in between the bound long wave frequency (f_1–f_2 = 0.035 Hz) and the lowest frequency of the primary wave components (f_2 = 0.155 Hz).

2.3. Validation Method

The validation of the numerical model OF to the large-scale experiment EXP is done both qualitatively and quantitatively. The qualitative validation entails a comparison of the time series of the main measured parameters. However, it is recommended to apply model performance statistics as well for a more quantified and objective validation [73]. Therefore, general numerical model performance will be evaluated by applying a skill score or dimensionless measure of average error, such as Willmott's refined index of agreement d_r [74]:

$$d_r = \begin{cases} 1 - \frac{MAE}{cMAD}, \; MAE \le cMAD \\ \frac{cMAD}{MAE} - 1, \; MAE > cMAD \end{cases} \tag{2}$$

where c is a scaling factor and is taken equal to 2 to obtain a balance between the number of deviations evaluated within the numerator and within the denominator of the fractional part of d_r; MAE is the mean absolute error defined by:

$$MAE = \frac{1}{N} \sum_{i=1}^{N} |P_i - O_i| \tag{3}$$

with N the number of samples in the time series, and P the predicted time series together with the pair-wise-matched observed time series O (for i = 1,2, ... ,N), and MAD is the mean-absolute deviation:

$$MAD = \frac{1}{N} \sum_{i=1}^{N} |O_i - \overline{O}| \tag{4}$$

where the overbar represents the mean of the time series. This model performance index d_r is bounded by [−1.0, 1.0] and, in general, more rationally related to model accuracy than other existing model performance indices or skill scores. For the purposes of this paper, d_r is used as a general measure of the model performance, and a d_r value of 0.5 is already considered to be a poor model performance. Since it is a single measure of model performance, it can be more easily used to evaluate, for example, the spatial model performance over the length of the wave flume.

Because a repetition of the selected experimental test is available (REXP), d_r can be evaluated between REXP and EXP as well. This can serve as a limit above which a d_r value of the numerical model signifies that the numerical model performance cannot be improved beyond the experimental model uncertainty due to model effects, etc. Therefore, similar to the relative errors as defined by van Rijn et al. [75], a relative refined index of agreement d'_r is proposed here which provides the performance of the numerical model relative to the experimental model uncertainty:

$$d_r^r = \begin{cases} 1 - \frac{MAE_{num} - MAE_{rexp}}{cMAD} = 1 - (d_{r,num} - d_{r,rexp}), MAE_{num} - MAE_{rexp} \le cMAD \\ \frac{cMAD}{MAE_{num} - MAE_{rexp}} - 1 = (d_{r,num} - d_{r,rexp}) - 1, MAE_{num} - MAE_{rexp} > cMAD \end{cases} \tag{5}$$

where the subscripts *num* and *rexp* indicate that the statistic is evaluated for the respective numerical and repeated experimental data, and c is again taken equal to 2. When the numerator MAE_{num}–MAE_{rexp} is negative (i.e., <0), the numerical error compared to the experiment is smaller than the experimental uncertainty, which means that the numerical model performance cannot be improved. In that case MAE_{num}–MAE_{rexp} = 0 is forced, so that d'_r = 1. A classification of model performance based on ranges of d'_r values and corresponding rating terminology is proposed in Table 2.

Table 2. Proposed classification of the relative refined index of agreement d'_r and corresponding rating.

d'_r Classification [-]	Rating
0.90–1.00	Excellent
0.80–0.90	Very Good
0.70–0.80	Good
0.50–0.70	Reasonable/Fair
0.30–0.50	Poor
(−1.00)–0.30	Bad

To obtain more insight into where the error of the model originates from, pattern statistical parameters are considered as well. They are here explained in terms of what they represent for a time series of η. The first additional statistical parameter is the standard deviation σ, which is a measure of the wave energy or wave height of a η time series. The normalised standard deviation is given by:

$$\sigma^* = \frac{\sigma_p}{\sigma_o} \tag{6}$$

where σ_p and σ_o are the standard deviations of the predicted and observed time series, respectively. Another important statistical parameter is the bias B, given by:

$$B = \overline{P} - \overline{O} \tag{7}$$

The bias indicates whether the model under- or over-predicts the observation, but provides no further assurances on the accuracy of the model result. The bias represents the difference in wave setup between two η time series. It is normalised by the standard deviation of the observed time series:

$$B^* = \frac{B}{\sigma_o} \tag{8}$$

The correlation coefficient R, is defined by:

$$R = \frac{\frac{1}{N}\sum_{i=1}^{N}\left(P_i - \overline{P}\right)\left(O_i - \overline{O}\right)}{\sigma_p \sigma_o} \tag{9}$$

which is a measure of the phase similarity between two time series and the wave periods in the case of η time series.

The length of the time series used for the analysis is based on the duration of the generated bichromatic waves including tapering (i.e., 125 s), beginning at the first time step when the baseline is first significantly exceeded (i.e., indicating arrival of the first wave). Since the experimental and numerical time series have different sampling frequencies, the time series with the highest sampling frequency was interpolated to the time steps of the time series with the lowest sampling frequency.

For some locations where wetting and drying occurs (i.e., on the dike, promenade, and vertical wall), the measurement regularly returned to the baseline or zero-line meaning that as a bore passed by, reflected against the wall and ran back down the dike slope, intervals were created in the time series of (near-) zero values. Including these "non-event" times in the statistical analyses would bias the statistics by:

- unnecessarily penalising the numerical model performance for an experimental measurement error. For example, in the experimentally measured and processed time series of p and F_x, often some residual instrumental noise or oscillations persisted during such non-event (or "dry") times;
- unnecessarily rewarding the model performance towards (almost) perfect agreement. For example, during the time between impacts no water reaches the wall and model performance would be perfect during such times (disregarding measurement noise).

It was therefore decided to focus the analysis on the event instances when the values of the time series (either experimental or numerical, to penalise phase differences or impacts not modelled by the numerical model) are larger than a certain threshold above the baseline. The threshold for each such time series is chosen to be as low as possible, but higher than the residual noise in the experiment.

3. Results

3.1. Time Series

The numerical model results are first compared qualitatively in the time domain to the experimental measurements of test EXP. The surface elevations η are compared in Figure 4, the horizontal velocity U_x on the promenade in Figure 5, and the total horizontal force F_x and pressures p on the vertical wall in Figures 6 and 7, respectively.

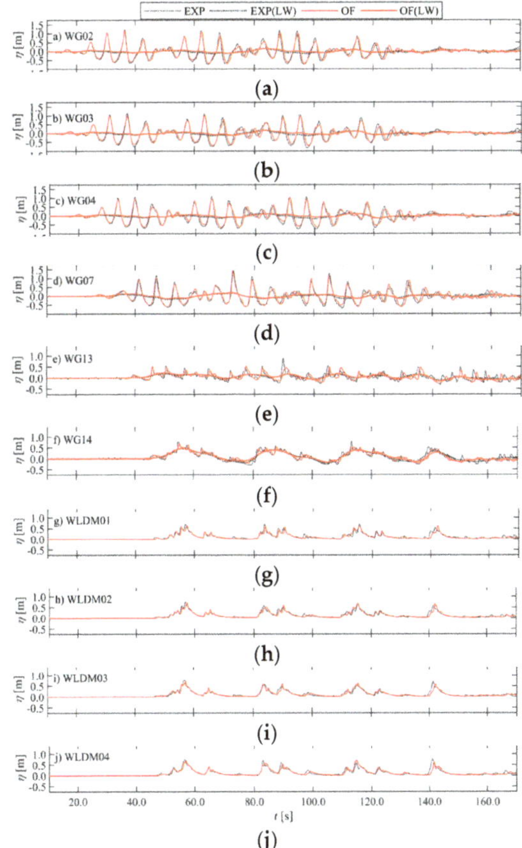

Figure 4. Comparison of the η time series at all sensor locations (**a–j**), including η_{LW} in (**a–f**) (bold lines). The zero-reference is the *SWL* for (**a–f**) and the promenade bottom for (**g–j**).

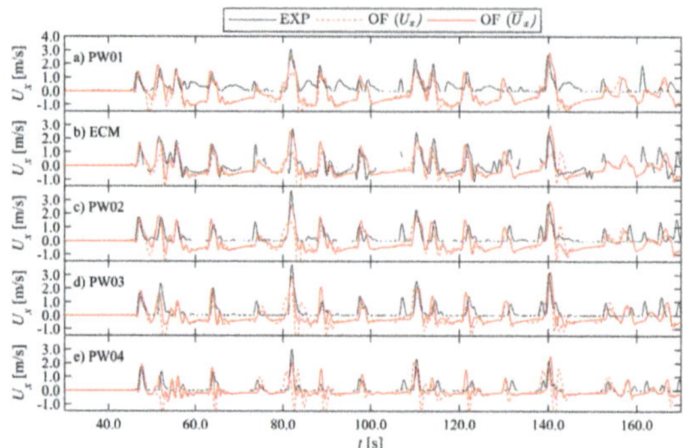

Figure 5. Comparison of U_x time series at all sensor locations (for the PWs in (**a**,**c**–**e**); for the ECM in (**b**)). The zero-reference is the promenade bottom at the sensor locations. For OF, both U_x at the measured height above the promenade and the depth-averaged $\overline{U}_x$ time series are shown.

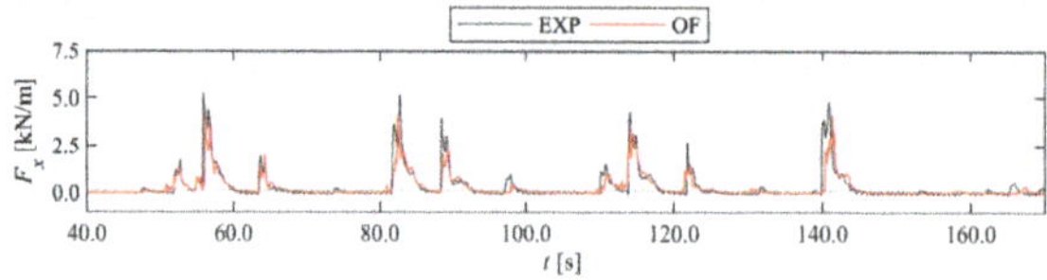

Figure 6. Comparison of F_x time series for the vertical wall. The experiment is the LC force measurement.

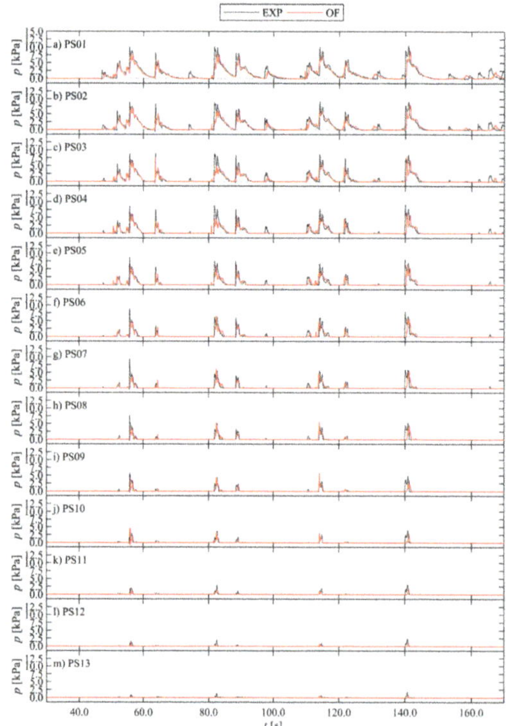

Figure 7. Comparison of p time series for all vertical pressure sensor locations (for PS01–13 in (**a**–**m**)), PS01 being the bottom PS and PS13 the top-most PS.

The η time series compare very well between OF and EXP (Figure 4), especially at the beginning of the simulation, but more discrepancies start to show over time and further along the flume. Overall, frequency dispersion, the non-linear wave transformation processes (i.e., SW shoaling (Figure 4d), breaking (Figure 4e,f), energy transfer to the subharmonic bound LW (Figure 4d–f)), overtopping (Figure 4g), bore interactions, and reflection processes (Figure 4g–j) seem to be well-represented by OF.

The simulated U_x on top of the promenade appears to significantly underestimate the experimental measurements (Figure 5). This underestimation mostly disappears when using the OF depth-averaged velocity $\overline{U}_x$ instead, which is done for the remainder of the validation. In addition, OF shows much better correspondence to the ECM than the PWs during return flow of a reflected bore ($U_x < 0$). This confirms that the PWs did not measure correct velocities during those instances (e.g., 57 s $\leq t \leq$ 63 s in Figure 5b–c).

In terms of F_x and p on the vertical wall, OF generally reproduces the timing of the impact events, including the evolution over time (Figures 6 and 7). However, the EXP time series peak values appear to be underestimated by OF for both F_x and p, and for a few impacts, the first dynamic impact peak is not entirely captured either (e.g., t = 82 s and 140 s). In the experiment, the lowest PSs were loaded more often than the PSs positioned higher up the vertical wall, because of different bore impact run-up heights. The lowest PSs also registered the highest values, indicating a mostly hydrostatic pressure distribution along the vertical wall [76]. Both these observations were reproduced by OF. Validation of the pressure distribution along the vertical wall is further investigated in Section 3.4.

3.2. Wave Characteristics

Based on the η time series the root mean square wave height H_{rms} is calculated in the time domain and represents a characteristic wave height and measure of the wave energy. The evolution of H_{rms}, the short- and long-wave components (i.e., $H_{rms,sw}$ and $H_{rms,lw}$), and the mean surface elevation $\overline{\eta}$ or wave setup over the wave flume up to the toe of the dike are displayed in Figure 8. The experimental repeatability of H_{rms} appears to be near-perfect, since the EXP and REXP data points are almost indistinguishable. The OF results for these wave characteristics are available along the complete distance from the wave paddle until the toe of the dike location. The numerical results seem to follow the experiments very well, although some discrepancies can be seen. The total and SW wave heights (H_{rms} and $H_{rms,sw}$, respectively, in Figure 8) decrease in the OF result from the wave paddle up to the toe of the foreshore and underestimate the EXP wave height along this distance. Over the foreshore, the SWs start to shoal until their steepness becomes too high and, according to OF, start to break about 11 m from WG07 towards the dike. The location of incipient wave breaking (or decrease in H_{rms}), x_b, cannot be validated with the experiment, because of insufficient wave gauges in the wave breaking zone. In any case, the EXP wave height increase due to shoaling (WG07) and decrease due to breaking (WG13–14) are reproduced well by OF. However, over the foreshore, OF slightly underestimates the wave amplitude. The experimental LW wave height ($H_{rms,lw}$ in Figure 8) is slightly underestimated by OF in front of the wave paddle (WG02–WG04), and at the dike toe (WG14).

In terms of the wave setup $\overline{\eta}$, the wave set-down observed in the experiment offshore from the foreshore toe is not reproduced by OF ($\overline{\eta}_{OF}$ remains close to zero). Further along the flume in the surf zone, however, $\overline{\eta}$ is better predicted by OF, showing a smaller overestimation.

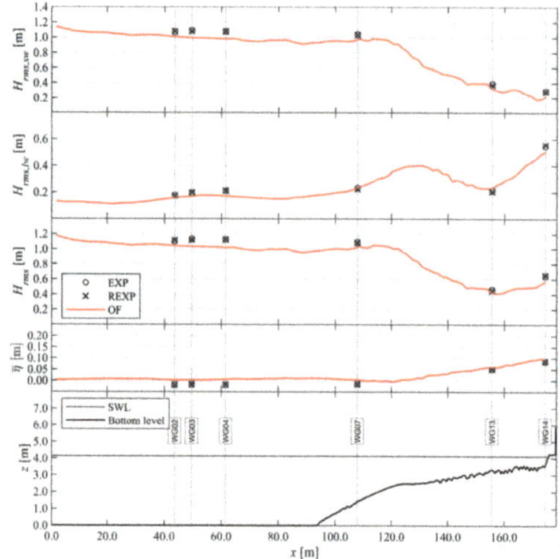

Figure 8. Comparison of H_{rms} between OF and (R)EXP up to the dike toe. From top to bottom: $H_{rms,sw}$ for the SW components, $H_{rms,lw}$ for the LW components, H_{rms} for the total η, the wave setup $\overline{\eta}$, and finally an overview of the sensor locations, SWL, and bottom profile.

3.3. Model Performance and Pattern Statistics

In this section, the model performance and pattern statics introduced in Section 2.3 are applied to obtain a quantitative numerical model performance evaluation. Tables 3 and 4 provide the pattern and model performance statistics for all sensor locations along the flume up to the vertical wall. The evolution of d_r and R at the WG locations along the wave flume up to the toe of the dike is visualised for η_{SW} ($d_{r,sw}$ and R_{sw}), η_{LW} ($d_{r,lw}$ and R_{lw}), and η ($d_{r,tot}$ and R) in Figures 9 and 10 respectively, and of d_r for η and U_x on the promenade in Figure 11.

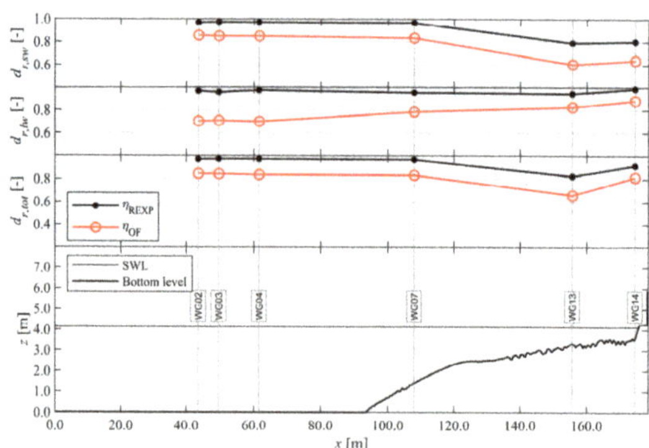

Figure 9. Index d_r of REXP and OF with EXP up to the dike toe. From top to bottom: $d_{r,sw}$ for η_{SW}, $d_{r,lw}$ for η_{LW}, $d_{r,tot}$ for η, and finally an overview of the sensor locations, SWL, and bottom profile.

Table 3. Pattern and model performance statistics for all η measurement locations.

Location	REXP				OF					
	B^* [-]	σ^* [-]	R [-]	d_r [-]	B^* [-]	σ^* [-]	R [-]	d_r [-]	d'_r [-]	Rating [-]
WG02	−0.01	1.01	1.00	0.97	0.06	0.94	0.96	0.85	0.88	Very Good
WG03	−0.01	0.99	1.00	0.97	0.05	0.92	0.95	0.85	0.87	Very Good
WG04	−0.01	1.00	1.00	0.97	0.06	0.91	0.95	0.84	0.87	Very Good
WG07	0.01	1.00	1.00	0.97	0.06	0.94	0.94	0.84	0.87	Very Good
WG13	0.00	0.97	0.94	0.83	0.04	0.95	0.73	0.66	0.83	Very Good
WG14	0.00	1.00	0.98	0.92	0.05	0.89	0.91	0.82	0.90	Very Good
WLDM01	−0.02	0.99	0.99	0.92	−0.08	1.00	0.89	0.80	0.88	Very Good
WLDM02	−0.02	1.01	0.99	0.92	−0.05	1.01	0.91	0.82	0.89	Very Good
WLDM03	0.00	0.98	0.99	0.92	−0.03	0.98	0.90	0.82	0.90	Very Good
WLDM04	0.01	0.97	0.98	0.92	−0.00	1.00	0.87	0.79	0.87	Very Good

Table 4. Pattern and model performance statistics for all U_x measurement locations on the promenade.

Location	REXP				OF					
	B^* [-]	σ^* [-]	R [-]	d_r [-]	B^* [-]	σ^* [-]	R [-]	d_r [-]	d'_r [-]	Rating [-]
PW01	0.02	0.96	0.91	0.80	−1.24	1.55	0.58	−0.10	0.10	Bad
ECM	−0.02	1.05	0.87	0.81	−0.25	0.94	0.73	0.63	0.82	Very Good
PW02	−0.05	0.99	0.88	0.82	−0.66	1.22	0.65	0.29	0.48	Poor
PW03	−0.02	1.00	0.92	0.86	−0.57	1.06	0.68	0.40	0.54	Reasonable/Fair
PW04	−0.03	1.02	0.88	0.77	−0.42	0.88	0.58	0.37	0.61	Reasonable/Fair

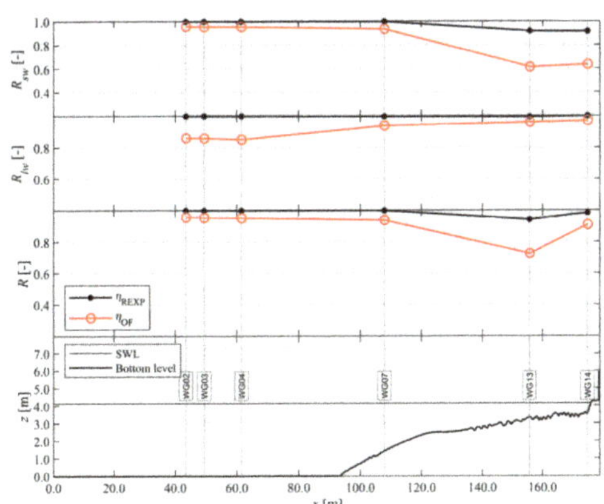

Figure 10. Comparison of R for η of REXP and OF with EXP up to the dike toe. From top to bottom: R_{sw} for η_{SW}, R_{lw} for η_{LW}, R for η, and finally an overview of the sensor locations, SWL, and bottom profile.

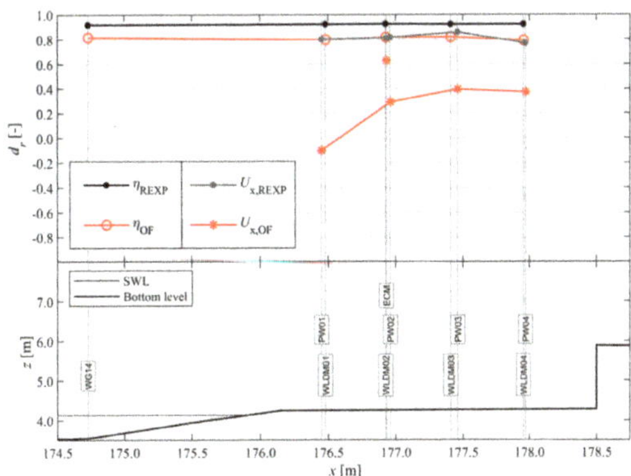

Figure 11. Index d_r of REXP and OF with EXP from the dike toe up to the vertical wall. From top to bottom: d_r for η and U_x, and finally an overview of the sensor locations, SWL, and bottom profile.

The evolution of $d_{r,tot}$ along the flume is very similar for both REXP and OF (Figure 9 and Table 3): it remains constant until the shoaling zone (WG02–WG07), decreases over the surf zone (WG07–13), and increases back up to the dike toe (WG13–14). This indicates that the decreased experimental model repeatability of the surface elevation in the surf zone is at least part of the cause of the decreased numerical model performance. The relative model performance d'_r for η is consequently fairly constant, corresponding to a model performance rating of Very Good, which remains consistently so up to the last sensor location in front of the vertical wall. Considering η_{SW} and η_{LW} separately reveals that $d_{r,sw}$ mostly follows the same trend as $d_{r,tot}$, and that $d_{r,lw,OF}$ clearly has a different behaviour: $d_{r,lw,OF}$ is not as high as $d_{r,sw,OF}$ in front of the wave paddle (i.e., $d_{r,lw,OF}$ = ~0.70 and $d_{r,sw,OF}$ = ~0.85 at WG02–WG04), but steadily increases towards the dike toe, while $d_{r,lw,rexp}$ remains relatively constant, causing d'_r to slightly increase as well.

The pattern statistics B^* and σ^* represent the accuracy of the respective wave setup and wave height from offshore until the dike toe and confirm the qualitative observations made in Section 3.2. However, spatial information about the accuracy of the numerical wave phase modelling was not included previously, and is shown separately here in Figure 10. The SW phase accuracy of OF decreases significantly over the surf zone (R = ~0.90 to ~0.60), while it increases for the LWs (R = ~0.85 to ~0.97). The total wave phase prediction accuracy of OF decreases at WG13 because it is located at a node of the standing long waves in front of the dike (Figure 8), thus R_{sw} has a higher weight in R there. Conversely, the dike toe (WG14) is located at an antinode, and therefore R_{lw} has higher weight in R than R_{sw}, leading to an increase of R again at the dike toe.

Along the promenade, the d_r for η and U_x is shown in Figure 11 and at first sight seems to indicate that the OF model performance for U_x is much worse than that for η, primarily for comparisons with the PW measurements, but also for the ECM measurement. Taking into account the experimental uncertainty, however, the model performance rating for U_x of ECM is actually Very Good ($d'_{r,ECM}$ in Table 4), which is the same as the OF model performance rating for η on the promenade ($d'_{r,WLDM01\text{-}04}$ in Table 3). For the PW measurements, the OF rating for U_x is still worse (Reasonable/Fair to Bad), but was explained before by the fact that the PW's had faulty positive U_x measurements during return flow (Section 3.1).

Although the wave setup at the dike toe is overestimated by OF ($B^*_{WG14} > 0$), η on the promenade is on average underestimated ($B^*_{WLDM01-04} < 0$) as is U_x ($B^* < 0$). Conversely, the bore wave height is well-represented on the promenade ($\sigma^*_{WLDM01-04}$ = ~1.00), while the wave height is underestimated at the dike toe ($\sigma^*_{WG14} = 0.89$). The surface elevation phase difference between OF and EXP observed at the dike toe ($R_{WG14} = 0.91$) is carried over on the promenade ($R_{WLDM01\text{-}04}$ = ~0.90), but higher phase differences are detected for U_x ($R_{ECM} = 0.73$).

Finally, the model performance in terms of p and F_x is evaluated at the vertical wall (Figure 12 and Table 5). Both REXP and OF show the highest model performance at the lowest pressure sensor location and a more or less linear decreasing model performance at PS locations higher along the vertical wall. The relative difference between the d_r of REXP and OF increases more along the vertical wall, leading to a numerical model performance rating from Very Good for PS01–PS06, to Good for PS05–PS11, and finally to Reasonable/Fair at the highest PS locations (PS12–PS13) (Table 5). Considering that the bottom PSs registered the highest p values and are therefore the most determinative in the calculation of F_x, it follows that the numerical model performance for F_x is rated Very Good as well. The pattern statistics in Table 5 reveal the remaining numerical errors to be that p and F_x are generally underestimated by OF (i.e., $B^* < 0.00$ and $\sigma^* < 1.00$) and that the impact events still slightly mismatch in time between OF and EXP ($R < 1.00$).

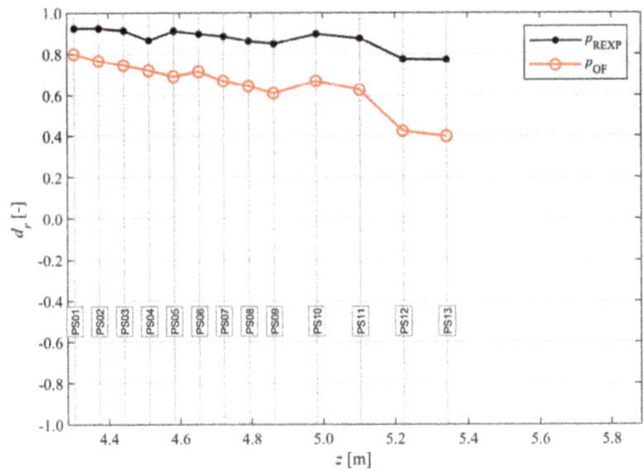

Figure 12. Index d_r of REXP and OF with EXP, for p at the vertical wall (horizontal axis).

Table 5. Pattern and model performance statistics for all p (PS) and F_x (LC) measurement locations.

Location	REXP				OF					
	B^* [-]	σ^* [-]	R [-]	d_r [-]	B^* [-]	σ^* [-]	R [-]	d_r [-]	d'_r [-]	Rating [-]
PS01	0.00	1.00	0.98	0.92	−0.14	0.84	0.84	0.80	0.88	Very Good
PS02	−0.01	0.99	0.97	0.92	−0.10	0.82	0.77	0.77	0.84	Very Good
PS03	0.00	1.00	0.96	0.91	−0.13	0.75	0.71	0.75	0.83	Very Good
PS04	0.02	0.99	0.94	0.87	−0.13	0.74	0.66	0.72	0.85	Very Good
PS05	0.01	1.00	0.96	0.91	−0.11	0.75	0.61	0.69	0.78	Good
PS06	−0.01	0.97	0.96	0.90	−0.13	0.78	0.61	0.72	0.82	Very Good
PS07	−0.01	0.93	0.95	0.89	−0.17	0.76	0.53	0.67	0.78	Good
PS08	−0.05	0.86	0.94	0.86	−0.20	0.74	0.46	0.65	0.78	Good
PS09	−0.07	0.88	0.93	0.85	−0.25	0.78	0.39	0.61	0.76	Good
PS10	−0.04	0.93	0.94	0.90	−0.24	0.77	0.48	0.67	0.77	Good
PS11	−0.04	0.91	0.94	0.88	−0.33	0.57	0.37	0.63	0.75	Good
PS12	−0.20	0.79	0.89	0.78	−0.55	0.53	−0.05	0.42	0.65	Reasonable/Fair
PS13	−0.15	0.57	0.92	0.77	−0.59	0.33	0.12	0.40	0.63	Reasonable/Fair
LC	0.00	0.97	0.90	0.90	−0.12	0.74	0.73	0.76	0.85	Very Good

3.4. Bore Interactions and Impact

To explain some of the numerical successes and failures encountered in the reproduction of the experimental bore impacts on the vertical wall, a detailed analysis is done of a selection of individual impact events and the bore interactions leading up to them. The analysis is based on an investigation of snapshots at important time instants during the first two largest impact events in the modelled time series (Figure 7). The first (t = ~56 s) and second (t = ~82 s) main impact events are chosen because they are good examples of a respective successful and less successful numerical reproduction of the experimental impacts.

Numerical snapshots of the flow on the dike, including the velocity distribution along the vertical cross-section at the ECM location or the pressure distribution along the vertical wall, are compared in Figures 13 and 14 to the equivalent experimental data and snapshots based on side and top view video images. Key time instants of overtopped bore behaviour are selected during these two main impacts and are listed chronologically in Table 6. Some of the key time instants occur at slightly different times in each model (due to slight wave phase differences). In those cases, the key time instants were selected from each model result based on identifiable features in the bore interaction images, the U_x time series or the F_x time series (e.g., peaks, troughs, …), making sure a relevant comparison is made of the bore interaction and the velocity or pressure profile.

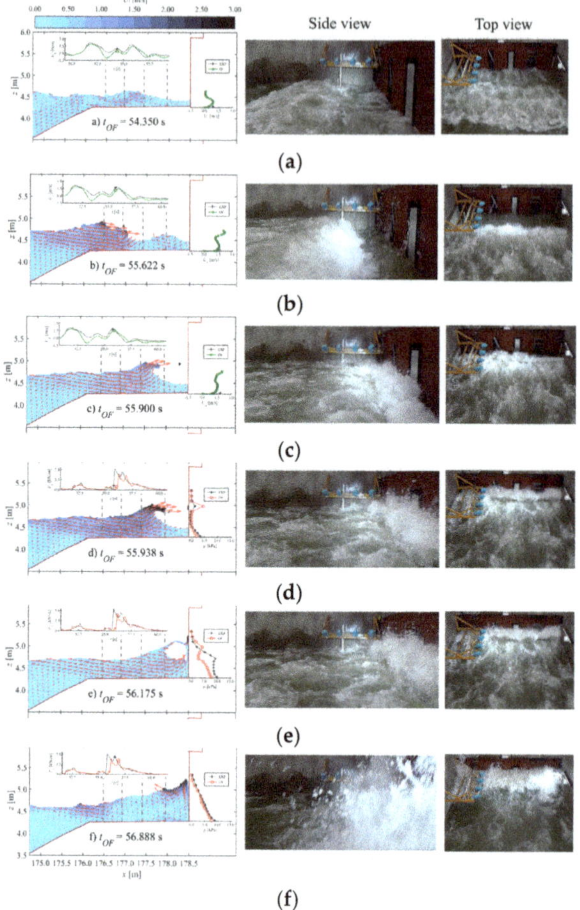

Figure 13. Snapshots of selected key time instants chronologically over the first main impact (**a**–**f**). The OF snapshot (left) is compared to the equivalent EXP snapshot from the side view (centre) and top view (right) cameras. In the OF snapshots, the colours of the water flow indicate the velocity magnitude $|U|$ according to the colour scale shown at the top. The red arrows are the velocity vectors, which are scaled for a clear visualisation. Each OF snapshot has two inset graphs: at the top is a time series plot of U_x (for EXP and $\overline{U_x}$ for OF) (**a**–**c**) or F_x (**d**–**f**), in which a circle marker (o) and a plus marker (+) indicate the time instant of the numerical and experimental snapshot, respectively. Along the vertical wall, U_x (**a**–**c**) or p (**d**–**f**) is plotted at the respective ECM sensor location or each PS location (the vertical axis is z [m]). Along the promenade, four vertical grey dashed lines indicate the sensor locations on the promenade, of which the WLDM gauges are also visible in the experimental snapshots (topped by blue plastic bags). The location of the ECM is at the second vertical grey dashed line from the left. The time instant of the numerical snapshot is provided by t_{OF}.

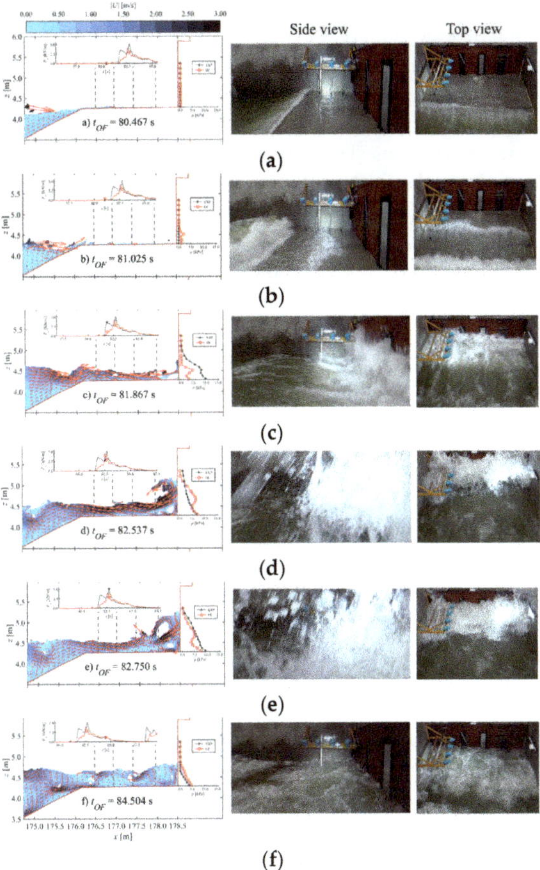

Figure 14. Snapshots of selected key time instants chronologically over the second main impact (**a**–**f**). See the caption of Figure 13 for further descriptions.

Table 6. Description of the snapshots shown in Figure 13 (main impact 1) and Figure 14 (main impact 2).

	Figure	Description
Main Impact 1	Figure 13a	Pre-impact of small overtopped wave.
	Figure 13b	Pre-collision of large overtopped bore and small wave reflected from vertical wall.
	Figure 13c	Collision of large overtopped bore and reflected small wave.
	Figure 13d	Impact on vertical wall of high velocity spray from overturned bore.
	Figure 13e	Dynamic impact of overturned bore on vertical wall.
	Figure 13f	Quasi-static impact of overturned bore on vertical wall.
Main Impact 2	Figure 14a	Very small overtopped bore.
	Figure 14b	Impact of small overtopped bore on vertical wall.
	Figure 14c	Impact of large overtopped bore on vertical wall.
	Figure 14d	Impact of large overtopped bore on vertical wall, continued.
	Figure 14e	Impact of large overtopped bore on vertical wall, continued.
	Figure 14f	Return flow of large bore reflected from vertical wall.

The first series of impacts mainly occurred while the LWs overtopped and reflected on the dike-wall structure for the first time. A good indication of this time period is when η at the dike toe (Figure 4f) was larger than the freeboard (i.e., 47 s $\leq t \leq$ 70 s). During the LW overtopping/reflection, several SWs propagated on top of the LW crest, overtopped the dike, and impacted the vertical wall along with the LWs; after a very small first overtopped bore (t = ~48 s in Figure 6), a second larger bore impacted and reflected on the vertical wall (t = ~52.5 s). While the reflected second bore returned seawards, a third small wave overtopped and headed towards the vertical wall (Figure 13a, termed "sequential overtopping bore pattern" by Streicher et al. [75]). This small wave then reflected against the vertical wall, while a very large turbulent bore overtopped the dike crest (Figure 13b). At that moment the small wave and large bores were propagating in opposite directions on the promenade. Eventually they collided, and the larger incident turbulent bore was forced to overturn (Figure 13c). This collision also caused spray to be ejected at a high velocity from the overturning wave tongue (see (x, z) = (178.3 m, 4.9 m) in Figure 13c). This airborne water volume hit the vertical wall first and separately from the main overturning wave tongue (see (x, z) = (178.5 m, 4.95 m) in Figure 13d), causing a local pressure peak at the location of PS10 (see the p-profile in Figure 13d). Subsequently, the main overturning wave hit the wall, causing a dynamic force peak $F_{x,1}$ (Figure 13e), and ran vertically up the wall temporarily reducing F_x during maximum run-up (not shown). The following run-down and reflection from the wall correspond to a second force peak $F_{x,2}$, this time of quasi-static nature (Figure 13f). This type of bore interaction was called a "plunging breaking bore pattern" by Streicher et al. [75], which (in this case) caused a quasi-static impact ($F_{x,1}/F_{x,2}$ < 1.20, according to Streicher et al. [75]). This is valid for both the experiment and the numerical model result, indicating that OF was able to reproduce these processes leading to a very similar shape of the pressure distribution along the vertical wall (see pressure profiles in Figure 13d–f) and time evolution of F_x (see time series graph insets in Figure 13d–f). Comparing $U_{x,ECM}$ from EXP with the velocity profile from OF at the ECM location (see velocity profiles in Figure 13a–c) reveals that OF locally, but consistently underestimated U_x at the vertical measurement position of the ECM, which was also observed in Figure 5b.

The second series of impacts occurred during the second LW overtopping and reflection event (Figure 4f–j: 74 s $\leq t \leq$ 100 s). Again, SWs propagated on top of the LW crest, bringing bore interactions to the promenade. This time, however, the bore interaction pattern modelled by OF that caused the main impact was different from the pattern observed in EXP. First, a very small bore overtopped the dike crest and was immediately followed by a much larger bore. In EXP, the smaller bore was overtaken by the larger bore (Figure 14b–c, termed "catch-up bore pattern" by Streicher et al. [75]), leading to a quasi-static impact. In the result from OF, however, the very small wave overtopped sooner (Figure 14a), so that it had time to reflect against the wall (Figure 14b) before colliding with the incoming larger bore (not shown). OF therefore modelled a collision bore pattern instead of a catch-up bore pattern, greatly reducing the first impact force peak of the main impact (by ~65% compared to EXP, Figure 14c). This also clearly affected the pressure profiles along the vertical wall: during the first F_x peak, p is severely underestimated, but the distribution is still similar, with a local peak at PS04. The p-profiles differentiate more at the F_x peak of the OF result (Figure 14d) and at the quasi-static F_x peak in the EXP result (Figure 14e). In the experiment, a quasi-hydrostatic pressure profile was measured, at both those time instants. In the OF result, however, a pressure peak is found at PS06, caused by a vortex formed at the foot of the vertical wall upon which a strong flow impinged on the wall at that location. After reflection of the bore, both models correspond again, showing a hydrostatic pressure profile along the wall (Figure 14f).

4. Discussion

4.1. Wave Transformation Processes Until the Dike Toe

In Sections 3.1 and 3.2 it was already established that OF is capable of reproducing the wave shoaling and breaking processes in terms of evolutions in η and H_{rms}. This section discusses the processes related to the LW transformations over the foreshore as modelled by OF and their correspondence to observations in EXP.

The modulation factor β_m of the SWs is high for the considered bichromatic wave conditions (Table 1), indicating that the incident-bound LW amplitude was relatively high as well. Furthermore, the normalised bed slope parameter β_b can be calculated [37]:

$$\beta_b = \frac{h_x}{\omega}\sqrt{\frac{g}{h_b}} \tag{10}$$

where h_x is the foreshore slope (= 1:35), ω is the radial frequency of the bound LW (= $2\pi(f_1\text{–}f_2)$), g the gravitational acceleration, and h_b a characteristic breaking depth (= 2.12 m at x_b = 115 m). A value of 0.28 is obtained, which means that the bound LW shoaling had a mild slope regime ($\beta_b <$ 0.3), so that the growth rate of the incoming LWs was much higher than that given by Green's Law (conservative shoaling), indicating significant energy transfer from the primary SWs to the bound LW [77]. Additionally, in a mild-slope regime, LW shoreline dissipation and shoreline reflection are high and low, respectively [37]. However, the beach considered here is not a beach by itself, but acts as a foreshore to a steep-sloped dike. Consequently, no such expected decrease in LW energy towards the shoreline is observed (i.e., $H_{rms,lw}$ in Figure 8). Indeed, the dike was positioned in the shoaling zone of the long waves, thereby preventing the LWs from breaking. Instead, LWs reflected against the dike, indicated by the oscillations of $H_{rms,lw}$ towards the dike in the OF result, which implies the presence of a (partial) standing wave system. Wave gauges WG13 in the inner surf zone and WG14 at the dike toe were positioned at a node and anti-node of this standing wave system. This is also clearly visible in the η time series plot, where η_{LW} is much closer to zero at WG13 (Figure 4e) than at WG14 (Figure 4f). In the surf zone the LW previously bound to the wave group became a free wave, traveling at its own wave celerity. Due to first-order wave generation at the boundary, other spurious free LWs were generated as well at the wavemaker and propagated as free waves towards the dike [78]. During a standing LW crest at the dike toe, the LWs themselves overtopped the dike (i.e., when $\eta >$ freeboard R_c = 0.117 m, Figure 4f) thereby temporarily aiding several breaking SWs to overtop the crest of the dike (the wave length of the free LWs was more than five times longer than the primary SW components in the inner surf zone). These results have illustrated OF's ability to reproduce the wave energy transfer to the subharmonics and LW transformations over the foreshore until the dike toe. All these observations also confirm that the contribution of LWs to the processes on the dike, including the wave impact loading on the vertical wall, is very important in the case that is considered here.

4.2. Importance of Differences in Wave Generation Methods

Although the overall OF model performance was rated to be Very Good, a few differences between the OF and EXP results remain to be explained. One of the largest OF inaccuracies was an underestimation of the wave height, primarily observed at the offshore WG locations (WG02–WG04, see Figure 8 and Table 3), suggesting an underestimation of the incident wave energy and/or numerical diffusion. The underestimation was likely caused by differences between the numerical wave generation method with static boundary in OF and the physically moving wave paddle in the EXP [69]. The wave boundary condition by olaFlow allows for a tuning factor to be applied to U_x and η at the boundary, to overcome a possible underestimation of the incident wave height. Such a calibration of the OF model (with a tuning factor of 1.13) was found to solve the underestimation of the wave height (not shown), but introduced or exacerbated other errors, finally leading to lower values of d_r and decreased model performance ratings for $U_{x,ECM}$ and F_x.

Another remaining discrepancy between OF and EXP is found in $\overline{\eta}$, which was primarily overestimated by OF in the offshore region (Figure 8). Also, after calibration of the incident wave height to EXP, H_{rms} (and consequently $\overline{\eta}$) increased in the surf zone, exacerbating the $\overline{\eta}$ overestimation there (not shown). The root cause of this difference is likewise related to the different wave generation methods applied in EXP and OF. In the experimental wave flume, the finite body of water and conservation of mass caused water mass to be redistributed from offshore to the surf zone during build-up of the wave setup, thereby causing a lowering of the mean water level in the offshore region. This process developed differently in OF because of the static boundary condition including AWA. The AWA assures a constant mean water level at the boundary [8,53], meaning that a net water mass is added to the computational domain until a quasi-steady state is achieved when wave setup is fully developed [55]. In this case, OF's method is closer to the field condition, where generally a large enough body of water is available to supply water mass for the wave setup to develop without noticeably lowering the offshore mean water level. Nevertheless, in the context of the validation, this difference in $\overline{\eta}$ is the cause of many of the remaining inaccuracies in the OF result compared to EXP, because the waves propagated in slightly different mean water depths, which affected the non-linear wave-wave interactions and wave phases in the surf zone. Consequently, it is believed to be the root cause of the strong decrease of R_{sw} observed in the surf zone (i.e., locations WG13–14 in Figure 10).

These two remaining inaccuracies in the OF results compared to EXP (i.e., underestimation of H_{rms} and overestimation of $\overline{\eta}$) are both attributable to the differences in wave generation methods applied. Although an overall Very Good model performance rating was achieved by OF, it is expected that even better results can be obtained by applying a closed dynamic wave boundary condition in OF, which mimics the EXP wave paddle movement. However, application of the dynamic boundary condition of olaFlow proved to be highly unstable for the present case, and no result was achieved to confirm this hypothesis.

4.3. OF Model Performance for Impacts on a Dike-Mounted Vertical Wall

The accuracy of a numerical wave model to reproduce wave overtopping over a dike with a very shallow foreshore depends on the quality of the incident waves at the dike toe location [10]. The same should therefore hold true for impacts on a dike-mounted vertical wall by such overtopped waves.

The overall Very Good model performance of OF in terms of p and F_x at the vertical wall can be explained by a generally correct reproduction of bore interactions over the promenade of the dike. Conversely, discrepancies (even small ones) in bore interactions between OF and EXP can lead to significant differences in the impact type on the vertical wall, and consequently in p and F_x (Section 3.4). In addition, the much lower values of B^*_{OF} and R_{OF} compared to B^*_{REXP} and R_{REXP} for $U_{x,ECM}$ (i.e., $B^*_{REXP} = -0.02$ and $R_{REXP} = 0.87$, $B^*_{OF} = -0.25$ and $R_{OF} = 0.73$ in Table 4) indicate an important contribution of the underestimation of U_x and of phase differences in U_x between OF and EXP to the remaining errors in the impact prediction by OF. The bore interactions on their part depend on the wave conditions at the dike toe location. This is illustrated by the calibrated OF model results, which were found to improve the wave height reproduction at the dike toe compared to the OF model (Section 4.2), while errors increased for the wave setup and wave phases at the dike toe location, leading to a lower model performance for the processes on the dike (not shown).

Even when the incident wave conditions at the dike toe would be perfectly reproduced, other model limitations would still contribute to residual errors in the numerical results for the wave impacts on the vertical wall:

- 3D effects in EXP (i.e., irregular and oblique wave fronts, wave breaking-induced 3D vortex formation), which are unreproducible by a 2DV RANS model;
- Water-air mixing in bores and air pressure fluctuations in entrained air pockets by overturning wave impacts on the wall, which are both processes not resolved by a multiphase numerical model of two incompressible and immiscible fluids.
- The applied VOF method, which is known to smear the water-air interface over several grid cells and to cause high spurious velocities in the air phase [45]. These limitations may be (partially) overcome by applying the following recent developments:
    - o An alternative geometric VOF method, isoAdvector, has been developed to obtain a sharper interface [79,80], specifically with applications for marine science and engineering

in mind [46]. However, the sharper interface may lead to a larger error in the velocity near the water surface [45] and the method is currently mainly tested and validated for wave propagation, but not yet for wave breaking, overtopping, and wave impact, all processes essential for this study.

- o Spurious velocities may be avoided by implementing, e.g., the ghost fluid method [81], although such an implementation is currently not available in any of the open source OpenFOAM versions.

- The turbulence model, which has been carefully chosen as the state-of-the-art (Section 2.2.1), but is still limited by its inherent assumptions.
- Douglas and Nistor [82] have shown that (compared to a dry-bed condition) a bore propagating over a thin layer of water on the bed (i.e., wet-bed condition) can substantially increase the steepness and depth of the bore-front and consequently affect the impact of the bore on the wall. The near-bed resolution of the OF grid along the promenade might not have been able to correctly reproduce wet-bed bore propagation in cases of a very thin layer of water, possibly even modelling a dry-bed bore propagation instead.
- Differences between OF and EXP in the treatment of friction on the bed of the promenade. The no-slip boundary condition and applied wall function in OF modelled a boundary layer, which lowered U_x close to the bed more than was measured in EXP. On average, U_x was underestimated by OF at the measurement locations of the PWs and ECM close to the promenade bed (Figure 5, B^* in Table 4 and Figure 13a–c).

Errors in the reproduction of the impact type and the first two model limitations listed above are also apparent in the numerical reproduction of the pressure distribution along the vertical wall: higher up the wall a decreasing OF model performance rating of p was observed (Figure 12, Table 5). The highest PS locations are the most sensitive to errors in the impact and run-up patterns along the vertical wall and to overly simplified water-air mixture modelling.

5. Conclusions

A RANS multiphase solver for two incompressible and immiscible fluids (water and air), interFoam of OpenFOAM® with olaFlow wave boundary conditions (OF), was applied in 2DV for bichromatic wave transformations over a cross-section of a hybrid beach-dike coastal defence system, consisting of a steep-sloped dike with a mildly-sloped and very shallow foreshore, and finally wave impact on a vertical wall. OF was not validated before in this context, where (prior to impact) waves undergo many nonlinear transformations and interact with a dike slope and promenade. A large-scale experiment of bichromatic waves and its repetition were selected for this validation. The repeated test allowed us to assess the accuracy of the measurements, uncertainty due to model effects, and variability due to stochastic processes in the experiment.

The validation consisted of both qualitative and quantitative comparisons. Pattern and model performance statistics were employed for the quantitative validation. Based on Willmott's refined index of agreement d_r, calculated for OF and the repeated test REXP with reference to the first test EXP, a relative refined index of agreement d_r' was proposed, which takes the experimental uncertainty, derived from REXP, into account in the numerical model performance evaluation. Based on value ranges of d_r', a classification into model performance ratings was proposed as well.

After a convergence analysis of the most important numerical parameters (i.e., grid resolution and CFL number), and without calibration of the numerical model, a model performance rating of Very Good was achieved by OF compared to the experiment for all relevant design parameters (i.e., η, U_x, p, and F_x), which demonstrates OF's applicability for the design of such hybrid coastal defence systems. Remaining discrepancies were found to be mainly caused by the different wave generation methods applied in OF (static boundary) and EXP (moving wave paddle), which caused an underestimation of the incident wave energy and an overestimation of the wave setup in OF compared to EXP. Consequently, when applying OF for a design of a hybrid coastal defence system, the incident wave energy is recommended to be calibrated, while the wave setup development for a static boundary condition with AWA in OF is actually closer to the field condition compared to EXP (finite water mass).

A detailed comparison of snapshots at key time instants of bore interactions leading up to two selected bore impacts on the vertical wall revealed that slight errors in wave phases can lead to very different bore interaction patterns on the promenade and finally to different bore impact types on the wall.

Future work includes a detailed inter-model comparison between the OF model presented here, a weakly compressible SPH model (DualSPHysics), and a non-hydrostatic wave model (SWASH) for the same case [65].

Author Contributions: Conceptualization, V.G., C.A., T.S. and P.T.; methodology, V.G.; validation, V.G. and M.S.; formal analysis, V.G.; investigation, V.G., C.A., T.S., M.S. and L.C.; resources, P.T.; data curation, V.G., M.S. and L.C.; writing—original draft preparation, V.G.; writing—review and editing, V.G., C.A., T.S., M.S., L.C., A.K. and P.T.; visualization, V.G. and M.S.; supervision, P.T. and A.K.; project administration, P.T.; funding acquisition, P.T. and A.K. All authors have read and agreed to the published version of the manuscript.

Funding: This research was part of the CREST (Climate REsilient CoaST) project (http://www.crestproject.be/en), funded by the Flemish Agency for Innovation by Science and Technology, grant number 150028. The work described in this section was supported by the European Community's Horizon 2020 Research and Innovation Programme through the grant to HYDRALABPLUS, Contract no. 654110. C.A. acknowledges funding from the European Union's Horizon 2020 research and innovation programme under the Marie Sklodowska-Curie, grant number 792370.

Conflicts of Interest: The authors declare no conflict of interest. The funders had no role in the design of the study; in the collection, analyses, or interpretation of data; in the writing of the manuscript, or in the decision to publish the results.

Appendix A. Numerical Convergence Analysis

The OF result is influenced by many of its settings, of which the spatial discretisation of the model domain and time stepping are the most important [45]. Their convergence analysis is presented here. The numerical model convergence analysis is based on η at the experimental wave gauge locations over the wave flume up to the dike toe, since it is the most important driver of model performance of the subsequent processes on the dike. The wave force at the vertical wall is not suitable as reference for the grid convergence analysis, because relatively small differences in wave phase can cause very different types of bore interactions on the promenade and therefore very different resulting bore impacts (Section 3.4).

Appendix A.1. Model Convergence Statistics

For the convergence analysis, four customised statistical error indicators are considered, among which the first three are defined to reflect several aspects of the η time series considered (i.e., wave setup, wave height and wave phase):

- Freeboard normalised bias, *NB*:

$$NB = \frac{B}{R_c} \tag{A1}$$

in which R_c is the freeboard, and B is the bias defined by (7). The bias or difference in the wave setup is normalised with the freeboard which is one of the governing parameters for waves overtopping a dike [83].

- Residual error of the normalised standard deviation, *RNSD*:

$$RNSD = 1 - \sigma^* \tag{A2}$$

in which σ^* is given by (6) and in which the observed time series is the reference time series and the predicted time series is the considered time series. A positive RNSD signifies a higher wave height and a negative RNSD signifies a lower wave height compared to the reference.

- Residual error of the correlation coefficient, *RCC*:

$$RCC = 1 - R \tag{A3}$$

in which *R* is the correlation coefficient, given by (9), between the reference time series and time series of interest. Lower RCC values indicate better phase correspondence of the considered time series to the reference.

- Normalised mean absolute error, *NMAE*, given by:

$$NMAE = \frac{MAE}{O_{max} - O_{min}} \times 100\% \tag{A4}$$

in which *MAE* is the mean absolute error, given by (3), and O_{max} and O_{min} are the respective maximum and minimum values of the reference time series.

The closer these statistics are to zero, the lower the difference between the considered and reference time series.

Appendix A.2. Convergence Analyses

The grid convergence analysis varies the refinement level in the surface elevation zone β_{sez} up to the dike toe (i.e., β_{sez} = 0, 1, 2, 3; Figure 3) and uses the mesh with the highest level (i.e., β_{sez} = 3 or Δx = Δz = 0.0225 m) as the reference to which the other (coarser) resolution simulations are compared to. Convergence is achieved when no other significant changes are observed compared to a finer grid resolution model. The time stepping convergence analysis uses the run with the lowest *maxCo* number (i.e., *maxCo* = 0.15) as the reference to which other temporally coarser simulations (i.e., *maxCo* = 0.45, 0.25) are compared to. The statistical error indicators from Section A1 are provided in Figures A1 and A2. All errors are close to or less than 5% at the toe of the dike for β_{sez} = 2 (i.e., Δx = Δz = 0.045 m) and *maxCo* = 0.25. Even though *maxCo* = 0.45 does not show much higher errors than a value of 0.25, *maxCo* = 0.25 was preferred, because higher *maxCo* simulations were found to be prone to numerical instabilities. In any case, as long as the *maxCo* number cannot be defined separately for the air and water phases, the time step is mostly determined by the high spurious velocities that occur at the water-air interface. Because these spurious velocities are much higher (2–3 times) than the velocities in the water phase, much lower Courant numbers are actually obtained in the water phase [46]. This also explains why only limited differences between the tested *maxCo* values are observed here.

Moreover, the *NMAE* shows in both cases a similar value at the toe of the dike (WG14) to that of the ~3% obtained between EXP and REXP. The remaining numerical error is therefore assumed acceptable, and the mesh resolution and time stepping can be considered sufficiently converged for those settings (β_{sez} = 2; *maxCo* = 0.25).

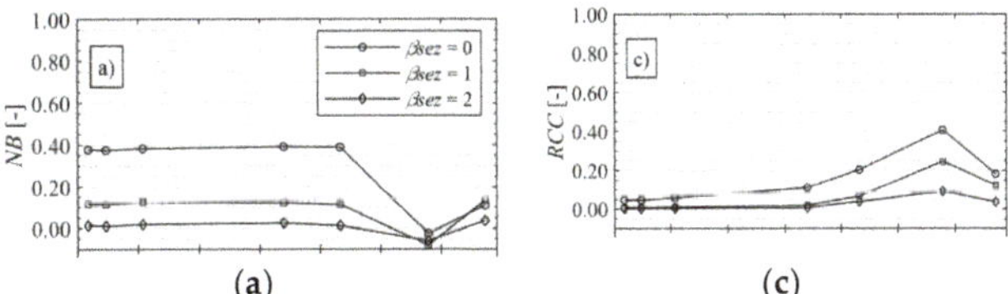

(a) (c)

Figure A1. *Cont.*

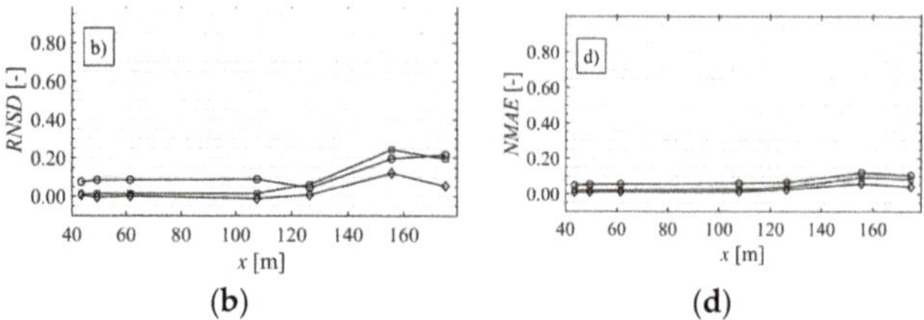

Figure A1. OF model grid resolution convergence analysis of the η time series at the WG locations along the flume up to the dike toe (WG14) based on (**a**) the normalised bias, (**b**) the residual normalised standard deviation, (**c**) the residual correlation coefficient, and (**d**) the normalised mean-absolute-error. The reference is the finest mesh with a refinement level in the surface elevation zones β_{sez} of 3.

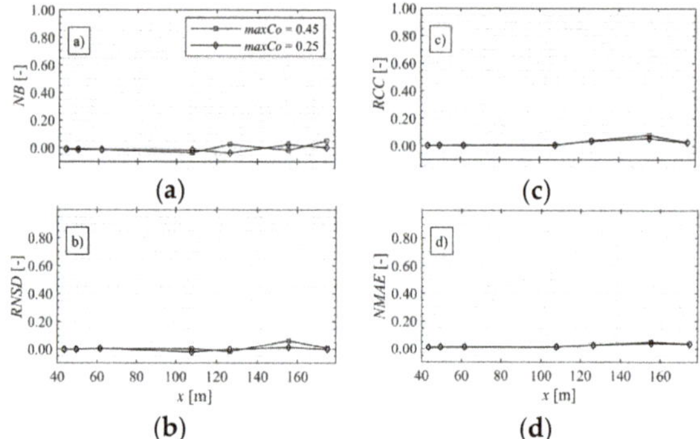

Figure A2. OF model time step convergence analysis based on *maxCo* for the mesh with β_{sez} = 2. The reference is the lowest maximum Courant number applied (*maxCo* = 0.15). See caption of Figure A1 for the description of (**a–d**).

References

1. IPCC. The Ocean. and Cryosphere in a Changing Climate. In *IPCC Special Report on the Ocean and Cryosphere in a Changing Climate*; IPCC: Geneva, Switzerland, 2019.
2. Van Doorslaer, K.; Romano, A.; De Rouck, J.; Kortenhaus, A. Impacts on a storm wall caused by non-breaking waves overtopping a smooth dike slope. *Coast. Eng.* **2017**, *120*, 93–111. [CrossRef]
3. De Finis, S.; Romano, A.; Bellotti, G. Numerical and laboratory analysis of post-overtopping wave impacts on a storm wall for a dike-promenade structure. *Coast. Eng.* **2020**, *155*, 103598. [CrossRef]
4. Chen, X.; Hofland, B.; Uijttewaal, W. Maximum overtopping forces on a dike-mounted wall with a shallow foreshore. *Coast. Eng.* **2016**, *116*, 89–102. [CrossRef]
5. Streicher, M.; Kortenhaus, A.; Gruwez, V.; Hofland, B.; Chen, X.; Hughes, S.; Hirt, M. Prediction of Dynamic and Quasi-static Impacts on Vertical Sea Walls Caused by an Overtopped Bore. *Coast. Eng.* **2018**. [CrossRef]
6. Gruwez, V.; Vandebeek, I.; Kisacik, D.; Streicher, M.; Verwaest, T.; Kortenhaus, A.; Troch, P. 2D overtopping and impact experiments in shallow foreshore conditions. In Proceedings of the 36th Conference on Coastal Engineering, Baltimore, MD, USA, 30 July–3 August 2018; pp. 1–13.
7. Xiao, H.; Huang, W.; Tao, J. Numerical modeling of wave overtopping a levee during Hurricane Katrina. *Comput. Fluids* **2009**, *38*, 991–996. [CrossRef]
8. Torres-Freyermuth, A.; Lara, J.L.; Losada, I.J. Numerical modelling of short- and long-wave transformation on a barred beach. *Coast. Eng.* **2010**, *57*, 317–330. [CrossRef]
9. Altomare, C.; Crespo, A.J.C.; Domínguez, J.M.; Gómez-Gesteira, M.; Suzuki, T.; Verwaest, T. Applicability of Smoothed Particle Hydrodynamics for estimation of sea wave impact on coastal structures. *Coast. Eng.* **2015**, *96*, 1–12. [CrossRef]

10. Suzuki, T.; Altomare, C.; Veale, W.; Verwaest, T.; Trouw, K.; Troch, P.; Zijlema, M. Efficient and robust wave overtopping estimation for impermeable coastal structures in shallow foreshores using SWASH. *Coast. Eng.* **2017**, *122*, 108–123. [CrossRef]
11. Altomare, C.; Tagliafierro, B.; Dominguez, J.M.; Suzuki, T.; Viccione, G. Improved relaxation zone method in SPH-based model for coastal engineering applications. *Appl. Ocean. Res.* **2018**, *81*, 15–33. [CrossRef]
12. Violeau, D. *Fluid Mechanics and the SPH Method: Theory and Applications*; Oxford University Press: New York, NY, USA, 2012; ISBN 978-0-19-965552-6.
13. Oñate, E.; Celigueta, M.A.; Idelsohn, S.R.; Salazar, F.; Suárez, B. Possibilities of the particle finite element method for fluid–soil–structure interaction problems. *Comput. Mech.* **2011**, *48*, 307. [CrossRef]
14. Didier, E.; Neves, D.R.C.B.; Martins, R.; Neves, M.G. Wave interaction with a vertical wall: SPH numerical and experimental modeling. *Ocean. Eng.* **2014**, *88*, 330–341. [CrossRef]
15. St-Germain, P.; Nistor, I.; Townsend, R.; Shibayama, T. Smoothed-Particle Hydrodynamics Numerical Modeling of Structures Impacted by Tsunami Bores. *J. Waterw. Port. Coast. Ocean Eng.* **2014**, *140*, 66–81. [CrossRef]
16. Domínguez, J.M.; Altomare, C.; Gonzalez-Cao, J.; Lomonaco, P. Towards a more complete tool for coastal engineering: Solitary wave generation, propagation and breaking in an SPH-based model. *Coast. Eng. J.* **2019**, *61*, 15–40. [CrossRef]
17. Subramaniam, S.P.; Scheres, B.; Schilling, M.; Liebisch, S.; Kerpen, N.B.; Schlurmann, T.; Altomare, C.; Schüttrumpf, H. Influence of Convex and Concave Curvatures in a Coastal Dike Line on Wave Run-up. *Water* **2019**, *11*, 1333. [CrossRef]
18. Mokos, A.; Rogers, B.D.; Stansby, P.K. A multi-phase particle shifting algorithm for SPH simulations of violent hydrodynamics with a large number of particles. *J. Hydraul. Res.* **2017**, *55*, 143–162. [CrossRef]
19. Torres-Freyermuth, A.; Losada, I.J.; Lara, J.L. Modeling of surf zone processes on a natural beach using Reynolds-Averaged Navier-Stokes equations. *J. Geophys. Res.* **2007**, *112*, C09014. [CrossRef]
20. Higuera, P.; Lara, J.L.; Losada, I.J. Simulating coastal engineering processes with OpenFOAM®. *Coast. Eng.* **2013**, *71*, 119–134. [CrossRef]
21. Ingram, D.M.; Gao, F.; Causon, D.M.; Mingham, C.G.; Troch, P. Numerical investigations of wave overtopping at coastal structures. *Coast. Eng.* **2009**, *56*, 190–202. [CrossRef]
22. An, M.H.; Jiang, Q.; Zhang, C.K. Simulation of wave propagation on sloping seadike. In Proceedings of the Proceedings 2013 International Conference on Mechatronic Sciences, Electric Engineering and Computer (MEC), Shengyang, China, 20–22 December 2013; pp. 2505–2509.
23. Kleefsman, K.M.T.; Fekken, G.; Veldman, A.E.P.; Iwanowski, B.; Buchner, B. A Volume-of-Fluid based simulation method for wave impact problems. *J. Comput. Phys.* **2005**, *206*, 363–393. [CrossRef]
24. Wenneker, I.; Wellens, P.; Gervelas, R. Volume-of-fluid model comflow simulations of wave impacts on a dike. *Coast. Eng. Proc.* **2011**, *1*, 17. [CrossRef]
25. Vanneste, D.F.A.; Altomare, C.; Suzuki, T.; Troch, P.; Verwaest, T. Comparison of numerical models for wave overtopping and impact on a sea wall. *Coast. Eng. Proc.* **2014**, *1*, 5. [CrossRef]
26. Xie, P.; Chu, V.H. The forces of tsunami waves on a vertical wall and on a structure of finite width. *Coast. Eng.* **2019**, *149*, 65–80. [CrossRef]
27. Wemmenhove, R.; Luppes, R.; Veldman, A.E.P.; Bunnik, T. Numerical simulation of hydrodynamic wave loading by a compressible two-phase flow method. *Comput. Fluids* **2015**, *114*, 218–231. [CrossRef]
28. Liu, S.; Gatin, I.; Obhrai, C.; Ong, M.C.; Jasak, H. CFD simulations of violent breaking wave impacts on a vertical wall using a two-phase compressible solver. *Coast. Eng.* **2019**, *154*, 103564. [CrossRef]
29. Streicher, M.; Kortenhaus, A.; Altomare, C.; Gruwez, V.; Hofland, B.; Chen, X.; Marinov, K.; Scheres, B.; Schüttrumpf, H.; Hirt, M.; et al. WALOWA (WAve LOads on WAlls): Large-Scale Experiments in the Delta Flume. In Proceedings of the International Short Course and Conference on Applied Coastal Research (SCACR), Santander, Spain, 3–6 October 2017; pp. 69–80.
30. Streicher, M.; Kortenhaus, A.; Hughes, S.; Hofland, B.; Suzuki, T.; Altomare, C.; Marinov, K.; Chen, X.; Cappietti, L. Non-Repeatability, Scale- and Model Effects in Laboratory Measurement of Impact Loads Induced by an Overtopped Bore on a Dike Mounted Wall. In Proceedings of the ASME 2019 38th International Conference on Ocean, Offshore and Arctic Engineering, Glasgow, Scotland, UK, 9–14 June 2019; p. 10.

78. Barthel, V.; Mansard, E.P.D.; Sand, S.E.; Vis, F.C. Group bounded long waves in physical models. *Ocean Eng.* **1983**, *10*, 261–294. [CrossRef]
79. Roenby, J.; Bredmose, H.; Jasak, H. A computational method for sharp interface advection. *R. Soc. Open Sci.* **2016**, *3*, 160405. [CrossRef]
80. Roenby, J.; Jasak, H.; Bredmose, H.; Vukčević, V.; Heather, A.; Scheufler, H. isoAdvector. 2019. Available online: https://github.com/isoAdvector/isoAdvector (accessed on 14 October 2019).
81. Vukčević, V.; Jasak, H.; Gatin, I. Implementation of the Ghost Fluid Method for free surface flows in polyhedral Finite Volume framework. *Comput. Fluids* **2017**, *153*, 1–19. [CrossRef]
82. Douglas, S.; Nistor, I. On the effect of bed condition on the development of tsunami-induced loading on structures using OpenFOAM. *Nat. Hazards* **2015**, *76*, 1335–1356. [CrossRef]
83. Van der Meer, J.W.; Allsop, N.W.H.; Bruce, T.; De Rouck, J.; Kortenhaus, A.; Pullen, T.; Schüttrumpf, H.; Troch, P.; Zanuttigh, B. Eurotop, Manual on Wave Overtopping of Sea Defences and Related Structures. An Overtopping Manual Largely Based on European Research, but for Worldwide Application, 2nd ed. 2018. Available online: www.overtopping-manual.com (accessed on 26 April 2019).

Journal of
Marine Science and Engineering

Article

An Inter-Model Comparison for Wave Interactions with Sea Dikes on Shallow Foreshores

Vincent Gruwez [1,*], Corrado Altomare [2], Tomohiro Suzuki [3,4,5], Maximilian Streicher [1], Lorenzo Cappietti [6], Andreas Kortenhaus [1] and Peter Troch [1]

1 Department of Civil Engineering, Ghent University, 9000 Ghent, Belgium; maximilian.streicher@ugent.be (M.S.); andreas.kortenhaus@ugent.be (A.K.); peter.troch@ugent.be (P.T.)
2 Maritime Engineering Laboratory, Department of Civil and Environmental Engineering, Universitat Politecnica de Catalunya—BarcelonaTech (UPC), 08034 Barcelona, Spain; corrado.altomare@upc.edu
3 Flanders Hydraulics Research, 2140 Antwerp, Belgium; tomohiro.suzuki@mow.vlaanderen.be
4 Faculty of Civil Engineering and Geosciences, Delft University of Technology, 2628 CN Delft, The Netherlands
5 ECOH Corporation, Tokyo 110-0014, Japan
6 Department of Civil and Environmental Engineering, University of Florence, 50139 Florence, Italy; lorenzo.cappietti@unifi.it
* Correspondence: vincent.gruwez@ugent.be

Received: 5 November 2020; Accepted: 25 November 2020; Published: 3 December 2020

Abstract: Three open source wave models are applied in 2DV to reproduce a large-scale wave flume experiment of bichromatic wave transformations over a steep-sloped dike with a mildly-sloped and very shallow foreshore: (i) the Reynolds-averaged Navier–Stokes equations solver interFoam of OpenFOAM® (OF), (ii) the weakly compressible smoothed particle hydrodynamics model DualSPHysics (DSPH) and (iii) the non-hydrostatic nonlinear shallow water equations model SWASH. An inter-model comparison is performed to determine the (standalone) applicability of the three models for this specific case, which requires the simulation of many processes simultaneously, including wave transformations over the foreshore and wave-structure interactions with the dike, promenade and vertical wall. A qualitative comparison is done based on the time series of the measured quantities along the wave flume, and snapshots of bore interactions on the promenade and impacts on the vertical wall. In addition, model performance and pattern statistics are employed to quantify the model differences. The results show that overall, OF provides the highest model skill, but has the highest computational cost. DSPH is shown to have a reduced model performance, but still comparable to OF and for a lower computational cost. Even though SWASH is a much more simplified model than both OF and DSPH, it is shown to provide very similar results: SWASH exhibits an equal capability to estimate the maximum quasi-static horizontal impact force with the highest computational efficiency, but does have an important model performance decrease compared to OF and DSPH for the force impulse.

Keywords: inter-model comparison; wave modelling; shallow foreshore; dike-mounted vertical wall; wave impact loads; OpenFOAM; DualSPHysics; SWASH

1. Introduction

Urban areas situated along low elevation coastal zones need to be protected against wave overtopping and flooding during storm conditions. Moreover, many existing sea dikes protecting such coastal urban areas need to be adapted to be prepared for the effects of sea level rise, which is one of the most challenging problems currently facing coastal communities around the world. A hybrid soft-hard coastal defence system is a promising solution in this context [1]. Such a coastal defence system consists of a soft mildly-sloped—usually nourished—beach that acts as a shallow foreshore to a hard steep-sloped sea dike. In many cases, structures on top of the sea dike (e.g., storm walls and buildings fronted by a promenade) are still in danger of being loaded by overtopping storm waves. In the design of these structures, such wave loading needs to be considered. However, this is a challenging task, because along the cross section of a hybrid beach–dike coastal defence system, storm waves are forced to undergo many transformation processes before they reach the structures on the dike. These hydrodynamic processes include shoaling, wave dissipation by breaking (turbulent bore formation) and bottom friction, energy transfer from the sea/swell or short waves (hereafter SW) to their super- and subharmonics (or long waves: hereafter LW) by nonlinear wave–wave interactions, reflection, wave run-up and overtopping on the dike, bore impact on a wall or building, and finally reflection back towards the sea interacting with incoming bores on the promenade.

Therefore, typically experimental modelling is done for this specific case [2]. However, numerical modelling has become a possibility as well by applying computational fluid dynamics (CFD) techniques. In particular, Gruwez et al. [3] have already shown that numerical modelling with a Reynolds-averaged Navier–Stokes (RANS) model (i.e., interFoam of OpenFOAM®) can provide very similar results to large-scale experiments of overtopped wave impacts on coastal dikes with a very shallow foreshore (i.e., from the WAve LOads on WAlls (WALOWA) project [4]). Yet, such Eulerian numerical methods require often expensive mesh generation and need to implement conservative multi-phase schemes to capture the nonlinearities and rapidly changing geometries. Conversely, meshless schemes can efficiently handle problems characterised by large deformations at interfaces, including moving boundaries and do not require special tracking to detect the free surface. Methods such as smoothed particle hydrodynamics (SPH) [5] and the particle finite element method (PFEM) [6] are examples, of which SPH is the most commonly applied in coastal engineering applications [7]. In SPH, the continuum is replaced by particles, which are calculation nodal points free to move in space according to the governing dynamics laws. Although, differently from Eulerian grid-based methods, multiphase air–water SPH models are still quite scarce and have a high computational cost [8,9]. Several studies on coastal engineering applications based on single-phase SPH have been published during the last decades, for example, wave propagation over a beach [10], solitary waves [11], modelling of surf zone hydrodynamics [12], wave run-up on dikes [13], tsunamis forces [14] and wave forces on vertical walls and storm walls [15,16]. Still, single-phase SPH is also inherently expensive computationally, therefore high-performance computing is required. In particular, graphics processing units (GPUs) are employed to accelerate the computations, as, for example, in GPU-SPH [17] and DualSPHysics [18].

Up to now, it has been assumed that numerical models based on the full Navier–Stokes (NS) equations (i.e., RANS and SPH) are necessary for the current case, particularly for the complex two dimensional vertical (2DV) flows occurring on the dike and promenade in front of the structure (or vertical wall) on top of the dike. However, bores impacting vertical walls have also been modelled before with more simplified numerical models. Overtopped bores propagating over a promenade and impacting a vertical wall show many similarities to tsunami bore impacts on vertical walls. Tsunami bore impacts on vertical walls have been numerically modelled by Xie and Chu [19] using shallow-water-hydraulics (SWH) equations, with a hydrostatic pressure assumption, and have shown results consistent with experiments. The nonlinear shallow water (NLSW) equations have been applied before for the modelling of wave overtopping on steep-sloped impermeable dikes [20] and for violent overtopping of steep-sloped seawalls [21], but the lack of frequency dispersion was identified as the limiting factor to be able to correctly reproduce wave grouping. Beyond simple bore propagation, wave frequency dispersion has been added to the NLSW equations in several ways. One such example is OXBOU, a one-dimensional horizontal (1DH) hybrid Boussinesq-NLSW model [22], in which the Boussinesq equations treat the frequency dispersion prior to wave breaking. Whittaker et al. [23] applied this model for the propagation of a transient focussed wave group, wave breaking, overtopping and loading on an inclined seawall with a steep foreshore. They found that when the hydrodynamic contributions are sufficiently small, the perturbed hydrostatic pressure force gives an accurate approximation to the experimentally measured horizontal force. Another example is SWASH, a non-hydrostatic NLSW equations model, where frequency dispersion is resolved by approximation of the vertical gradient of the non-hydrostatic pressure together with a vertical terrain-following grid in a multi-layer approach [24]. It has been shown to be a very efficient and accurate method for the simulation of wave transformations over a (barred) beach [25], including transfer of wave energy to the LWs [26], and mean overtopping discharge [27] and maximum individual overtopping volume [28] over dikes with a very shallow foreshore. However, SWASH has never been validated before for (overtopped) bore impact loading on vertical walls.

To help identify and highlight the (dis)advantages of different types of numerical models relative to each other, inter-model comparisons are typically performed. Vanneste et al. [29] compared a RANS model (FLOW-3D) with DualSPHysics and SWASH for the application of wave overtopping and impact on a quay wall with berm and storm wall on top. They found a qualitatively good correspondence of the wave overtopping and impact on the storm wall for FLOW-3D and DualSPHysics, without an attempt to assess the hydrostatic pressure result from SWASH. They showed that a one-layer SWASH model was not able to accurately predict the wave overtopping in such a case. Buckley et al. [30] compared SWASH with SWAN [31] and XBeach [32] for application to irregular wave transformations in reef environments, where the formation of LWs was also found to be important and was reproduced by SWASH and XBeach. St-Germain et al. [33] compared DualSPHysics and SWASH for monochromatic wave transformation and overtopping on a dike with a shallow foreshore based on numerical snapshots and included a visual validation of surface elevation time series with a physical experiment. In addition, they made a numerical comparison of an irregular wave case. Both models appeared to give similar results for the bulk parameters. However, their comparison was mostly qualitative and therefore not quantified with model performance statics. Park et al. [34] did a laboratory validation and inter-model comparison of two RANS models: a single-phase model (ANSYS-Fluent) and a two-phase model (IHFOAM, part of the open source CFD toolbox OpenFOAM® [35–37]). They investigated non-breaking, impulsive breaking, and broken monochromatic wave interactions with elevated coastal structures, and found that the numerical accuracy of wave shoaling and breaking processes played a key role for the accuracy of the forces and pressures on the structure. Both models provided similarly good results, but validation was again mostly limited to a qualitative visual comparison of time series. One exception was the model performance in terms of force and pressure, which was quantified by calculating a normalised residual impulse of force/pressure. González-Cao et al. [38] both validated and inter-compared DualSPHysics and IHFOAM to experiments of breaking monochromatic waves impacting a vertical sea wall with a hanging horizontal cantilever slab, placed on a steep foreshore. They applied model performance and pattern statistics and showed that both models provide comparable results, with IHFOAM narrowly obtaining higher skill scores for low and medium resolutions, whereas for high spatial resolutions both models provided a similar level of accuracy. Finally, Lashley et al. [39] applied a broad range of wave models, including both SWASH and OpenFOAM®, to irregular wave overtopping on dikes with shallow mildly sloping foreshores (similar to the case considered in this paper). They found that accurate modelling of the LWs was essential to obtain accurate results for the mean overtopping discharge q and that the most computationally expensive model is not always necessary to obtain an accurate result. However, the analysis was strictly limited to the bulk parameters of wave transformation until the dike toe and q, and did not consider time series nor individual wave related events.

Therefore, clearly there is still a lack of literature about inter-model comparisons of numerical wave modelling for the combined processes leading to wave impacts on sea dikes and dike-mounted walls in presence of a very shallow foreshore, which also includes a detailed quantitative analysis based on both model performance and pattern statistics. The main goal of this paper is to investigate which type of numerical model is most accurate and most applicable in practice for the considered case. Three open source wave models are selected for standalone application, each representing one of the most popular in its category: (i) a RANS model (i.e., interFoam of OpenFOAM®), (ii) a weakly compressible SPH model (i.e., DualSPHysics) and (iii) a non-hydrostatic NLSW equations model (i.e., SWASH). We chose to investigate the performance of each model as standalone for the present work in order to provide a detailed overview of model capabilities and limitations applied to wave–structure interaction phenomena in very shallow water conditions. The RANS model is the same one as presented by Gruwez et al. [3], which was validated with large-scale experiments of overtopped wave impacts on coastal dikes with a very shallow foreshore from the WALOWA project [4]. In this paper, the same experiment and RANS model are used as a basis for the inter-model comparison with the (until now untested for this case) DualSPHysics and SWASH models. The analysis is done both (i) qualitatively, based on a comparison of the time series of the main measured parameters and snapshots of bore interactions and impacts on the dike, and (ii) quantitatively, based on model performance and pattern statistics, to critically assesses the performance of all three models to reproduce the large-scale experiment. The computational cost of each numerical model is also evaluated in terms of computational and model setup time. Finally, the results are discussed by comparing to results of the numerical models for the individual processes in other available literature, and the applicability of each numerical model for a design case is investigated.

2. Methods

2.1. Large-Scale Laboratory Experiments

Experimental tests to study overtopped wave loads on walls were undertaken in the Deltares Delta Flume. The model, at Froude length scale 1/4.3, consisted of a sandy foreshore and a concrete sea dike with promenade (Figure 1). On top of the promenade a vertical wall was positioned. The water surface elevation η was measured using wave gauges (WG) positioned over the wave flume and promenade, the horizontal velocity U_x with an electromagnetic current meter (ECM) positioned on the promenade [40], and the horizontal wave impact force F_x and pressures p by load cells (LC) and pressure sensors (PS), respectively. Both bichromatic and irregular wave tests were conducted with active reflection compensation (ARC), of which the repeated bichromatic wave test Bi_02_6 (Table 1) was chosen for the inter-model comparison. The test included mostly plunging breakers on the 1:10 transition slope and spilling breakers on the 1:35 foreshore slope in front of the dike. All other relevant details of the tests and the processing of the experimental data used in this paper for the inter-model comparison are provided by Gruwez et al. [3]. For further information on the experimental model setup, the reader is referred to the work in [4]. The WALOWA experimental dataset is available open access [41].

Table 1. Hydraulic parameters for the WALOWA bichromatic wave test (EXP) and its repetition (REXP): h_o is the offshore water depth, h_t the water depth at the dike toe, $H_{m0,o}$ the incident offshore significant wave height, R_c the dike crest freeboard, f_i the SW component frequency, a_i the SW component amplitude and δ ($=a_2/a_1$) the modulation factor. Reproduced from the work in [3], with permission from the authors, 2020.

TestID [-]	Duration [s]	h_o [m]	h_t [m]	$h_t/H_{m0,o}$ [-]	R_c [m]	f_1 [Hz]	a_1 [m]	f_2 [Hz]	a_2 [m]	δ [-]
Bi_02_6 (EXP) & Bi_02_6_R (REXP)	209	4.14	0.43	0.33	0.117	0.19	0.45	0.155	0.428	0.951

2.2. Numerical Models

2.2.1. OpenFOAM

The OpenFOAM® model and model setup as described by Gruwez et al. [3] is used. To summarise, and citing the work in [3], the solver interFoam of OpenFOAM® v6 [42] is applied, "where the advection and sharpness of the water–air interface are handled by the algebraic volume of fluid (VOF) method [43] based on the multidimensional universal limiter with explicit solution (MULES)" [44–46]. The boundary conditions for wave generation and absorption are managed by olaFlow [47], while "the turbulence is modelled by the k-ω SST turbulence closure model" that was "stabilised in nearly potential flow regions by Larsen and Fuhrman [48], with their default parameter values [49]". Hereafter, OF refers to the OpenFOAM® numerical model as presented by the authors of [3].

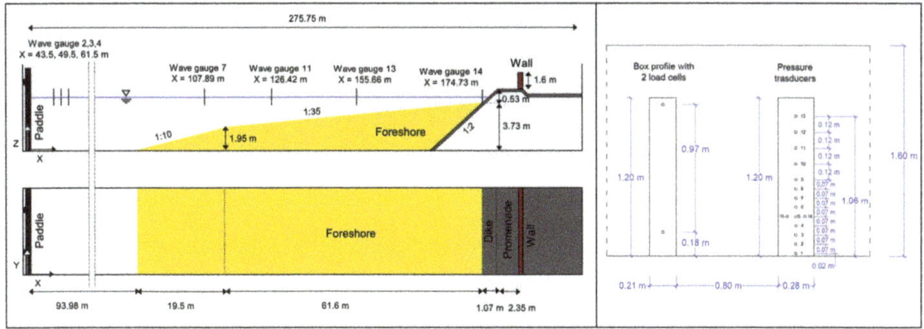

Figure 1. (**a**) Overview of the geometrical parameters of the wave flume and WALOWA model setup, with indicated WG locations. Reproduced from the work in [4], with permission from the authors, 2020. (**b**) Front view of the vertical wall on the promenade with indication of the LCs and PS array.

2.2.2. DualSPHysics

In the present study, DualSPHysics v5.0 [50], based on the weakly compressible SPH method [18], is applied, with the Wendland kernel [51] which controls the interactions between the particles based on a smoothing length h_{SPH}; and with artificial viscosity [52], tuned with parameter α_{av}, which represents the fluid viscosity, prevents particles from interpenetrating, and provides numerical stability for free surface flows [53]. Moreover, employing the artificial viscosity scheme has been shown to exhibit interesting features related to the turbulence field under breaking waves [12]. The weakly compressible SPH method requires that the speed of sound is usually maintained at least 10 times higher than the maximum velocity in the system (managed by the empirical coefficient $coeff_{sound}$). One consequence is that numerical pressure noise tends to develop [54]. To combat this, a density diffusion term (DDT) was introduced in the continuity equation [54]. This so-called δ-*SPH* approach increases the smoothness and the accuracy of pressure profiles. The δ-*SPH* method is applied in this study, by using the improved DDT of Fourtakas et al. [55] where the dynamic density is substituted with the total one. The modified Dynamic Boundary Conditions (mDBC) are employed for the fluid–boundary interactions [56]. Waves are generated by means of moving boundaries that mimic the movement of a laboratory wavemaker. The model also has its own embedded wave generation and absorption system capable for generation of random sea states, monochromatic waves and multiple solitary waves [11,57]. Hereafter, the DualSPHysics numerical model as presented here is simply referred to as DSPH.

The DSPH 2DV model domain extends from the wave paddle, over the foreshore and dike up to the vertical wall on top of the promenade (Figure 2). Some distance behind the vertical wall is also included to allow limited wave or splash overtopping without recirculation of the overtopped water. The boundary of the model domain and vertical wall consists of fixed particles. The water area, bounded by the still water level (SWL) and the fixed bottom, consists of particles that are allowed to move freely according to the SPH governing equations. The particles of the wave paddle move back and forth in the x-direction to reproduce the realised wave piston motion of the experiment including ARC. All fixed or wave paddle prescribed moving particles provide a boundary for the fluid particles according to the mDBC.

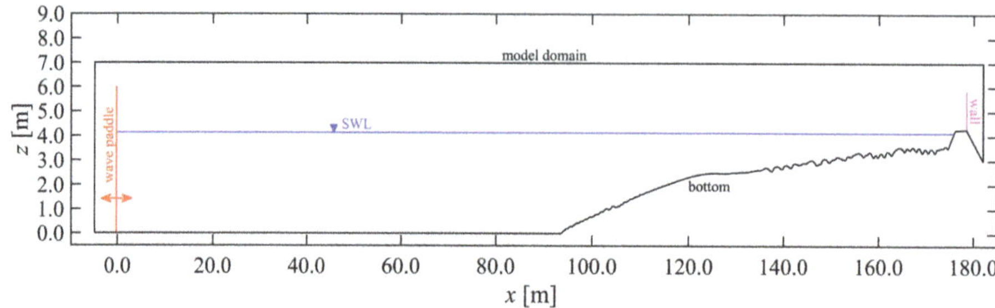

Figure 2. Definition of the DSPH 2DV computational domain, with coloured indication of the model fixed and movable boundaries. The still water level (SWL) is indicated in blue (z = 4.14 m). Note: The axes are in a distorted scale.

Unlike OF, where a variable grid resolution over the studied domain is used, no likewise or adaptive refinement is implemented in DSPH yet, being still one of the unresolved questions in SPH, also defined within the main SPHERIC Grand Challenges [58]. The resolution in DSPH is determined by the initial particle spacing dp. Previous experience has shown that at least ten particles per regular wave height (i.e., $H/dp \geq 10$) are necessary to obtain an accurate regular nonlinear wave profile and propagation [59]. However, to resolve the thin layer flows on the promenade, the water phase particles are assigned an initial particle spacing of $dp = 0.02$ m, leading to a total of 1,309,056 particles in the model domain. This choice is confirmed by a grid convergence analysis in Appendix A.

All simulations were carried out employing $\alpha_{av} = 0.01$, which is most commonly used for sea wave modelling [16], and $h_{SPH}/dp = 2.12$, where the smoothing length is calculated in DSPH according to the initial interparticle distance as $h_{SPH} = \text{coefh}\ \sqrt{2}dp$ in 2DV. In the present calculations, coefh = 1.8 was assumed (usually in the range 1.2 to 1.8 [59]). The recommended and default density diffusion parameter value of 0.1 was chosen. The results of a sensitivity analysis of these parameters showed negligible influence (not shown). The so-calculated kernel size is equal to 0.051 m, which can be considered as the effective model resolution since, citing Lowe et al. (2019), "the kernel size effectively reduces the model resolution by smoothing the results over the length-scale h_{SPH}". It is therefore twice the finest resolution used on the promenade in the OF model (i.e., dx = dz = 0.0225 m).

The symplectic position Verlet time integrator scheme was employed for time integration, with a variable time step dependent on the Courant–Friedrich–Lewy ($CFL = 0.18$) condition, the forcing terms and the diffusion term of Monaghan and Kos [10]. The DSPH simulations were run on a NVidia GeForce GTX TITAN Black with 2880 CUDA cores and FP64 (double) performance of 1.882 TFLOPS.

2.2.3. SWASH

In this study, SWASH v5.01 is applied. SWASH (Simulating WAves till SHore) is based on the nonlinear shallow water equations with addition of non-hydrostatic terms. It employs an implementation of the equations of mass and momentum conservation similar to incompressible RANS models, but with a significantly reduced vertical resolution. In the x-direction, the computational domain is discretised in equally sized grid cells and in z-direction the water column is divided into a fixed number of vertical layers K, each with a thickness of $\Delta z = h/K$ (where h is the local water depth). During wave breaking, SWASH does not model air inclusions and simulates it in a more simplified way, without overturning waves and turbulent vortices, applying a shock-capturing scheme. Moreover, for low vertical resolutions ($K < 10$), the bore front is approximated in a hydrostatic front approximation (HFA), by analogy of the turbulent bore to a hydraulic jump and by ensuring conservation of mass and momentum [25]. For a complete model description and numerical implementation reference is made to the works in [24–26]. Hereafter, the SWASH numerical model as presented here is referred to as SW1L or SW8L depending on the amount of vertical layers K applied (respectively, $K = 1$ and $K = 8$).

In SWASH, no boundary condition exists that replicates the wave paddle motion from an experiment. The SWASH 2DV model domain therefore starts at the first experimental wave gauge (i.e., WG02, x = 43.5 m) so that the incident wave time series obtained from a reflection analysis can be applied as a boundary condition for the incident waves. The domain extends further horizontally up to some distance past the top of the vertical wall where overtopped water is allowed to exit the domain (Figure 3). The model domain is vertically bounded by the free surface ($z = \eta(x,t)$) and a fixed bottom.

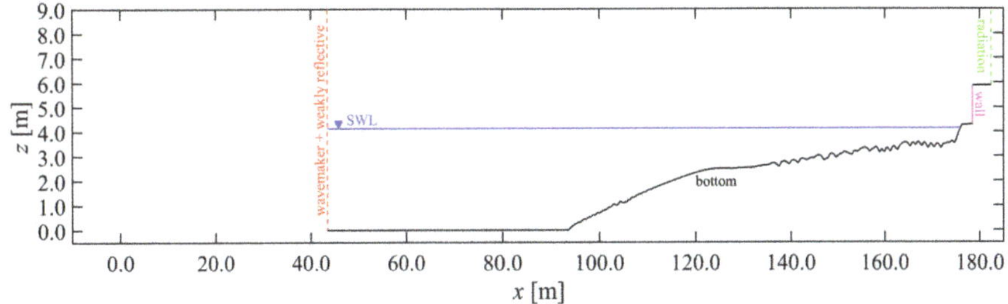

Figure 3. Definition of the SWASH computational domain, with coloured indication of the model boundaries. The wavemaker and weakly reflective boundary is positioned at the most offshore wave gauge WG02 location (x = 43.5 m). The SWL is indicated in blue (z = 4.14 m). Note: The axes are in a distorted scale.

In the horizontal direction, rectilinear cell sizes are used. For relatively high waves, 100 grid cells per wave length are recommended [60]. The shortest of the two primary components of the bichromatic wave group has a wave length of approximately 30 m (or $\Delta x = 0.30$ m) for the water depth at the wave paddle and about 12 m (or $\Delta x = 0.12$ m) for the water depth at the toe of the dike. A grid size of 0.2 m is therefore assumed, which is confirmed by a grid convergence analysis in Appendix A. To obtain a correct total water depth in each cell in the vicinity of the steep dike slope and the vertical wall, a bottom level in the cell centres is necessary, which is taken equal to the upper-right corner of each computational cell (by activating BOTCel SHIFT mode in SWASH), thereby preventing errors due to bottom interpolation.

The number of vertical layers is determined by the value of kh_0 (where k is the wavenumber) [61]. For both primary wave components, kh_0 is below 1.0 and a one-layer approach (or depth-averaged, $K = 1$) is acceptable with respect to frequency dispersion. Although a one-layer approach also appeared to be sufficient in terms of accuracy of water surface elevation for the wave–structure interactions with the dike and vertical wall, a second SWASH simulation was done as well using eight layers ($K = 8$) to resolve more the flow on top of the dike and in front of the wall. This allows a comparison of the velocity field in the snapshot comparisons with the other two numerical models (Section 3.5) and an evaluation of the model performance of a multi-layer model. Discretisation of the vertical pressure gradient is done by the implicit Keller-box scheme for SW1L, while the explicit central differences layout was applied for SW8L to ensure robustness.

The input at the wavemaker boundary of SWASH is the incident η time series obtained at WG02 by a reflection analysis using the three offshore wave gauges (WG02, 03 and 04), and following the method of Mansard and Funke [62] as it is implemented in WaveLab [63]. Note that the inter-distances of the three wave gauges were not optimised for the considered bichromatic waves, but still the reflection analysis was found to provide a reasonable incident time series. In addition, a calibration factor of 0.95 is applied to the incident surface elevation time series to match the amount of wave energy with the experiment in WG02 (see Section 3.2) introduced into the computational domain. The wavemaker boundary has a weakly reflective boundary condition, which is a numerical active wave absorption system emulating the effects of the experimental ARC. At the outlet boundary past the top of the vertical wall, a Sommerfeld radiation condition is applied, which allows overtopped water to leave the domain. A Manning's roughness coefficient n value of 0.019 is applied for the entire domain (default value [61]), for both the sand bottom and dike.

2.3. Data Sampling and Processing

Data sampling and processing of the OF model results and synchronisation of the numerical results to the experimental data were discussed by Gruwez et al. [3] and remain valid here. The same methods were applied to the DSPH and SW1L/SW8L model results. Of interest to repeat here is that a 3rd-order Butterworth low-pass filter with a cut-off frequency of 6.22 Hz (i.e., 3.0 Hz at prototype scale which is larger than the natural frequency of 1.0 Hz of typical buildings along the Belgian coast) was applied to the F_x and p time series of both the experiment and numerical model results. This removed the high frequency oscillations caused by stochastic processes during dynamic or impulsive impacts, so that the experimental signal can be reproduced by the deterministic numerical models [3,64].

For the water surface elevation measurement, both the OF and DSPH methods have uncertainties in a breaking region, where the free surface is complex and air/void inclusions are present. However, experimental instruments, such as the resistive wave gauges applied here, can suffer from similar uncertainties [3,12]. In the case of SW1L/SW8L, no air or void inclusions are modelled and η is available explicitly from the governing equations.

The pressure was sampled in OF at the PS locations along the vertical wall, while F_x was calculated by integration of p along the height of the LC (by using the OpenFOAM® library "libforces.so"). In DSPH, p is calculated by interpolating the fluid particle pressure at a distance from the wall equal to h_{SPH} and forces are calculated as the summation of the acceleration values (solving the momentum equation) multiplied by the mass of each boundary particle belonging to the vertical wall. In SWASH, the total pressure, including both the hydrostatic and non-hydrostatic pressures, exhibited strong oscillations in the grid cells closest to the vertical wall (not shown). Contrary to OF and DSPH, the (numerical noise) oscillations were not removed completely by the applied filtering, with significant residual—and in some cases even exacerbated—spurious oscillations. No immediate explanation was found to their root cause. In any case, it was found that these oscillations are attributable to the non-hydrostatic part of the pressure. Therefore, they disappeared entirely when only considering the hydrostatic pressure. The SW1L/SW8L p and F_x time series are therefore limited to the hydrostatic part in further analyses. For SW1L, the hydrostatic pressures at the pressure sensor locations were then calculated by $\rho g(\eta\text{-}z_{PS})$, where ρ is the water density (1000 kg/m^3), g the gravitational acceleration (9.81 m/s^2), η is taken from the grid cell closest to the vertical wall (which represents most closely the bore run-up height against the vertical wall) and z_{PS} is the z-coordinate of the considered pressure sensor. For SW8L, the hydrostatic pressure was interpolated between the 8 vertical grid cell values closest to the PS locations. The horizontal impact force F_x was obtained by integration of the hydrostatic pressure along the vertical wall.

Furthermore, citing the work in [3], "to investigate the model performance for the SW and LW components separately, the η time series were separated into η_{SW} and η_{LW} by applying a 3rd order Butterworth high- and low-pass filter, respectively. A separation frequency of 0.09 Hz was employed, which is in between the bound long wave frequency ($f_1 - f_2$ = 0.035 Hz) and the lowest frequency of the primary wave components (f_2 = 0.155 Hz)."

2.4. Inter-Model Comparison Method

The inter-model comparison is done qualitatively by comparing the time series of the main measured parameters between the numerical model results and the experimental data. The same model performance and pattern statistics used in the detailed OF model validation by Gruwez et al. [3] (Appendix B) are applied here for the quantitative model performance of DSPH and SW1L/SW8L and the numerical inter-model comparison.

For easier and faster selection of the best model performance between different models, pattern statistics can be visualised in one graph. The Taylor diagram [65] is such an example, which makes use of the relation between the normalised standard deviation σ^* (Equation (A9)), the correlation coefficient R (Equation (A11)) and the root mean square error ($RMSE$). While this diagram allows a straightforward comparison of the performance for amplitude (represented by σ^*) and phase (represented by R) between different models, no information about the bias is provided. Moreover, the Taylor diagram relies heavily on the (centred) $RMSE$, which is known to be misleading, because it is biased by extremes or outliers in the dataset and is dependent on the data sample size [66]. Minimising the $RMSE$ therefore does not always lead to an improved model performance [67]. Alternatively, Jolliff et al. [67] therefore proposed a skill target diagram, based on the normalised bias B^* (Equation (A10)) and the model skill score S:

$$S = 1 - \left(e^{-\frac{(\sigma^*-1)^2}{0.18}}\right)\left(\frac{1+R}{2}\right), \tag{1}$$

which has a scale between 0 and 1 (lower values present a better model prediction), and allows to independently move σ^* and R closer to 1. In a skill target diagram, B^* is taken as the Y-axis of the target diagram and $S\sigma_d$ as the X-axis, with σ_d being the sign of the standard deviation difference:

$$\sigma_d = sign\left(\sigma_p - \sigma_o\right). \tag{2}$$

Positive and negative X-axis values therefore indicate respectively a higher or a lower standard deviation (or wave height when the considered variable is η) of the modelled time series compared to the observed time series. The closer the model point is to the diagram origin, the better the model performance is to represent the observation. The total model skill score based on this diagram can then be summarised as the distance ST from the origin of the target diagram or

$$ST = \sqrt{(B^*)^2 + S^2} \tag{3}$$

which is bounded by [0, 1]. The closer ST is to zero, the better the skill of the model is to reproduce the pattern of the experimental measurements. The main advantages of a skill target diagram are that it clearly visualises the pattern statistics, and that it provides more insight into different aspects of the model performance than a general numerical model performance statistic, such as Willmott's refined index of agreement d_r (Equation (A5)).

So far, none of the statistics mentioned provide specific information on the model performance of the peak forces and duration of the wave-induced force on the vertical wall. However, both are of high significance to structural damage [68], and their model performance should be assessed as well. The model performance to reproduce the experimental peak forces of each independent wave impact event during the test is evaluated by a d_r-value between predicted and observed maximum horizontal force per impact event $F_{x,max}$ (i.e., $d_{r,Fx,max}$). The duration of the wave impact can be evaluated by the impulse of the total horizontal force I:

$$I = \int_0^{t_N} F_x(t)dt \tag{4}$$

where t_N is the total duration of the test. To evaluate the model performance, a normalised predicted impulse is considered:

$$I^* = \frac{I_p}{I_o} \tag{5}$$

where I_p and I_o are the predicted and observed force impulses calculated by Equation (4), respectively. The observed total horizontal force impulse is overestimated, equal to or underestimated by the prediction when $I^* > 1$, $I^* = 1$ or $I^* < 0$, respectively. Note that I^* is evaluated for the complete F_x time series, so that phase differences are disregarded by this parameter. Therefore, I^* purely evaluates the correspondence of the total impulse on the vertical wall during the complete test.

3. Results

3.1. Time Series Comparison

The three numerical model results are first compared qualitatively in the time domain to each other and to EXP. For the sake of brevity, not all measured locations, but a selection of sensor locations is presented here. Sensor locations were selected to be representative for different areas along the flume with clearly different physical behaviours of the waves. In Figure 4, the time series of η are compared at measurement location WG04, representing the offshore waves between the wave paddle and foreshore toe; WG07, representing the wave shoaling and incipient breaking area; WG13, representing the surf zone; WG14, representing the inner surf zone and toe of the dike location; and WLDM02, representing the bore interaction area on the promenade. For clarity, the η_{LW} time series are shown separately in Figure 5. The time series of U_x are compared in Figure 6 at the ECM location on the promenade. For the numerical models, actually the depth-averaged horizontal velocity $\overline{U}_x$ is shown instead, as it was shown to deliver a better correspondence to EXP than U_x for OF [3]. The same was found to be the case for DSPH (not shown), and SW1L only provides $\overline{U}_x$, since it is a depth-averaged model. In Figure 7, the time series of F_x are compared to the LC measurements, and in Figure 8 the time series of p are compared at the PS locations selected at approximately equidistant positions along the array (i.e., 0.28 m from PS01 up to PS09 and 0.24 m up to PS13).

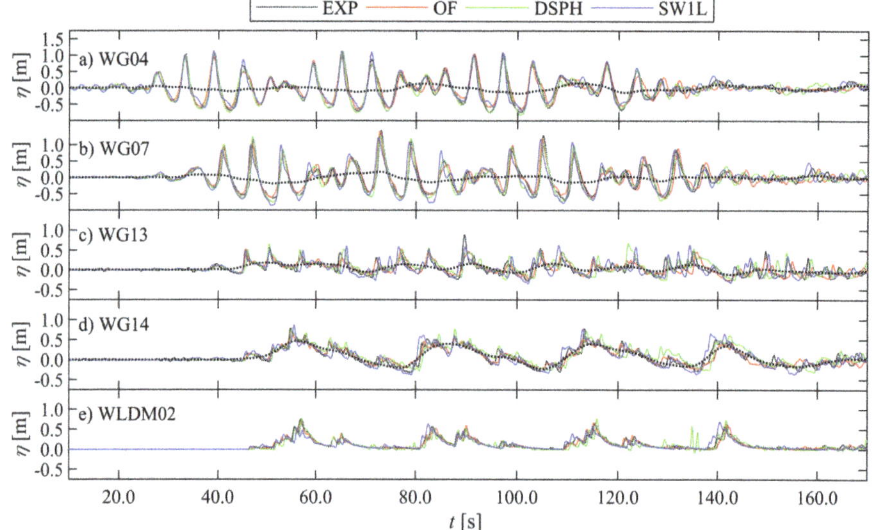

Figure 4. Comparison of the η time series at selected sensor locations (**a–e**). The zero-reference is the SWL for panels (**a–d**) and the promenade bottom for panel (**e**). Note: η_{LW} is shown as well, but only for EXP (bold dotted lines in panels (**a–d**)). Adapted from the work in [3], with permission from the authors, 2020.

From these figures, it is immediately clear that all three numerical models provide results that are very close to EXP. Especially for η, differences appear to be very small with more significant differences in the surf zone (Figure 4c,d) and on top of the dike (Figure 4e). Further differences are revealed when comparing the η time series of the LW components only. OF does not correspond as well to EXP in the offshore zone (Figure 5a) compared to the other two numerical models. In the surf zone, however,

OF shows better correspondence to EXP together with SW1L, while DSPH starts to diverge more from EXP for t greater than approximately 120 s (Figure 5c,d).

Reproducing U_x appears to be more challenging than η for all numerical models. Most of the positive U_x peak values (i.e., flow towards the vertical wall) are reproduced, while some of the return flow durations (i.e., t for $U_x < 0$) are modelled longer by OF than SW1L, with DSPH being in between (Figure 6). Unfortunately, return flow velocities were often not captured by the ECM measurements in EXP, mostly by too thin flow layers (no data).

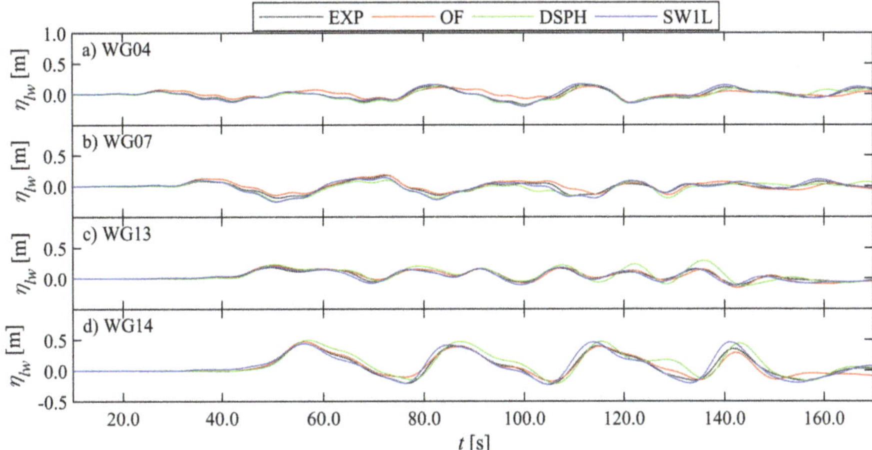

Figure 5. Comparison of the η_{LW} time series at selected sensor locations (**a–d**). The zero-reference is the SWL. Adapted from the work in [3], with permission from the authors, 2020.

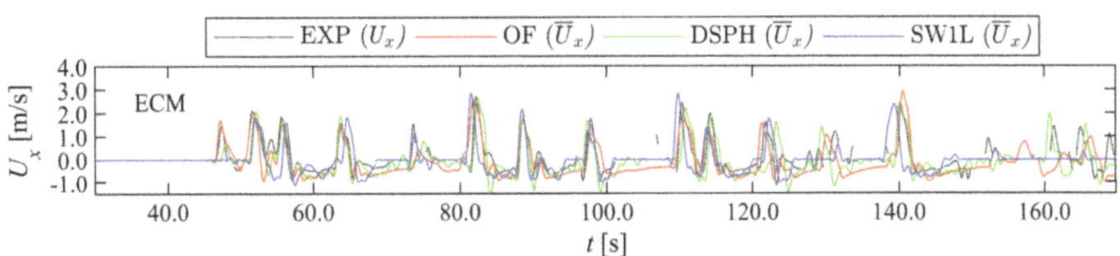

Figure 6. Comparison of the U_x time series at the ECM location. The zero-reference is the promenade bottom. Adapted from the work in [3], with permission from the authors, 2020.

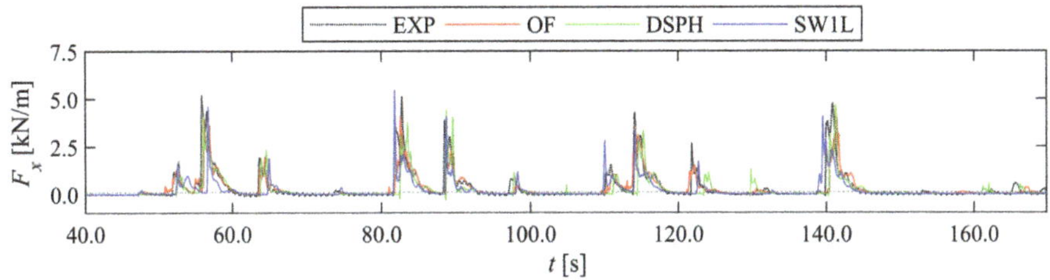

Figure 7. Comparison of the F_x time series at the vertical wall. The experiment is the load cell force measurement. Adapted from the work in [3], with permission from the authors, 2020.

For the F_x (Figure 7) and p time series (Figure 8), differences become more distinctive. DSPH shows (small) negative or sub-atmospheric p peaks, not observed in EXP, that occur before some of the dynamic impact peaks and mostly for the lowest PSs (Figure 8a). Both OF and DSPH appear to underestimate most peak values for both F_x and p, while phase differences with EXP are most apparent for DSPH and SW1L.

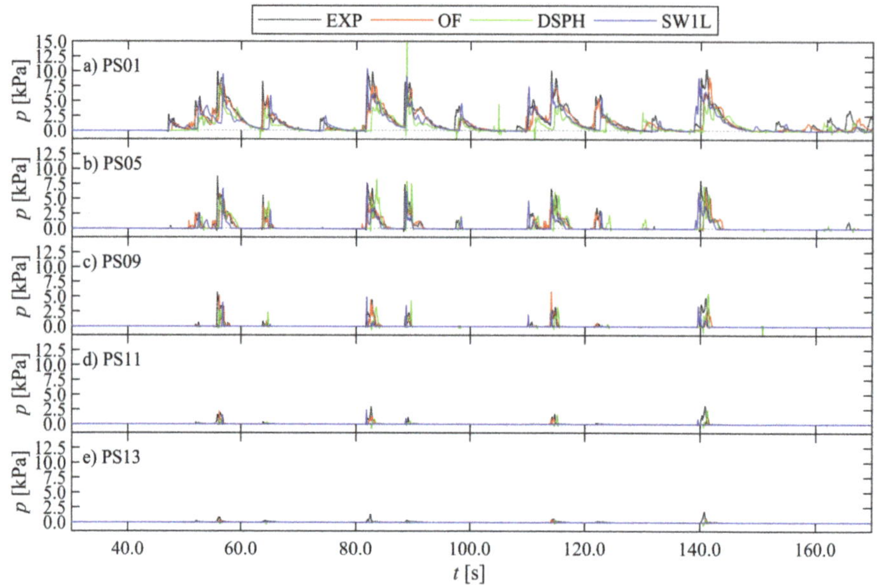

Figure 8. Comparison of the p time series at selected sensor locations (**a**–**e**), PS01 being the bottom PS (**a**) and PS13 the top-most PS (**e**). Adapted from the work in [3], with permission from the authors, 2020.

3.2. Spatial Distribution of Wave Characteristics

The evolution of the root mean square wave height H_{rms}, the SW and LW components (i.e., $H_{rms,sw}$ and $H_{rms,lw}$), and the mean surface elevation $\overline{\eta}$ (wave setup) over the wave flume up to the toe of the dike are compared in Figure 9. All models agree on the general evolution of the H_{rms} curves along the flume. The wave height slightly decreases from the wave paddle up to the toe of the foreshore, more so in case of OF than DSPH (with DSPH closest to EXP): from the wave paddle location to the foreshore toe, H_{rms} decreases about 10% more for OF than DSPH. SW1L shows a similar behaviour from its offshore boundary (i.e., at WG02) until the foreshore toe and corresponds most with DSPH. On the other hand, SW shoaling is overestimated by both DSPH and SW1L ($H_{rms,sw}$ at WG07 in Figure 9). In the surf zone, DSPH reproduces the energy loss due to SW breaking best of all three numerical models (i.e., closest result to (R)EXP for $H_{rms,sw}$ at WG13 and WG14 in Figure 9), while SW1L overestimates and OF underestimates $H_{rms,sw}$ there.

The LW wave height $H_{rms,lw}$ evolution along the flume in the experiment is also reproduced by all three numerical models. An unexpected peak appears in the DSPH result near x = 126 m, which is not found in the results by OF and SW1L. Moreover DSPH significantly overestimates $H_{rms,lw}$ at WG13, while OF underestimates $H_{rms,lw}$ at the dike toe (WG14).

Overall, DSPH provides the best correspondence with (R)EXP for H_{rms} followed by SW1L, while OF clearly underestimates it for all measured locations. In terms of the wave setup $\overline{\eta}$, however, SW1L shows the best correspondence with (R)EXP, while OF and DSPH, respectively, over- and underestimate it until at least WG07. In the inner surf zone, OF corresponds better with (R)EXP for $\overline{\eta}$ together with SW1L, while DSPH significantly overestimates it ($\overline{\eta}$ at WG13–14 in Figure 9).

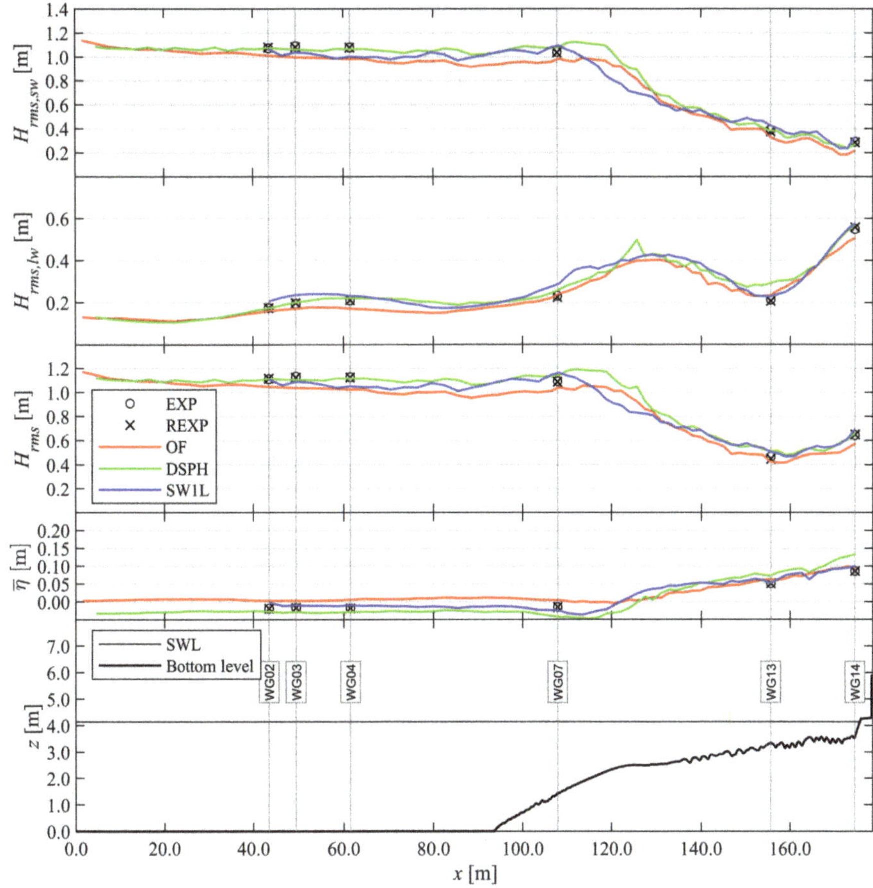

Figure 9. Comparison of the root mean square wave height H_{rms} between each numerical model and the (repeated) experiment up to the toe of the dike. From top to bottom: $H_{rms,sw}$ for the short wave components, $H_{rms,lw}$ for the long wave components, H_{rms} for the total surface elevation, mean surface elevation or wave setup $\overline{\eta}$ and an overview of the sensor locations, SWL and bottom profile. Adapted from the work in [3], with permission from the authors, 2020.

3.3. Model Performance and Pattern Statistics

More than a qualitative validation and evaluation of inter-model performance is not possible with the time series comparison of Section 3.1, especially when visually almost no discernible and consistent trends of distinction between the model results can be made. The model performance and pattern statistics, provided in Appendix B and Section 2.4, then become very useful for a quantitative evaluation. As d_r provides a single dimensionless measure of average error, it is suitable to provide insight into the spatial variation of model error in the flume. In addition to the d_r of each numerical model, the d_r of the repeated experiment (REXP) is also included in this analysis. A numerical model d_r higher than the d_r of REXP means that the numerical error cannot be reduced further compared to the experimental repeatability error and a near "perfect" model performance would be achieved with regard to the experiment [3]. Therefore, a relative refined index of agreement d'_r (Equation (A8)) and a corresponding rating (Table A1) was defined by Gruwez et al. [3] which provides the performance of the numerical model relative to the experimental model uncertainty. Tables 2 and 3 provide the

d_r-values and the pattern statistics at key locations (dike toe and on the promenade, respectively). It is noted that the statistics for η reported in Table 3 were averaged over the four measured locations (WLDM01—WLDM04), for the sake of brevity and because it better represents the statistics for the processes on the promenade. The evolution of d_r and R at the WG locations along the wave flume up to the toe of the dike is shown, respectively, in Figures 10 and 11, for η_{SW} ($d_{r,sw}$ and R_{sw}), for η_{LW} ($d_{r,lw}$ and R_{lw}) and for η ($d_{r,tot}$ and R), and of d_r for η and U_x on the dike in Figure 12.

Offshore, DSPH has the best model performance (WG02-04 in Figure 10, rated Excellent) followed by OF (rated Very Good), and this continues to be so up to the shoaling zone (WG07), although the rating for DSPH drops slightly to Very Good. On the other hand, while SW1L starts offshore with a (relative) model performance similar to OF (rated Very Good), a notable decrease in (relative) model performance occurs in the shoaling zone (WG07, rated Good). All models show a generally decreasing trend of $d_{r,tot}$ over the surf zone (WG07-13) and increases back up to the dike toe (WG13-14). Over the surf zone, DSPH gradually becomes the least performing numerical model (WG13-14, rated Good) followed by SW1L (WG13, rated Good). The relative model performance of SW1L increases back to Very Good at the dike toe (WG14). The performance of OF is not as good as DSPH in the offshore area (WG02-04), but becomes the highest of all three numerical models in the surf zone (WG13-14, rated Very Good) and continues to perform the best on the dike as well (Figure 12), where the d_r of η remains more or less constant for all models, with exception of DSPH which increases slightly back to a rating of Very Good.

Separating η into the SW and LW components reveals that $d_{r,sw}$ mostly follows the same trend as $d_{r,tot}$, with the exception that DSPH performs better than SW1L at the dike toe for $d_{r,sw}$ (Figure 10). On the other hand, $d_{r,lw}$ clearly has a different behaviour: OF does not reproduce the incident LWs as good as DSPH and SW1L, but its LW performance steadily increases towards the dike toe (Figure 10), where the LW energy increases (Figure 9). SW1L shows the overall best LW performance as it shows similar $d_{r,lw}$ values offshore to DSPH and similar values to OF in the surf zone. It is also revealed that the increase in SW1L error at WG07 is mostly caused by a decrease in η_{sw} performance. Even though SW1L shows the least performance in modelling η_{sw} over the foreshore, that does not seem to affect its capability of reproducing the LW shoaling and energy transfer from the SW to LW components, with a similar accuracy to OF for modelling η_{lw} in the surf zone. Increased accuracy of SWASH of the SW modelling can be obtained however, with increased vertical resolution: the SW8L model exhibits much better performance in the shoaling zone (i.e., SW1L: $d_{r,sw,WG07} = 0.73$, SW8L: $d_{r,sw,WG07} = 0.86$, not shown), and attains the same model performance for η at the toe of the dike as DSPH ($d_{r,sw,WG14} = 0.53$). However, because of the smaller wave height of the SW components at the dike toe compared to the LW components (Figure 9), this improvement only slightly increases the overall model performance at the dike toe (Table 2, SW1L: $d_{r,sw,WG14} = 0.85$, SW8L: $d_{r,sw,WG14} = 0.86$).

Table 2. Model performance and pattern statistics evaluated for η of REXP, OF, DSPH, SW1L and SW8L at measured location WG14 (dike toe location). Adapted from the work in [3], with permission from the authors, 2020.

Model [-]	B^* [-]	σ^* [-]	R [-]	d_r [-]	d'_r [-]	Rating [-]
REXP	0.00	1.00	0.98	0.92	1.00	Excellent
OF	0.05	0.89	0.91	0.82	0.90	Very Good
DSPH	0.21	1.02	0.85	0.71	0.79	Good
SW1L	0.04	1.09	0.87	0.77	0.85	Very Good
SW8L	0.08	1.00	0.88	0.78	0.86	Very Good

The pattern statistics B^* and σ^* in Table 2 represent, respectively, the accuracy of the wave setup and wave amplitude at the toe of the dike [3], and spatial information of these errors could already be derived implicitly from the $\overline{\eta}$ and H_{rms} graphs in Figure 9. Both were already discussed in Section 3.2. The result is that at the toe of the dike, DSPH has the best result of the three numerical models in terms of reproduction of the wave height (Table 2, σ^* = 1.02), but the worst result in terms of the wave setup (Table 2, B^* = 0.21). OF has the worst result for the wave height (Table 2, σ^* = 0.89), while delivering a close second-best result with SW1L for the wave setup (Table 2, B^* = 0.05). SW1L provides the lowest wave setup error (Table 2, B^* = 0.04).

Table 3. Model performance and pattern statistics evaluated for η of REXP, OF, DSPH, SW1L and SW8L averaged over all measured locations on the promenade (WLDM01—WLDM04) and for U_x at the measured location ECM. Adapted from the work in [3], with permission from the authors, 2020.

Model [-]	Parameter [-]	B^* [-]	σ^* [-]	R [-]	d_r [-]	d'_r [-]	Rating [-]
REXP	η	−0.01	0.99	0.99	0.92	1.00	Excellent
	U_x	−0.02	1.05	0.87	0.81	1.00	Excellent
OF	η	−0.04	1.00	0.89	0.81	0.89	Very Good
	U_x	−0.25	0.94	0.73	0.63	0.82	Very Good
DSPH	η	−0.04	1.01	0.81	0.72	0.80	Very Good
	U_x	−0.26	0.92	0.68	0.62	0.81	Very Good
SW1L	η	−0.03	0.96	0.78	0.74	0.82	Very Good
	U_x	−0.22	0.84	0.51	0.55	0.74	Good
SW8L	η	−0.14	0.91	0.82	0.75	0.83	Very Good
	U_x	−0.09	0.86	0.62	0.59	0.78	Good

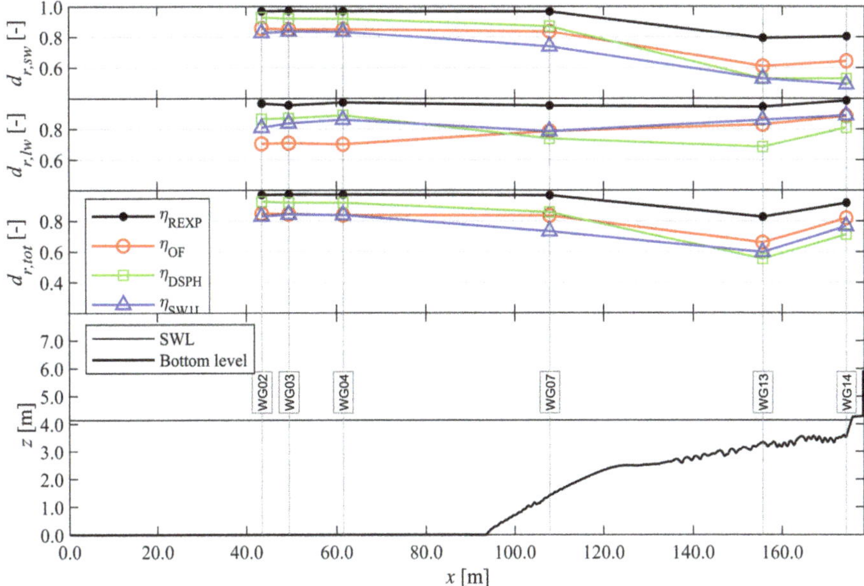

Figure 10. Comparison of d_r, evaluated for REXP, OF, DSPH and SW1L with reference to EXP, up to the toe of the dike. From top to bottom: $d_{r,sw}$ for η_{SW}, $d_{r,lw}$ for η_{LW}, $d_{r,tot}$ for η, and finally an overview of the sensor locations, SWL, and bottom profile. Adapted from the work in [3], with permission from the authors, 2020.

However, in the previous, spatial information about the accuracy of the wave phase modelling is missing and is shown separately in Figure 11. From this figure it is clear that DSPH introduces the largest error in wave phases over the surf zone up to the dike toe, and that the error is mostly due to phase errors in the SWs. Additionally, an important contribution of phase error is present in the η_{LW}

result of DSPH as well, which is not observed in the other numerical model results. Consequently, at the toe of the dike the phase error is largest for DSPH (i.e., lowest R value in Table 2). However, at the dike toe the difference with SW1L is small, while OF provides the best phase correspondence with EXP.

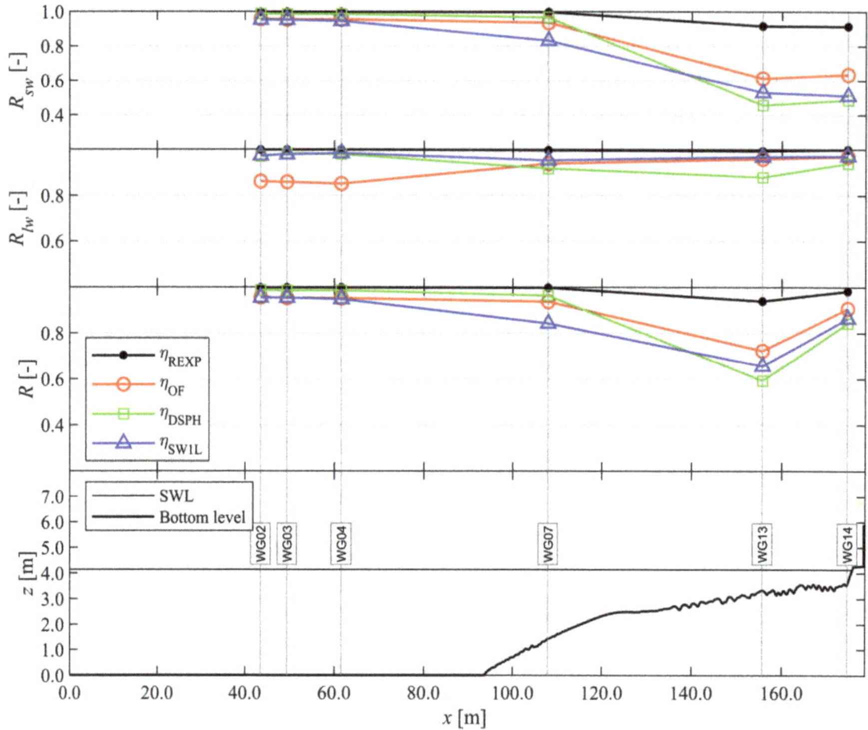

Figure 11. Comparison of R evaluated for η (of REXP, OF, DSPH and SW1L with reference to EXP) up to the dike toe. From top to bottom: R_{sw} for η_{SW}, R_{lw} for η_{LW}, R for η, and finally an overview of the sensor locations, SWL, and bottom profile. Adapted from the work in [3], with permission from the authors, 2020.

On top of the dike, the d_r of U_x is provided in Figure 12 and Table 3, and indicates a lower model performance for all three numerical models than obtained for η. However, for the relative model performance d'_r this difference significantly reduces, so that the same rating is obtained for U_x as for η in case of DSPH and OF (Table 3, rated Very Good). SW1L (and SW8L) has the lowest d'_r for U_x and rating (Table 3, rated Good). Although the wave setup at the dike toe is overestimated by each numerical model (Table 2, $B^* > 0$), η on the promenade is generally underestimated and U_x as well (Table 3, $B^* < 0$). The bore wave height is best represented by OF (indicated by σ^*), closely followed by DSPH and SW1L. Phase differences are observed for all numerical models (i.e., $R < 1.00$ in Table 3), but are lowest for OF, followed by DSPH and SW1L.

Next the d_r-values of the pressures at the vertical wall are compared in Figure 13 and the statistics in Table 4. Again, all models show a lower model performance than REXP, and OF obtained the highest value followed by SW1L and DSPH, both of which have very similar model performance along the PS array. Model performances of all models tend to decrease and converge to each other towards higher sensor locations on the vertical wall.

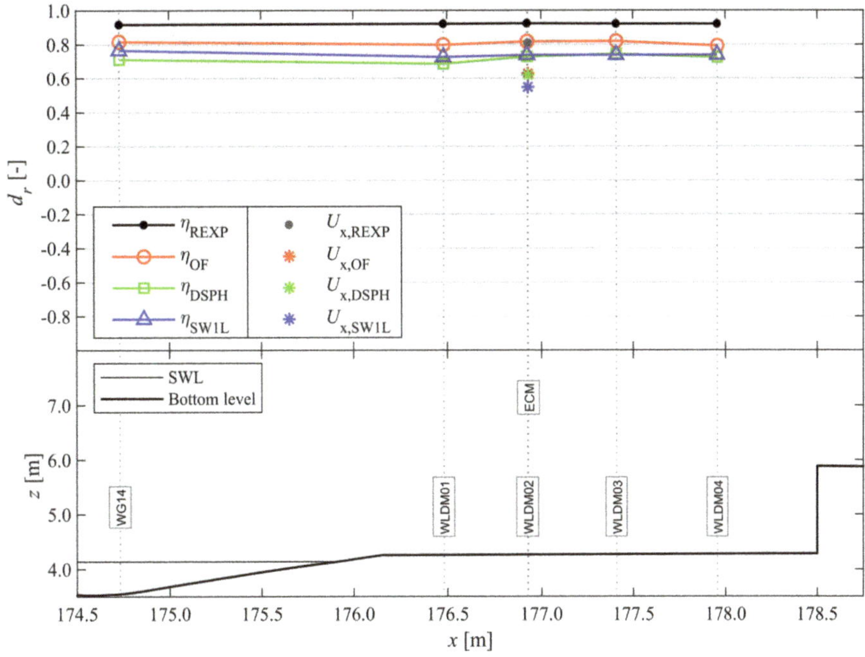

Figure 12. Comparison of d_r, evaluated for η and U_x (of REXP, OF, DSPH and SW1L with reference to EXP) from the toe of the dike up to the vertical wall. From top to bottom: d_r for η and U_x, and finally an overview of the sensor locations, SWL, and bottom profile. Adapted from the work in [3], with permission from the authors, 2020.

The d_r value of F_x for each model is shown in an overview table (Table 4) together with the other pressure and force related statistics. Even though OF has the highest model performance in terms of the F_x time series (Table 4, $d_{r,Fx}$ = 0.76), the force peaks are better estimated by SW1L (Table 4, $d_{r,Fx,max,OF}$ = 0.85 and $d_{r,Fx,max,SW1L}$ = 0.88), while it has the largest errors in the F_x time series (Table 4, $d_{r,Fx}$ = 0.64). Moreover, SW1L underestimates the total impulse much more than OF does (Table 4, I^*_{OF} = 0.85 and I^*_{SW1L} = 0.62). DSPH has a similar model performance as OF for the force peaks $F_{x,max}$, while I^* is in between OF and SW1L and its overall model performance for F_x is similar to SW1L. Consequently, the relative model performance for F_x is rated Very Good for OF and Good for DSPH and SW1L. While the model performance slightly increases for SW8L compared to SW1L at the dike toe (Table 2) and on the promenade (Table 3), this does not translate into a better model performance for the wave impact on the vertical wall; in fact, almost every F_x statistic is lower for SW8L than for SW1L (Table 4).

Pattern statistics (B^*, σ^* and R) are included as well in Table 4. They indicate that all numerical models underestimate the wave impact force and exhibit phase differences. OF shows the least overall underestimation (i.e., B^* closest to zero) and the least phase differences (i.e., highest R), while DSPH has the highest σ^* value. The results for F_x are slightly worse for the multi-layer model SW8L compared to the single layer model SW1L, except for σ^*.

3.4. Skill Target Diagrams

After the spatial inter-model comparisons of Sections 3.1–3.3 based on the model performance and pattern statistics, the pattern statistics are visualised here together in a skill target diagram as described in Section 2.4 (Figure 14). The selection of observed locations that is considered for these diagrams, is the same as that for the time series plots (Section 3.1). This is to prevent as much as possible a biased evaluation of the model skill in the target diagram, because a particular area had more sensors (i.e., the offshore area for η and the lower half of the pressure sensor array for p). One exception is η on the promenade, for which the pattern statistics were averaged over the four measured locations (WLDM01–WLDM04), and therefore the values from Table 3 are used here as well. The general model performance is visualised by a circle with a radius equal to the mean of the ST skill scores (Equation (3)) or distances from the origin of each model data point in the target diagram of either η (Figure 14a) or p (Figure 14b). The repeated experiment (REXP) is visualised in the skill target diagrams as well, but only by the mean skill circle for η and p. This circle is included to have a reference of the experimental repeatability error. For both U_x and F_x only one representative observed location is available. They are visualised as pentagrams together with the model data points for η and p, respectively.

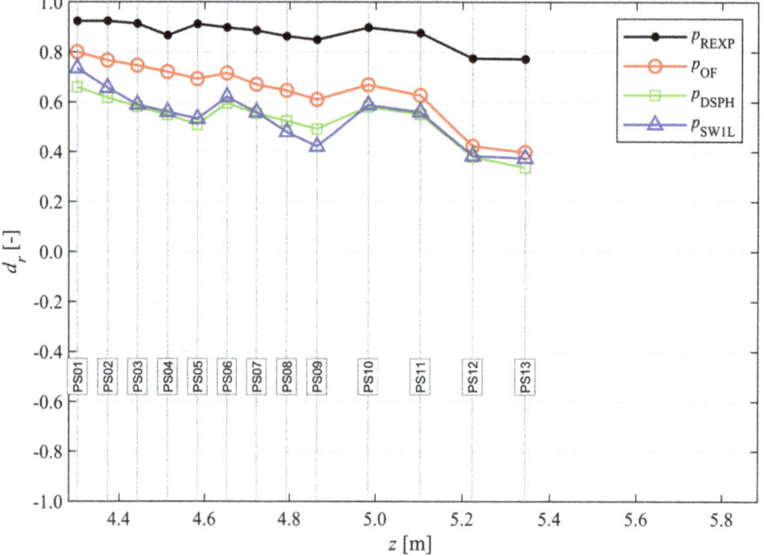

Figure 13. Comparison of d_r evaluated for p (of REXP, OF, DSPH and SW1L with reference to EXP) at the vertical wall (horizontal axis). Adapted from the work in [3], with permission from the authors, 2020.

Table 4. Model performance and pattern statistics evaluated for p and F_x of REXP, OF, DSPH, SW1L and SW8L at the respective measured locations PS05 and LC. Adapted from the work in [3], with permission from the authors, 2020.

Model [-]	Variable [-]	B^* [-]	σ^* [-]	R [-]	I^* [-]	$d_{r,Fx,max}$ [-]	d_r [-]	d'_r [-]	Rating [-]
REXP	p	0.01	1.00	0.96	-	-	0.91	1.00	Excellent
	F_x	0.00	0.97	0.90	0.99	0.92	0.90	1.00	Excellent
OF	p	−0.11	0.75	0.61	-	-	0.69	0.78	Good
	F_x	−0.12	0.74	0.73	0.85	0.85	0.76	0.86	Very Good
DSPH	p	−0.21	0.93	0.30	-	-	0.51	0.60	Reasonable/Fair
	F_x	−0.21	0.78	0.55	0.73	0.84	0.66	0.76	Good
SW1L	p	−0.43	0.74	0.43	-	-	0.53	0.62	Reasonable/Fair
	F_x	−0.31	0.66	0.55	0.62	0.88	0.64	0.74	Good
SW8L	p	−0.56	0.81	0.25	-	-	0.46	0.55	Reasonable/Fair
	F_x	−0.37	0.71	0.40	0.54	0.87	0.60	0.70	Reasonable/Fair

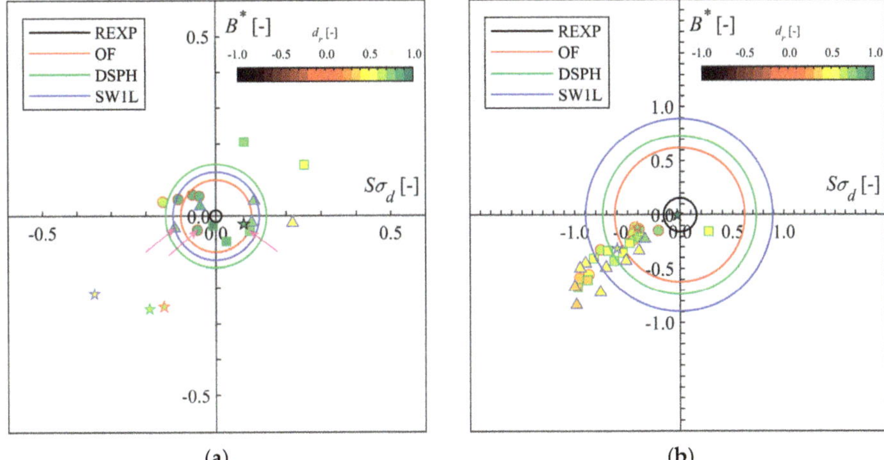

Figure 14. Numerical model skill target diagrams (target: EXP) for selected sensor locations along the flume. All markers are colour-filled according to the d_r colour scale. The circles represent the mean value of all markers for a specific model. The data points of REXP are not plotted for clarity, but only the mean (black circle). (**a**) Target diagram with pattern statistics for η at locations WG04, 07, 13, 14 and averaged pattern statistics over WLDM01-WLDM04 for the promenade (OF: circles, DSPH: squares; and SW1L: triangles) and U_x at the ECM location (pentagrams). The magenta arrows indicate the markers representing the model performance of η on the promenade (WLDM02). (**b**) p at approx. equidistant locations PS01, 03, 05, 07, 09, 10–13 (OF: circles, DSPH: squares; and SW1L: triangles) and F_x (pentagrams). Note: the target diagrams have different axis ranges.

None of the mean skill circles of the three numerical models have a smaller radius than the REXP circle, which means that none of the models has an ideal model skill relative to the experimental repeatability. In case of η, the model skill circle of DSPH is largest, and therefore DSPH has the lowest overall model skill, followed by the smaller circles of SW1L and OF (highest model skill), respectively. However, the U_x pentagrams suggest a better U_x model skill of DSPH than SW1L. For p, OF remains the numerical model with the highest model skill, followed by DSPH and SW1L. The F_x pentagrams have the same ranking.

In the η skill target diagram, the numerical model skill for the location on the dike is indicated by an arrow. The remaining numerical model skill scores are those of the measured locations along the flume up to the dike toe. For OF, they are positioned in the top left quadrant, which means that the wave energy is underestimated ($\sigma_d < 0$) and the wave setup is slightly overestimated ($B^* > 0$) (both confirmed by Figure 9). For DSPH, the wave energy is mostly overestimated ($\sigma_d > 0$). The two points furthest removed from the origin are the measured locations in the surf zone (i.e., WG13-14), where high B^* and low R-values (or wave phase differences) are the largest contributors to the decreased DSPH model skill in this area. SW1L generally shows an overestimation of the wave energy ($\sigma_d > 0$) and increased wave phase differences (lower R) in the shoaling and surf zones. Generally, SW1L has the best wave setup results (lowest B^* values). The U_x skill of all numerical models is one of a clear underestimation (pentagrams in lower left quadrant), of both the mean value ($B^* < 0$) and standard deviation ($\sigma_d < 0$), and increased phase difference (low R-values). The same is valid for p and F_x, where all numerical model skill points are also positioned in the lower left quadrant.

3.5. Snapshot Inter-Model Comparison

The numerical models applied in this paper typically have a higher spatial resolution of the physical parameters of interest (e.g., η, U, p) than an experiment. This allows a comparison of snapshots between the numerical models. To allow a comparison of the velocity field, the multi-layered model SW8L is used instead of SW1L. The first main impact series is most appropriate for such a comparison, because accumulation of errors is lowest at the beginning of the test. Snapshots of the flow on the dike and the pressure distribution along the vertical wall are compared in Figures 15–17. A few key time instants in the F_x and U_x time series were selected during this series of impacts and are listed chronologically in Table 5. These time instants were selected from each model result independent of time, because due to phase errors these key time instants have occurred at (slightly) different times between the models.

The first main impact was identified by Gruwez et al. [3] to be caused by a plunging breaking bore pattern impacting on the vertical wall. The overturning wave arose when a large incoming bore collided with a smaller bore that was reflected against the vertical wall only a few moments before. This collision occurred at different locations on the promenade for each model and explains the timing mismatch of the F_x impact peak with EXP (see F_x graph insets in Figure 15c). The timing of the smaller bore impact corresponds with EXP in case of OF and SW8L, but is late for DSPH (time instant 1, Figure 15a). For all numerical models, the large incoming bore arrived later than was observed in EXP. This means that the collision between the reflected small bore and incoming large bore was timed differently for each model, with repercussions for the subsequent impact of the overturned wave on the vertical wall.

The best result is obtained by the OF model, which modelled a correct timing of the smaller bore reflection against the wall, but the late arrival of the larger incoming bore (i.e., by approximately 0.3 s) caused the collision to occur further from the wall than in the EXP (time instant 2, Figure 17a). Nevertheless, for the impact on the wall (time instants 3 and 4, Figure 15b,c), OF is able to—albeit mostly qualitatively—reproduce the shape of the pressure distribution which is distinctly different from a hydrostatic pressure distribution: both the pressure peak at PS10 for time instant 3 and the general shape of the pressure distribution at time instant 4 are captured by the model. We direct interested readers to the work in [3] for a more detailed description of this comparison. DSPH has a very similar result, but the timing of the smaller bore was late as well (i.e., by approximately 0.7 s), meaning that the collision with the larger bore occurred closer to the wall than observed in EXP (time instant 3, Figure 15b). Although the model did not capture the pressure peak at PS10 for time instant 3, it did manage to approach the pressure distribution qualitatively as well during the dynamic impact (time instant 4, Figure 15c). Although SW8L managed to get the timing of the smaller bore impact right (time instant 1, Figure 15a), the larger bore impacted the wall much later than in EXP (i.e., by approximately 1.2 s). This means that no interaction between the bores was modelled on the promenade. In any case, SWASH is a depth-integrated model, so it is inherently not able to model overturning waves explicitly. Moreover, only hydrostatic pressure distributions are provided by the model to avoid spurious numerical oscillations (Section 2.3). However, even with those limitations, SW8L (and SW1L) still managed to predict an F_x peak during the dynamic impact (time instant 4, Figure 15c), but the pressure distribution remains hydrostatic and therefore no local pressure peaks were captured at all.

After the dynamic impact (time instant 4), the bore ran up the vertical wall (time instant 5, Figure 15d) and reflected, causing a second quasi-static F_x peak (time instant 6, Figure 16a). In both the OF and DSPH results, a clockwise vortex formed near the bottom of the wall during the run-up process. However, in case of DSPH, this vortex was much stronger and lasted during the quasi-static F_x peak as well. The p distribution of EXP was mostly hydrostatic, except for a small local peak at PS06 (time instant 5), which seems to have been captured qualitatively by DSPH. On the other hand, the strong vortex modelled by DSPH also caused a non-hydrostatic p distribution during time instant 6, while it was mostly hydrostatic in EXP. In this time instant, again OF was closest to the EXP observation. SW8L was not successful in correctly predicting the wave run-up against the vertical wall during reflection of the large impacting bore (time instant 5) and consequently underestimated p and F_x more than the other numerical models for time instants 5 and 6. During return flow (time instant 8,

Figure 16b) the pressure distribution was mostly hydrostatic, and all numerical models were able to predict the p distribution well. Only DSPH shows a pressure decrease near the bottom of the wall (PS01), possibly caused by the persistent vortex modelled there, but now further removed from the wall compared to the previous time instants 5 and 6.

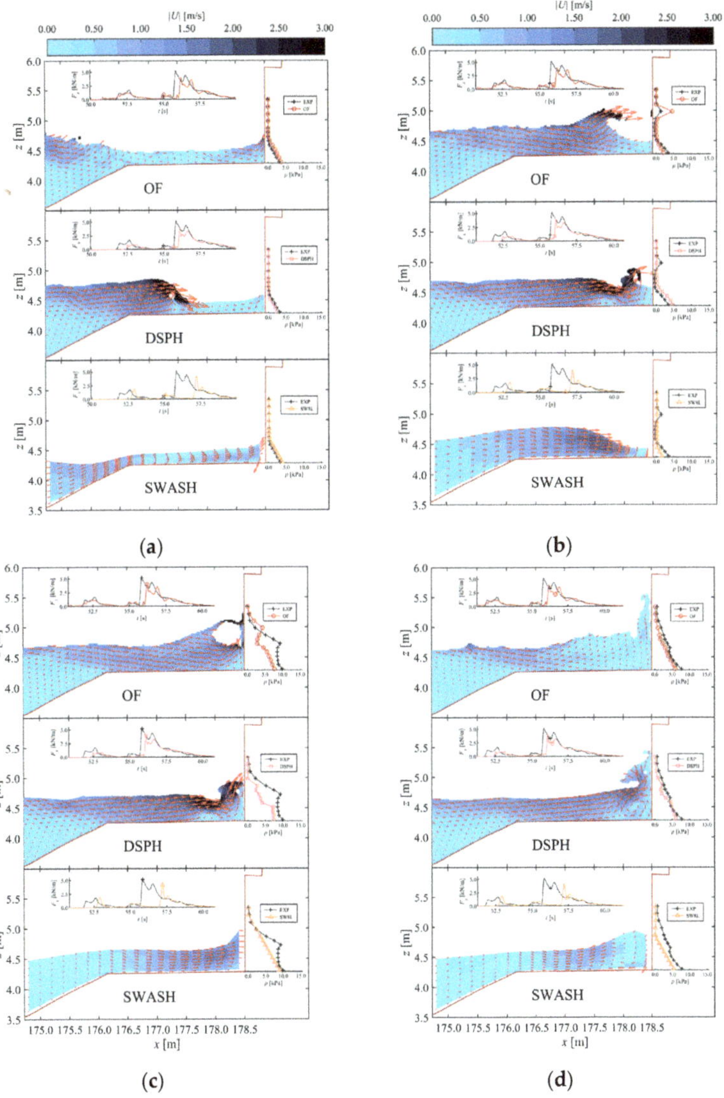

Figure 15. Numerical comparative snapshots of the water flow on the dike. Colours are the velocity magnitude $|U|$ according to the colour scale shown at the top of each figure. The red arrows are the velocity vectors, which are scaled for a clear visualisation. Each model snapshot has two inset graphs: at the top is a time series plot of F_x in which a marker indicates the time of the snapshot, and along the vertical wall p is plotted at each pressure sensor location (vertical axis is z [m]). Adapted from the work in [3], with permission from the authors, 2020. (**a**) Time instant 1; (**b**) time instant 3; (**c**) time instant 4; (**d**) time instant 5.

In terms of U, generally similar velocity field patterns are found for all three numerical models with differences mostly explained by (small) phase differences of the individual bores interacting on the promenade or limitations in the physics it can represent (in case of SWASH). Considering the velocity profile at the ECM location during a maximum incident U_x event (time instant 2, Figure 15b) followed by a maximum return flow U_x event (time instant 7, Figure 17b), it can be seen that all numerical models underestimate U_x of EXP close to the bottom for the incident bore (time instant 2), while OF predicts it very well, DSPH slightly underestimates it, and SW8L overestimates it during the return flow (time instant 7).

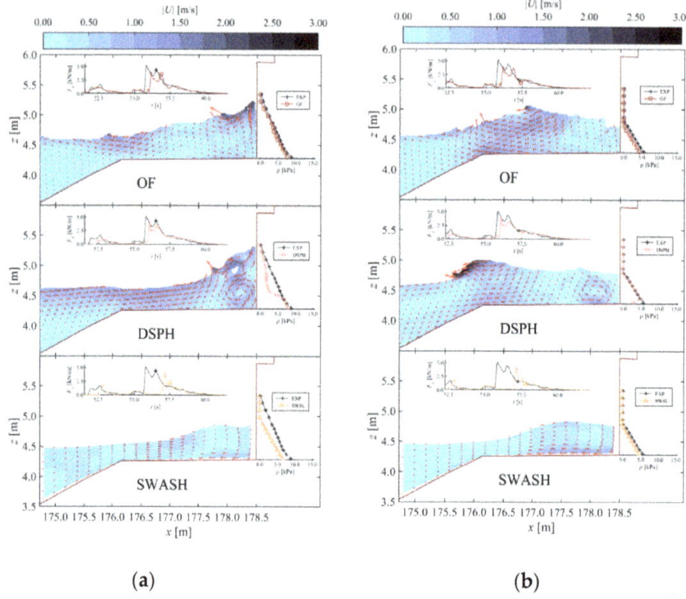

(**a**) (**b**)

Figure 16. Figure 15 continued: (**a**) time instant 6; (**b**) time instant 8.

3.6. Computational Cost

Not only the model performance is of importance in practical applications of numerical models, but also the computational cost that it requires. An overview is provided in Table 6 of the model resolution, total amount of grid cells/particles and corresponding computational cost for the computational hardware applied. The numerical convergence analyses in Appendix A showed that the main characteristics of η at the toe of the dike would not change more than 5% by increasing the grid or initial particle distance resolution beyond the values provided in Table 6. For OF, this was achieved by $\Delta x = \Delta z = H/20 = L_{m,t}/260$ (with H the wave height and $L_{m,t}$ the mean wave length of the SW components at the dike toe), for DSPH by $dp = H/50 = L_{m,t}/585$ and for SWASH by $\Delta x = L_{m,o}/170 = L_{m,t}/60$ (with $L_{m,o}$ the mean wave length of the SW components offshore).

Because of its Lagrangian description of the NS equations, DSPH has the advantage that it can be highly parallelised and is therefore able to make use of the many computational cores available in GPUs. On the other hand, OF and SWASH can be run in a parallelised way as well, but only on CPU cores, which are typically much less numerous. Different hardware and different amounts of cores are applied for each model, so only a qualitative comparison of the computational cost is possible. However, the applied hardware is in each case currently representative of what is typically available at research labs. OF and DSPH were run on multiple cores (CPU and GPU respectively) and SWASH on a single core.

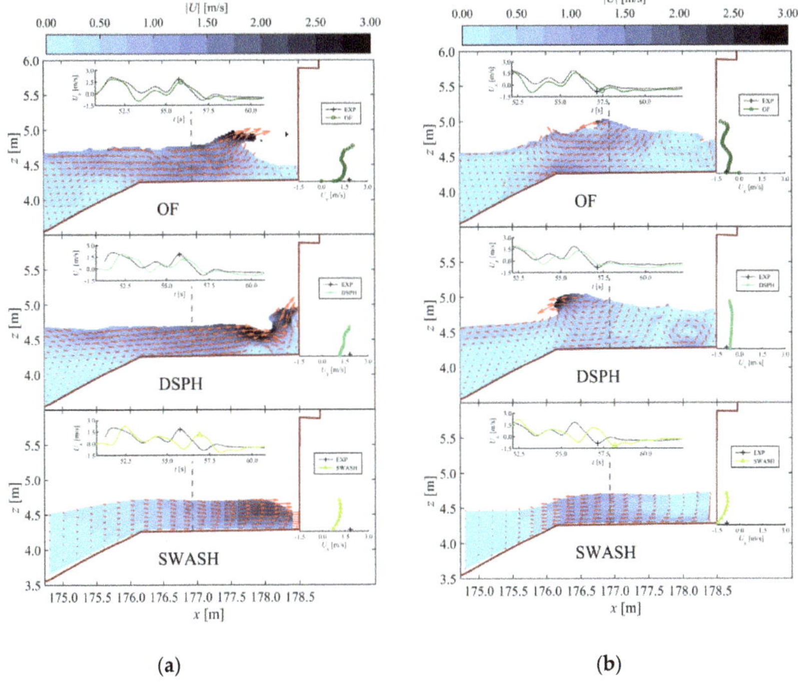

(**a**) (**b**)

Figure 17. Numerical comparative snapshots of the water flow on the dike. Colours are the velocity magnitude $|U|$ according to the colour scale shown at the top of each figure. The red arrows are the velocity vectors, which are scaled for a clear visualisation. Each model snapshot has two inset graphs of U_x at the ECM location on the promenade: at the top is a time series plot of U_x in which a marker indicates the time of the snapshot, and next to the vertical wall the numerical U_x-profile is plotted over the water column at the ECM location (vertical axis is z [m]) together with the single point U_x measurement by the ECM (+marker). Adapted from the work in [3], with permission from the authors, 2020. (**a**) Time instant 2; (**b**) time instant 7.

Table 5. Chronological list (in the EXP time frame) and description of the numerical model snapshots shown in Figures 15–17.

Time Instant	Figure	Description
1	Figure 15a	Dynamic F_x peak of the small wave impact preceding the main bore impact.
2	Figure 17a	Local positive U_x peak at the ECM location preceding the dynamic force peak of the main bore impact.
3	Figure 15b	Local F_x peak preceding the dynamic force peak of the main bore impact.
4	Figure 15c	Dynamic F_x peak of the main bore impact.
5	Figure 15d	Local F_x minimum between the dynamic and quasi-static force peaks of the main bore impact.
6	Figure 16a	Quasi-static F_x peak of the main bore impact.
7	Figure 17b	Local negative U_x peak at the ECM location after the quasi-static force peak of the main bore impact.
8	Figure 16b	Local F_x minimum after the quasi-static force peak of the main bore impact.

Because the flow in the vertical dimension is fully resolved by OF and DSPH in addition to the horizontal dimension and at a much higher resolution, their computational cost is significantly higher than for the depth-averaged SW1L model. The increase of calculation time compared to SW1L is about 5000 times in case of OF and 1000 times for DSPH even though DSPH has more than 4 times the computational points than the OF model grid. Adding eight vertical layers to the SWASH model (SW8L) leads to a factor 7 increase in calculation time (on a single core machine). Although it will not affect this conclusion, it should be noted, however, that the SWASH computational domain is about 25% shorter compared to OF and DSPH, because waves are generated at the WG02 location in the SWASH model (Figure 3).

Table 6. Grid resolution, number of cells and approximate computational times of each applied numerical model.

Parameter	OF	DSPH	SW1L	SW8L
Δx [m]	0.0225–0.18	n/a	0.20	0.20
Δz [m]	0.0225–0.18	n/a	-	-
dp [m]	-	0.02	-	-
K [-]	-	-	1	8
# computational points [-]	318,381	1,309,056	1032	8256
hardware [-]	24 cores (CPU)	2880 cores (GPU)	1 core (CPU)	1 core (CPU)
computational time [DD HH:MM:SS]	03 12:00:00	00 15:30:00	00 00:00:53	00 00:06:31

In the computational cost, the model setup time should also be included. This depends on the experience of the practitioner, so the model setup time is only discussed here in general terms based on experienced practitioners for each respective model. SWASH is the most straightforward in this respect and requires the least "hands-on" time. The model setup of DSPH is also quite straightforward, where SPH particles are initially created on nodes of a regular lattice in a few seconds. The model domain boundaries and the water volume are defined by particles in DSPH, while an intricate mesh is needed for Eulerian models such as OF. Fortunately, the mesh generation for OF has been made much smoother thanks to automatic mesh generation algorithms such as cfMesh as applied here. Still, OF is found to require the most model setup time (i.e., typically more than double). This is certainly the case when a variable grid resolution is used because the refinement zones need to be defined beforehand, and this can only be done accurately by having a reliable estimation of the location of the surface elevation. This is most easily obtained from a fast SWASH model result (at least for the wave propagation until the dike toe), or alternatively by introducing refinement zones iteratively between OF model runs.

4. Discussion

4.1. Inter-Model Comparison of Wave Transformation Processes

4.1.1. Wave Transformation Processes Until the Dike Toe

The main prerequisite for a numerical model to obtain accurate results of the processes on the dike (i.e., wave overtopping [27] and bore interactions [3]) is that the wave transformation and wave setup are predicted well towards and particularly at the dike toe.

Because DSPH was the only model able to reproduce the wave paddle motion exactly as in the experiment (note: OF has the functionality, but numerical instabilities prevented its successful application [3]), it is not surprising that it was the best performing model in terms of η from offshore up to the shoaling zone (Figures 9 and 11, WG02-WG07). In the EXP, H_{rms} decreased slightly (4%) between the offshore region (Figure 9, WG02-WG04) and the shoaling zone (WG07), indicating an energy loss greater than the H_{rms} gain due to shoaling, possibly due to bottom friction and/or model effects (e.g., lateral wall friction, sediment transport on the sandy slope). However, too few EXP measurement locations were available in this area to confirm this. In any case, all three applied numerical models agreed that H_{rms} first decreased towards the foreshore toe, and then increased due to SW shoaling (Figure 9). In case of both OF and SW1L, the H_{rms} decrease was more pronounced than DSPH and could be explained by wave energy losses due to bottom friction to a certain degree (i.e., in OF by the no-slip boundary and in SW1L by parameterised bottom friction). However, these energy losses cannot be explained by the bottom friction alone due to the relatively short propagation distance from the offshore model boundary to the foreshore toe of less than three wave lengths. Viscous and/or diffusive numerical schemes could have contributed as well, especially in the case of OF [45], which mostly used second order discretization schemes but also included a first order scheme (i.e., Euler time discretisation for the volume fraction of the VOF method) for numerical stability reasons [3]. Such numerical diffusion was limited as much as possible in each model by a careful selection of parameters and schemes, which is a balancing exercise between model accuracy and efficiency. DSPH is presently unable to model bottom friction and the numerical diffusion was least noticeable of all three models, causing the H_{rms} overestimation at WG07 by DSPH. SW1L/SW8L tended to overestimate the shoaling (mostly for the LWs) so that an overestimation similar to DSPH at WG07 was obtained. Clearly OF suffered most from (numerical) energy losses, as it started with approximately the same wave energy at the wave paddle as DSPH, but underestimated EXP at WG07. DSPH and OF did agree on the location of the mean breaking point (x_b = ~120 m, Figure 9) and simulated both spilling and plunging breakers, also observed in EXP [3]. Indeed, both models previously have been shown to predict the breaking point (and hydrodynamics) similarly well for both spilling and plunging breakers (i.e., OF [69] and DSPH [12]). SW1L/SW8L, on the other hand, predicted the breaking point to be located more offshore (by about 10 m). Contrary to OF and DSPH, SWASH does not explicitly model the turbulent wave front during the breaking process, but treats it at the sub-grid level instead by assuming similarity to a hydraulic jump [25]. SW1L/SW8L therefore did not reproduce the overturning of the wave front. Moreover, the vertical resolution (or K) for both SW1L and SW8L was too low to be able to model wave breaking without the use of the HFA, which has been shown before to cause the breaking point to be predicted too much offshore in case of plunging breakers (see their Figure 5 in [25]). Based on these observations, the breakpoint was most likely better predicted by OF and DSPH than by SW1L/SW8L. Although, it should be noted that SWASH has the potential to match the breakpoint location in H_{rms} by increasing K to 20 layers [25] (not tested). In the surf zone, DSPH and SW1L/SW8L predicted very similar wave heights until the dike toe and both models ended up slightly overestimating $H_{rms,EXP}$ at the dike toe, SW1L more than DSPH (σ^* in Table 2). The evolution of H_{rms} was also very similar for OF, but the values were lower because H_{rms} at the breakpoint was lower as well, with OF the only model that underestimated $H_{rms,EXP}$ at the dike toe.

Both wave set-down and set-up were overestimated by DSPH compared to EXP. This overestimation in the shoaling and the surf zone was mostly caused by the overestimation of the wave height and therefore radiation stresses in the same areas. The overestimation of the offshore wave set-down (Figure 9, WG02–04) was a result of the overestimation of the wave setup in the surf zone, and because of mass conservation and a finite water mass available in the computational domain. Over about the final 5 m towards the dike toe, the wave setup suddenly increased, causing locally an even higher overestimation of the wave setup and the worst correspondence there with EXP of all three numerical models. This is most likely related to the small water depth this area relative to the particle size dp. Indeed, a similar increase in wave setup was noted by Lowe et al. [12] in the shallowest water depths near the shoreline and exacerbated for increasing dp values (see their Figure S1b of [12] in their supplementary material). The mismatch of the mean water level between OF and EXP was already explained by Gruwez el al. [3] to be a result of the static wave boundary condition treatments and the general underestimation of the wave height. At each measurement location, the wave set-down/up was best predicted by the SW1L model.

The accuracy of the wave phases was best for DSPH and approximately equal for OF and SW1L in the offshore zone (Figure 11, WG02–WG04). However, in the surf zone towards the dike toe, DSPH had the most wave phase errors (Figure 11, WG13–WG14), mostly caused by both the SWs (Figure 4d,e) and LWs (Figure 5c,d) lagging behind those of EXP. In case of SW1L/SW8L, the wave phase accuracy decreased at WG07 (Figure 11), where the SWs were leading those of EXP near its predicted breaking point (Figure 4b). This trend mostly continued in the surf zone until the dike toe (Figure 4d,e), where SW1L/SW8L obtained a slightly better phase accuracy than DSPH (Figure 11 and R in Table 2). The simplified wave breaking modelling with the HFA is thought to be the main cause for the increased phase errors of the η_{sw} result in the surf zone observed for SW1L in Figure 11.

The observed numerical model accuracy for the reproduction of the wave amplitude, wave set-up and wave phase of the EXP at the dike toe, lead to OF having the highest model performance at the dike toe, followed by SW1L and DSPH (d_r in Table 2). Although DSPH achieved the best wave amplitude result, the errors in both wave setup and wave phase caused it to have the lowest model performance at the dike toe location. OF also achieved the best overall model skill in terms of η over the foreshore (and promenade), with the mean skill circle having the smallest radius in the target diagram (Figure 14, left), again, respectively, followed by SW1L and DSPH.

It should be noted, however, that the results of all three models remain very close to each other right up until the dike toe, which is reflected in the model performance rating for η: at the dike toe it was Very Good for both OF and SW1L, while it was Good for DSPH, but only because the d'_r value fell just below the limit for a Very Good rating (i.e., $d'_{r,WG14,DSPH} = 0.79 < 0.80$, Tables 2 and A1). This means that all three numerical models are shown to be able to represent frequency dispersion, and the nonlinear wave transformation processes: SW shoaling (Figure 4b), breaking (Figure 4c,d) and energy transfer to the subharmonic bound LW (Figure 4b–d). This is a confirmation of what has been proven before by Torres-Freyermuth et al. [70,71] for RANS modelling, by Lowe et al. [12] for DualSPHysics and Rijnsdorp et al. [26] for SWASH.

4.1.2. Bore Interactions on the Promenade and Impacts on the Vertical Wall

The numerical model performances obtained for η at the dike toe were maintained along the promenade, except for DSPH which showed an improved model performance to match with SW1L closer to the vertical wall (Figure 12). Because of SW breaking over the foreshore, transfer of energy to the LW, LW shoaling and reflection against the steep slope of the dike, the LW wave height became almost twice the SW wave height at the dike toe location (Figure 9, WG14). Indeed, Van Gent et al. [72] and Hofland et al. [73] have shown that the wave energy is dominated by LWs at the toe of a dike with a very or extremely shallow foreshore. Therefore, the LWs at the dike toe had a dominant effect on the wave overtopping processes on the dike [3], which is a confirmation of the same observation made by Lashley et al. [39]. During each crest of the LWs at the dike toe, the freeboard decreased significantly and the broken or breaking SWs were able to overtop the dike crest much more easily. Consequently, the bore interactions on the promenade depended mostly on the SWs instead. This is shown by the fact that the model performance of DSPH for η_{LW} at the dike toe was lower than SW1L (Figure 10), while it was higher than SW1L for η_{SW} at the same location, and the model performance for the total η increased on the promenade for DSPH relative to SW1L (Figure 12). This was mostly because of a better wave amplitude and phase accuracy by DSPH than SW1L/SW8L on the promenade (Table 3, σ^* and R, respectively). Moreover, in terms of U_x for the bore interactions on the promenade, DSPH had a higher model performance than SW1L and was comparable to OF. In any case, the highest model performance for both η and U_x on the promenade was achieved by OF. This is mostly indicated by a higher phase accuracy than both DSPH and SW1L/SW8L and a higher amplitude accuracy than SW1L/SW8L (Table 3, R and σ^*, respectively). OF obtained the highest phase accuracy on the promenade as a result of the highest wave phase accuracy achieved at the dike toe (Figure 11, WG14), especially because of the SW phase accuracy (i.e., R_{sw}) at the dike toe which was notably higher for OF than both DSPH and SW1L/SW8L. At the dike toe, DSPH had the highest error in $\overline{\eta}$ by overestimating it (Table 2, $B^* = 0.21$) more than OF and SW1L/SW8L. Nevertheless, on the promenade DSPH underestimated $\overline{\eta}$ with a very similar result as OF and SW1L (Table 3, $\overrightarrow{B}^*_{WLDM01-WLDM04} = -0.04$). A possible cause for this change in behaviour is that thin layered flows with a water depth on the promenade were not captured by DSPH because the water depth was smaller than a couple of particles high. The fact that SW1L/SW8L was also able to achieve a Very Good model performance for the overtopped flow layer thickness on the promenade (Table 3) indicates that a very good wave overtopping accuracy has been demonstrated as well, which is a confirmation of previous validation works on wave overtopping over a dike with a shallow foreshore modelled with SWASH [27,28].

Therefore, what seems to matter the most for the accuracy of the processes on the dike is to obtain a good model performance at the toe of the dike. A correct wave phase and wave setup at the dike toe is more important for a correct wave impact simulation than a correct wave amplitude at the dike toe. Indeed, OF had the best overall performance at the dike toe, with the best wave phase and close second best wave setup prediction (Table 2), while underestimating the wave amplitude, and showed the best model performance in terms of η and U_x on the promenade (Table 3), and consequently also for F_x and p at the vertical wall (Table 4, Figures 13 and 14b). Conversely, DSPH had the best wave amplitude, but the worst wave setup and phase prediction at the dike toe and showed a lower performance in terms of F_x at the vertical wall. In addition, the snapshot comparison of the first series of impacts at the vertical wall (Section 3.5) confirmed that the wave phase accuracy was critical to obtain the correct bore interaction pattern on top of the promenade, which was—in its turn—shown to be vital to the accuracy of the impact itself. Likewise, SW1L showed similar SW phase errors at the dike toe (Figure 11, R_{sw}), which consequently caused phase errors in the overtopped bores (R for SW1L/SW8L is mostly lower than OF and DSPH for both η and U_x in Table 3) and their impacts on the vertical wall (R for SW1L/SW8L is mostly lower than OF and DSPH for both p and F_x in Table 4). In Section 3.5, this manifested itself in the SW8L result by a clear phase difference in the U_x time series (Figure 17) and in the F_x time series of the main impact (Figure 15c), which were both a result of a phase error in η for the largest bore of the first wave group arriving at the dike toe (see t ~55 s in Figure 4d).

In the context of the design of vertical walls on top of the dike (e.g., storm walls and buildings), the accurate prediction of the force peak per impact event $F_{x,max}$ and the total horizontal force impulse I is of particular interest. The model performance for $F_{x,max}$ was found to be similar for all three applied numerical models (Table 4, $d_{r,Fx,max}$). However, for I^* important differences were noted, with the best total impulse prediction obtained by OF, followed by DSPH and SW1L/SW8L, respectively. For SWASH, the best result was obtained by the depth-averaged model SW1L. Adding vertical resolution (i.e., SW8L) generally improved the model performance for η along the wave flume (not shown, except at the dike toe in Table 2) and η and U_x along the promenade (Table 3). However, SW8L unexpectedly decreased the model performance for F_x compared to SW1L, with slightly worse performance for $F_{x,max}$ and a higher error in I^* (Table 4). The cause for this is unclear, but this might indicate that SW1L's good estimation of F_x could be—in part—caused by chance due to numerical errors (e.g., overestimation of the wave height at the toe of the dike, Table 2, $\sigma^* > 1.00$). Nevertheless, the SW1L/SW8L model results show that SWASH can provide similar or only slightly worse model performance for F_x compared to OF and DSPH, including the best estimation of $F_{x,max}$ per impact event, albeit with an important underestimation of I.

This comparison shows that SWASH, with a hydrostatic assumption for the calculation of F_x, is able to provide a similarly accurate prediction of F_x (especially $F_{x,max}$) compared to more complex and computationally expensive RANS or SPH models. Xie and Chu [19] already showed that with the hydrostatic pressure assumption, a tsunami bore impact force on a vertical wall can be obtained similar to OF (which includes non-hydrostatic effects as well) and experimental data. This is confirmed and extended here for overtopped bore impacts on dike-mounted vertical walls, determined from the SW1L/SW8L results based on the hydrostatic pressure only. Still, SW1L/SW8L had a lower model performance than OF and DSPH for F_x, most probably due to the hydrostatic pressure assumption. Indeed, while Whittaker et al. [23] found that the perturbed hydrostatic pressure gives an accurate approximation to the pulsating horizontal force on a gently sloped seawall, they expected the hydrodynamic contributions in F_x to increase in importance for steeper slopes (including vertical walls), particularly in the case of breaking waves. In addition, it is important to note that a low-pass filter was applied to both the experimental and numerical p and F_x time series that mostly removed high frequency oscillations in the dynamic impact peak of the double peaked signal caused by a bore impact (Section 2.3). Because these oscillations are stochastic by nature, not even REXP was able to reproduce them exactly and they cannot be reproduced by deterministic numerical models (which is why they were filtered out). However, in both the OF and DSPH results such high-frequency oscillations were present in the unfiltered signals (not shown). Although it is unclear whether they are caused by numerical errors or actual physical processes (or a combination thereof), it still might suggest that they are able to represent this phenomenon in some capacity. Conversely, in the case of SW1L/SW8L, even in the unfiltered F_x signal no such high-frequency oscillations were observed, because of the hydrostatic assumption.

Finally, OF is the only numerical model considered here that simulated both the water and air phases, albeit as incompressible and immiscible fluids. Although OF had the best overall model performance and some influence of the air was noted (especially for plunging wave impacts on the vertical wall), no immediate proof has been found that the air phase contributed positively to the model performance for the parameters considered in this paper. It is estimated that if there would have been a contribution (positive or otherwise), the effect would be small compared to the relatively larger errors in wave setup, height and phase. The fact that OF includes the air phase was actually more cumbersome to the simulation than an advantage. Indeed, spurious velocities in the air phase near the water–air interface caused significantly reduced time steps and consequently a large increase in computational cost [45,46]. Although, as discussed by Gruwez et al. [3], these spurious velocities can be overcome thanks to recent developments in OpenFOAM® [74] (however, presently not available open source).

4.2. Application Feasibility of the Numerical Models for a Design Case

This section provides recommendations on which of the three numerical models to use when, during the design of a structure on a dike with a shallow foreshore, a numerical model is used as an alternative to or complementary to experimental modelling in a wave flume. In such a case, typically a maximum expected wave impact force F_{max} and corresponding impulse I for given design conditions has to be predicted. Generally, two methods can be used to achieve this:

1. Modelling of a short-duration focused wave group. This method has been developed recently [23,75] based on the NewWave approach [76] and aims to reproduce the extreme event that causes F_{max} in an irregular wave train, thereby significantly reducing the test duration.
2. Modelling of an irregular wave train of sufficiently long duration (i.e., 1000 waves or more) to obtain statistically relevant results for F_{max}.

In this paper, bichromatic waves were used for the inter-model comparisons, which included many of the wave transformation processes found in irregular waves and focused wave groups (e.g., frequency dispersion, and transfer of wave energy to sub- and superharmonics), so that this case still provides a good indication of how each numerical model would perform in terms of accuracy in each of these design methods.

Disregarding for a moment the numerical model performances for the estimation of F_{max} (more on that later), the computational time necessary to achieve a result becomes the determinative factor in the choice between the three considered models. In the first method, a single wave group is modelled to obtain a certain focus location relative to the structure. Such a test duration has a similar length to the test considered here. For this design approach a standalone application of all three considered numerical models is therefore possible. However, many combinations of the focus location and phase at focus might be necessary to obtain the "true" F_{max} for a given offshore design wave state [23], which would be more challenging (in terms of runtime) for standalone application of especially OF, but also DSPH.

In the second method, an irregular wave train typically needs to be modelled for a much longer duration (i.e., 1000 waves or about a 55 times longer duration than the test considered here). The high computational time or resources that this currently would require, proves unpractical for both OF and DSPH (i.e., in terms of time: approximately 190 days and 35 days, respectively, based on Table 6), especially in case of coastal structures with a shallow foreshore where the lengthy foreshore needs to be included in the computational domain. Moreover, in case of DSPH an increase in errors in the η results was noted in the surf zone compared to EXP towards the end of the simulation (i.e., for $t > 120$ s in Figures 4c–e and 5c,d). For longer simulations it is possible that this accumulation of errors continues and exacerbates. A smaller particle size might solve this, but would increase the calculation time significantly. Therefore, similar to OF, DSPH would benefit from (adaptive) mesh refinement or variable resolution [77] to reduce the computational cost and increase the accuracy near and on top of the dike. Alternatively, coupling or hybridization with a computationally less demanding but accurate model regarding the wave transformations over the foreshore until the dike toe (e.g., SWASH [27]) is a possibility as well. Such a coupling would dramatically reduce the required computational time necessary to model this type of cross section, making longer simulations of 1000 irregular waves a practical possibility. For DSPH, such a hybridization with SWASH is already available [78]. Work in this direction for OF has been initiated by Vandebeek et al. [79,80]. The authors actively encourage further development and validation of these coupling approaches, especially since the coupling zone in current developments is located in the pre-breaking area, so that still a large area of the foreshore needs to be included in the model domain of DSPH and OF. Challenges remain in moving the coupling zone more towards the dike toe location, which would achieve the highest reduction in computational cost without losing too much accuracy for the processes on the dike. As opposed to both OF and DSPH, standalone application of SWASH for this method is definitely practically feasible for both a single layer and multi-layer approach, at least purely from a computational cost standpoint (i.e., for SW1L and SW8L approximately 1 h and 6 h, respectively, for an irregular wave train of 1000 waves, based on Table 6, and with even faster calculation times for parallel runs).

Another consideration in the choice between the three considered numerical models is of course the accuracy of the numerical model in the estimation of F_x itself. In terms of I, SWASH clearly underperformed compared to both DSPH and OF. Therefore, when I is important, OF is the recommended model to apply (i.e., highest value for I^*, Table 4). However, the inter-model comparison provided in this paper has shown that all three models have a similar model performance in terms of estimation of $F_{x,max}$ for individual impact events (Table 4, $d_{r,Fx,max}$). This means that for the estimation of $F_{x,max}$, SWASH is actually the most recommended model to apply, because of the lowest computational cost. Although, it is important to note that this conclusion is only valid for a relatively straightforward geometry of a dike slope, promenade and vertical wall. For more complex geometries of dikes (e.g., presence of roughness elements, small storm walls with or without parapets, etc.), all numerical models considered in this paper remain untested and especially the simplified model SWASH is expected to be insufficiently accurate, because the importance of vertical flows would increase. However, a meshless approach could potentially be the most capable to capture nonlinearities of fluid–structure interactions derived from extremely complex dike geometries.

Finally, the snapshot comparisons in Section 3.5 have shown that the p distributions modelled by OF resembled most closely to EXP during each stage of a bore impact on the vertical wall and is therefore recommended when pressure distributions are of particular interest. It is estimated that DSPH should be capable of the same when the bore interactions are modelled more correctly by lowering the wave phase error. On the other hand, because the non-hydrostatic effects in p needed to be disregarded due to numerical discrepancies, SW1L/SW8L is limited to hydrostatic p distributions.

5. Conclusions

Three open source CFD models were applied in 2DV to reproduce large-scale wave flume experiments of bichromatic wave transformations over a cross section of a hybrid beach-dike coastal defence system, consisting of a steep-sloped dike with a mildly-sloped and very shallow foreshore, and finally wave impact on a vertical wall: (i) the RANS solver interFoam of OpenFOAM® (OF), (ii) the weakly compressible SPH model DualSPHysics (DSPH) and (iii) the non-hydrostatic NLSW equations model SWASH (depth-averaged (K = 1): SW1L, and multi-layered (K = 8): SW8L). The inter-model comparison of those three numerical models to the experiment (EXP) demonstrated that they are all capable of modelling the dominant wave transformation (i.e., propagation, shoaling, wave breaking, energy transfer from the SW components to the bound LW via nonlinear wave–wave interactions) and the wave–structure interaction (i.e., individual wave overtopping, bore interactions, and reflection processes) processes involved leading up to the impacts on the vertical wall, albeit with a varying degree of accuracy. Based on a time series comparison, all three applied numerical models initially appeared to have a good correspondence of η, U_x, p and F_x to EXP. However, consistent differences between the models were hard to distinguish in this purely qualitative way. The accuracy was subsequently quantified more objectively by employing model performance statistics and the nature of the errors was exposed by pattern statistics. These statistics were plotted over the wave flume to provide spatial insight into the model performance, and the pattern statistics were plotted in a skill diagram, which visualised both the model performance and pattern statistics in a summarised way. In all statistics, the original EXP was used as the comparison reference, so that the repeated experiment (REXP) statistics could be used as reference for an ideal numerical model performance. While none of the numerical models managed to achieve such an ideal model performance, a rating of Good to Very Good was achieved by all three of them for most parameters and measured locations. The best overall model performance was achieved by OF, but required the highest computational cost. Although DPSH managed the best reproduction of the wave height until the dike toe, accumulation of errors in the wave setup and wave phase in the surf zone and near the dike toe caused a lower model performance than OF at the dike toe and for the processes on the dike. From this, it followed that accurate modelling of the wave setup and wave phases at the dike toe seem to be most important for accurate modelling of the bore interactions on the promenade. An analysis and comparison of snapshots of the numerical results on the dike revealed that these bore interactions are determinative for the impacts on the vertical wall. Even though SWASH is a much more simplified model than both OF and DSPH, it is shown to provide very similar results for the wave transformations until the dike toe and even for the processes on the dike and impacts on the vertical wall. When the impulse of the force on the structure is of lesser importance, SWASH is even most recommended for this application, because it is able to predict $F_{x,max}$ relatively accurate for each individual impact, with a significantly reduced computational cost, compared to OF and DSPH. However, SWASH is limited to hydrostatic pressure profiles for the impacts on the vertical wall, which is not always valid during more dynamic impact events. In addition, when the force impulse is of importance and more accurate and detailed wave–dike interactions are needed, OF is most recommended for this application. For future work, it is suggested to investigate whether the same conclusion is valid (particularly regarding applicability of SWASH) in the case of more complex dike geometries.

Author Contributions: Conceptualization, V.G., C.A., T.S., and P.T.; methodology, V.G.; validation, V.G. and M.S.; formal analysis, V.G.; investigation, V.G., C.A., T.S., M.S., and L.C.; resources, P.T.; data curation, V.G., M.S. and L.C.; writing—original draft preparation, V.G.; writing—review and editing, V.G., C.A., T.S., M.S., L.C., A.K., and P.T.; visualization, V.G. and M.S.; supervision, P.T. and A.K.; project administration, P.T.; funding acquisition, P.T. and A.K. All authors have read and agreed to the published version of the manuscript.

Funding: This research was part of the CREST (Climate REsilient CoaST) project (http://www.crestproject.be/en), funded by the Flemish Agency for Innovation by Science and Technology, grant number 150028. The work described in this section was supported by the European Community's Horizon 2020 Research and Innovation Programme through the grant to HYDRALABPLUS, Contract no. 654110. C.A. acknowledges funding from the European Union's Horizon 2020 research and innovation programme under the Marie Sklodowska-Curie, grant number 792370.

Conflicts of Interest: The authors declare no conflict of interest. The funders had no role in the design of the study; in the collection, analyses, or interpretation of data; in the writing of the manuscript; or in the decision to publish the results.

Appendix A. Numerical Convergence Analyses

The numerical model convergence analyses of the DSPH and SW1L/SW8L models was done following the same methodology as described by Gruwez et al. [3] for the OF model (see their Appendix A).

Appendix A.1. Model Convergence Statistics

For the convergence analysis, the four statistical error indicators as determined by [3] are considered:

- Freeboard normalised bias (NB):

$$NB = \frac{B}{R_c} \tag{A1}$$

in which R_c is the freeboard, and B is the bias between the considered and reference time series.

- Residual error of the normalised standard deviation ($RNSD$):

$$RNSD = 1 - \sigma^* \tag{A2}$$

in which σ^* is given by (A9) where the observed time series is the reference time series and the predicted time series is the considered time series.

- Residual error of the correlation coefficient (RCC):

$$RCC = 1 - R \tag{A3}$$

in which R is the correlation coefficient, given by (A11), between the reference time series and time series of interest.

- Normalised mean absolute error ($NMAE$) given by

$$NMAE = \frac{MAE}{O_{max} - O_{min}} \times 100\% \tag{A4}$$

in which MAE is the mean absolute error, given by (A6), and O_{max} and O_{min} are the maximum and minimum value of the reference time series.

For their interpretation, reference is made to the work in [3].

Appendix A.2. Convergence Analyses

Appendix A.2.1. DSPH

Rota Roselli et al. [59] identified the most important parameters that influence the model accuracy in terms of nonlinear wave propagation, among them the smoothing length h_{SPH}, the artificial viscosity parameter α_{av} and the initial particle distance dp. However, the current case also includes wave transformations over a beach with decreasing water depth towards the dike, for which dp is found to be the most important parameter. The convergence analysis for DSPH is therefore focused on dp (Figure A1).

Three initial particle distances (i.e., dp = 0.04 m, 0.03 m, 0.024 m) are compared to the finest resolution (i.e., dp = 0.02 m). In Figure A1 it is shown that for dp = 0.024 m, most statistical errors stay close to 5% at the toe of the dike, but approaches 8% for the NB, which is too high to be able to assume a converged state. This indicates that the resolution might still be too low to be able to sufficiently resolve the wave setup at the toe of the dike (~0.05 m based on the experimental result). To check convergence of NB an even higher resolution of dp = 0.01 m would be needed. However, this resolution was not practically feasible due to the huge amounts of data and computational costs involved. Therefore, the initial inter-particle distance dp of 0.02 m is assumed to be a sufficiently converged resolution and is used for the analysis in the paper.

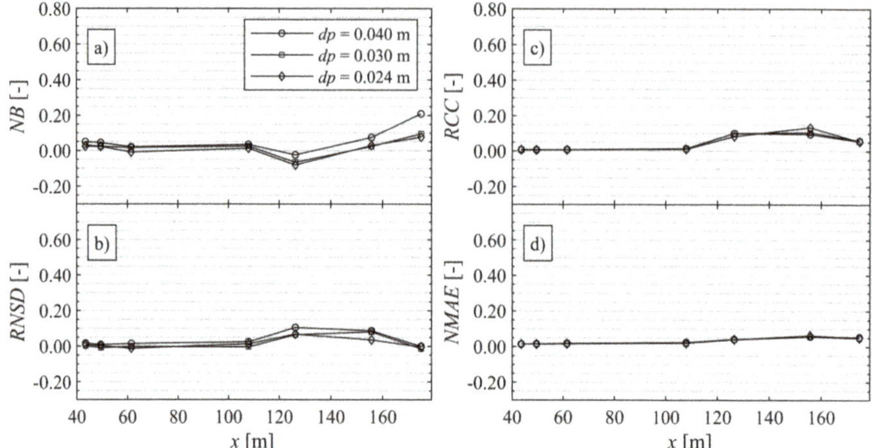

Figure A1. DSPH model inter-particle distance dp convergence analysis of the η time series at the WG locations along the flume up to the dike toe (WG14), based on (**a**) the normalised bias, (**b**) the residual normalised standard deviation, (**c**) the residual correlation coefficient, and (**d**) the normalised mean-absolute-error. The reference is the highest resolution simulation with $dp = 0.02$ m.

Appendix A.2.2. SWASH

In case of the SWASH model, the convergence analysis is focused on the grid resolution. Grid cell sizes $\Delta x = 0.2$, 0.4 and 0.8 m are considered, and the convergence statistics are applied with reference to the finest resolution ($\Delta x = 0.1$ m). The result in Figure A2 shows that the convergence errors of $\Delta x = 0.2$ m stay within +/−5% of the finest resolution. A resolution of $\Delta x = 0.2$ m is therefore used for the analysis in the paper. The same conclusion is valid for both SW1L and SW8L (not shown).

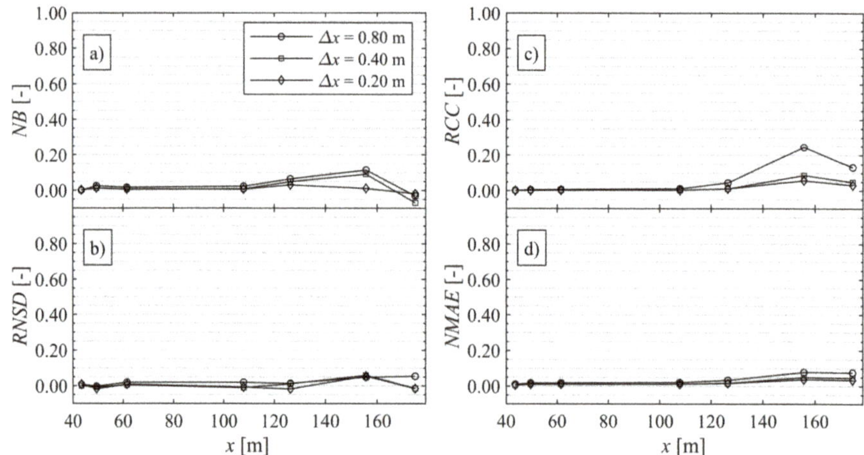

Figure A2. Same as Figure A1 but for the SW1L model grid resolution convergence analysis. See caption of Figure A1 for the description of (**a**–**d**). The reference is the finest grid model result with a horizontal grid cell resolution of $\Delta x = 0.10$ m.

Appendix B. Numerical Model Performance and Pattern Statistics

In this appendix, the equations are provided of the numerical model performance and pattern statistics used in this paper. For more descriptions and interpretations, reference is made to the work in [3]. The general numerical model performance is evaluated by applying Willmott's refined index of agreement d_r [81]:

$$d_r = \begin{cases} 1 - \frac{MAE}{cMAD}, & MAE \leq cMAD \\ \frac{cMAD}{MAE} - 1, & MAE > cMAD \end{cases} \tag{A5}$$

where d_r is bounded by [−1.0, 1.0], c is a scaling factor and is taken equal to 2, MAE is the mean absolute error defined by

$$MAE = \frac{1}{N}\sum_{i=1}^{N}|P_i - O_i| \tag{A6}$$

with N the number of samples in the time series, and P the predicted time series together with the pair-wise-matched observed time series O (for $i = 1, 2, \ldots, n$), and MAD is the mean absolute deviation:

$$MAD = \frac{1}{N}\sum_{i=1}^{N}\left|O_i - \overline{O}\right| \tag{A7}$$

where the overbar represents the mean of the time series.

In addition, a relative refined index of agreement d'_r was defined by [3] which provides the performance of the numerical model relative to the experimental model uncertainty (in case a repetition of the experiment is available):

$$d'_r = \begin{cases} 1 - \frac{MAE_{num} - MAE_{rexp}}{cMAD} = 1 - \left(d_{r,num} - d_{r,rexp}\right), & MAE_{num} - MAE_{rexp} \leq cMAD \\ \frac{cMAD}{MAE_{num} - MAE_{rexp}} - 1 = \left(d_{r,num} - d_{r,rexp}\right) - 1, & MAE_{num} - MAE_{rexp} > cMAD \end{cases} \tag{A8}$$

where the subscripts *num* and *rexp* indicate that the statistic is evaluated respectively for the numerical and repeated experimental data, and c is again taken equal to 2. When the numerator $MAE_{num} - MAE_{rexp}$ is negative (i.e., <0), the numerical error compared to the experiment is smaller than the experimental uncertainty, which means that the numerical model performance cannot be improved. In that case $MAE_{num} - MAE_{rexp} = 0$ is forced, so that $d'_r = 1$. A classification of d'_r and corresponding rating terminology as proposed by the authors of [3] is provided in Table A1.

Table A1. Proposed classification of the relative refined index of agreement d'_r and corresponding rating. Reproduced from [3], with permission from the authors, 2020.

d'_r Classification [-]	Rating
0.90–1.00	Excellent
0.80–0.90	Very Good
0.70–0.80	Good
0.50–0.70	Reasonable/Fair
0.30–0.50	Poor
(−1.00)–0.30	Bad

The pattern statistics are provided by:

- The normalised standard deviation σ^*:

$$\sigma^* = \frac{\sigma_p}{\sigma_o} \tag{A9}$$

where σ_p and σ_o are the standard deviations of the predicted and observed time series, respectively.

- The normalised bias B^*:

$$B^* = \frac{B}{\sigma_O} \tag{A10}$$

where B is the bias between the predicted and observed time series.

- The correlation coefficient R:

$$R = \frac{\frac{1}{N}\sum_{i=1}^{N}\left(P_i - \overline{P}\right)\left(O_i - \overline{O}\right)}{\sigma_P \sigma_O} \tag{A11}$$

References

1. IPCC. The Ocean and Cryosphere in a Changing Climate. In *IPCC Special Report on the Ocean and Cryosphere in a Changing Climate*; IPCC: Geneva, Switzerland, 2019.
2. Gruwez, V.; Vandebeek, I.; Kisacik, D.; Streicher, M.; Altomare, C.; Suzuki, T.; Verwaest, T.; Kortenhaus, A.; Troch, P. 2D overtopping and impact experiments in shallow foreshore conditions. In Proceedings of the 36th Conference on Coastal Engineering, Baltimore, MD, USA, 30 July–3 August 2018; pp. 1–13.
3. Gruwez, V.; Altomare, C.; Suzuki, T.; Streicher, M.; Cappietti, L.; Kortenhaus, A.; Troch, P. Validation of RANS modelling for wave interactions with sea dikes on shallow foreshores using a large-scale experimental dataset. *J. Mar. Sci. Eng.* **2020**, *8*, 650. [CrossRef]
4. Streicher, M.; Kortenhaus, A.; Altomare, C.; Gruwez, V.; Hofland, B.; Chen, X.; Marinov, K.; Scheres, B.; Schüttrumpf, H.; Hirt, M.; et al. *WALOWA (WAve LOads on WAlls)-Large-Scale Experiments in the Delta Flume*; SCACR: Santander, Spain, 2017.
5. Sibilla, S. Fluid mechanics and the SPH method: Theory and applications. *J. Hydraul. Res.* **2013**, *51*, 339–340. [CrossRef]
6. Oñate, E.; Celigueta, M.A.; Idelsohn, S.R.; Salazar, F.; Suárez, B. Possibilities of the particle finite element method for fluid-soil-structure interaction problems. *Comput. Mech.* **2011**, *48*, 307–318. [CrossRef]
7. Gotoh, H.; Khayyer, A. On the state-of-the-art of particle methods for coastal and ocean engineering. *Coast. Eng. J.* **2018**, *60*, 79–103. [CrossRef]
8. Colagrossi, A.; Landrini, M. Numerical simulation of interfacial flows by smoothed particle hydrodynamics. *J. Comput. Phys.* **2003**, *191*, 448–475. [CrossRef]
9. Mokos, A.; Rogers, B.D.; Stansby, P.K. A multi-phase particle shifting algorithm for SPH simulations of violent hydrodynamics with a large number of particles. *J. Hydraul. Res.* **2017**, *55*, 143–162. [CrossRef]
10. Monaghan, J.J.; Kos, A. Solitary waves on a Cretan beach. *J. Waterw. Port. Coastal, Ocean. Eng.* **1999**, *125*, 145–155. [CrossRef]
11. Domínguez, J.M.; Altomare, C.; Gonzalez-Cao, J.; Lomonaco, P. Towards a more complete tool for coastal engineering: Solitary wave generation, propagation and breaking in an SPH-based model. *Coast. Eng. J.* **2019**, *61*, 15–40. [CrossRef]
12. Lowe, R.J.; Buckley, M.; Altomare, C.; Rijnsdorp, D.; Yao, Y.; Suzuki, T.; Bricker, J. Numerical simulations of surf zone wave dynamics using Smoothed Particle Hydrodynamics. *Ocean. Model.* **2019**, *144*, 101481. [CrossRef]
13. Subramaniam, S.P.; Scheres, B.; Schilling, M.; Liebisch, S.; Kerpen, N.B.; Schlurmann, T.; Altomare, C.; Schüttrumpf, H. Influence of convex and concave curvatures in a coastal dike line on wave run-up. *Water* **2019**, *11*, 1333. [CrossRef]
14. St-Germain, P.; Nistor, I.; Townsend, R.; Shibayama, T. Smoothed-particle hydrodynamics numerical modeling of structures impacted by tsunami bores. *J. Waterw. Port. Coastal, Ocean. Eng.* **2014**, *140*, 66–81. [CrossRef]
15. Didier, E.; Neves, D.R.C.V.; Martins, R.; Neves, M.G. Wave interaction with a vertical wall: SPH numerical and experimental modeling. *Ocean. Eng.* **2014**, *88*, 330–341. [CrossRef]
16. Altomare, C.; Crespo, A.; Domínguez, J.M.; Gomez-Gesteira, M.; Suzuki, T.; Verwaest, T. Applicability of Smoothed Particle Hydrodynamics for estimation of sea wave impact on coastal structures. *Coast. Eng.* **2015**, *96*, 1–12. [CrossRef]

17. Hérault, A.; Bilotta, G.; Dalrymple, R.A. SPH on GPU with CUDA. *J. Hydraul. Res.* **2010**, *48*, 74–79. [CrossRef]
18. Crespo, A.; Dominguez, J.; Rogers, B.; Gómez-Gesteira, M.; Longshaw, S.; Canelas, R.; Vacondio, R.; Barreiro, A.; García-Feal, O. DualSPHysics: Open-source parallel CFD solver based on Smoothed Particle Hydrodynamics (SPH). *Comput. Phys. Commun.* **2015**, *187*, 204–216. [CrossRef]
19. Xie, P.; Chu, V.H. The forces of tsunami waves on a vertical wall and on a structure of finite width. *Coast. Eng.* **2019**, *149*, 65–80. [CrossRef]
20. Hu, K.; Mingham, C.; Causon, D. Numerical simulation of wave overtopping of coastal structures using the non-linear shallow water equations. *Coast. Eng.* **2000**, *41*, 433–465. [CrossRef]
21. Shiach, J.B.; Mingham, C.G.; Ingram, D.M.; Bruce, T. The applicability of the shallow water equations for modelling violent wave overtopping. *Coast. Eng.* **2004**, *51*, 1–15. [CrossRef]
22. Orszaghova, J.; Borthwick, A.G.; Taylor, P.H. From the paddle to the beach-A Boussinesq shallow water numerical wave tank based on Madsen and Sørensen's equations. *J. Comput. Phys.* **2012**, *231*, 328–344. [CrossRef]
23. Whittaker, C.; Fitzgerald, C.; Raby, A.; Taylor, P.; Borthwick, A.G.L. Extreme coastal responses using focused wave groups: Overtopping and horizontal forces exerted on an inclined seawall. *Coast. Eng.* **2018**, *140*, 292–305. [CrossRef]
24. Zijlema, M.; Stelling, G.; Smit, P. SWASH: An operational public domain code for simulating wave fields and rapidly varied flows in coastal waters. *Coast. Eng.* **2011**, *58*, 992–1012. [CrossRef]
25. Smit, P.; Zijlema, M.; Stelling, G. Depth-induced wave breaking in a non-hydrostatic, near-shore wave model. *Coast. Eng.* **2013**, *76*, 1–16. [CrossRef]
26. Rijnsdorp, D.P.; Smit, P.B.; Zijlema, M. Non-hydrostatic modelling of infragravity waves under laboratory conditions. *Coast. Eng.* **2014**, *85*, 30–42. [CrossRef]
27. Suzuki, T.; Altomare, C.; Veale, W.; Verwaest, T.; Trouw, K.; Troch, P.; Zijlema, M. Efficient and robust wave overtopping estimation for impermeable coastal structures in shallow foreshores using SWASH. *Coast. Eng.* **2017**, *122*, 108–123. [CrossRef]
28. Suzuki, T.; Altomare, C.; Yasuda, T.; Verwaest, T. Characterization of overtopping waves on sea dikes with gentle and shallow foreshores. *J. Mar. Sci. Eng.* **2020**, *8*, 752. [CrossRef]
29. Vanneste, D.F.; Altomare, C.; Suzuki, T.; Troch, P.; Verwaest, T. Comparison of numerical models for wave overtopping and impact on a sea wall. *Coast. Eng. Proc.* **2014**, *1*, 5. [CrossRef]
30. Buckley, M.; Lowe, R.; Hansen, J. Evaluation of nearshore wave models in steep reef environments. *Ocean. Dyn.* **2014**, *64*, 847–862. [CrossRef]
31. Booij, N.; Ris, R.C.; Holthuijsen, L. A third-generation wave model for coastal regions: 1. Model description and validation. *J. Geophys. Res. Space Phys.* **1999**, *104*, 7649–7666. [CrossRef]
32. Roelvink, D.; Reniers, A.; Van Dongeren, A.; Vries, J.V.T.D.; McCall, R.; Lescinski, J. Modelling storm impacts on beaches, dunes and barrier islands. *Coast. Eng.* **2009**, *56*, 1133–1152. [CrossRef]
33. St-Germain, P.; Nistor, I.; Readshaw, J.; Lamont, G. Numerical modeling of coastal dike overtopping using SPH and non-hydrostatic NLSW equations. In Proceedings of the 34th Conference on Coastal Engineering, Seoul, Korea, 15–20 June 2014.
34. Park, H.; Do, T.; Tomiczek, T.; Cox, D.T.; Van De Lindt, J.W. Numerical modeling of non-breaking, impulsive breaking, and broken wave interaction with elevated coastal structures: Laboratory validation and inter-model comparisons. *Ocean. Eng.* **2018**, *158*, 78–98. [CrossRef]
35. Higuera, P.; Lara, J.L.; Losada, I.J. Three-dimensional interaction of waves and porous coastal structures using OpenFOAM®. Part I: Formulation and validation. *Coast. Eng.* **2014**, *83*, 243–258. [CrossRef]
36. Higuera, P.; Lara, J.L.; Losada, I.J. Three-dimensional interaction of waves and porous coastal structures using OpenFOAM®. Part II: Application. *Coast. Eng.* **2014**, *83*, 259–270. [CrossRef]
37. IHFOAM Team. IHCantabria. Available online: https://ihfoam.ihcantabria.com/ (accessed on 5 November 2020).
38. González-Cao, J.; Altomare, C.; Crespo, A.; Domínguez, J.; Gómez-Gesteira, M.; Kisacik, D. On the accuracy of DualSPHysics to assess violent collisions with coastal structures. *Comput. Fluids* **2019**, *179*, 604–612. [CrossRef]
39. Lashley, C.H.; Zanuttigh, B.; Bricker, J.D.; Van Der Meer, J.; Altomare, C.; Suzuki, T.; Roeber, V.; Oosterlo, P. Benchmarking of numerical models for wave overtopping at dikes with shallow mildly sloping foreshores: Accuracy versus speed. *Environ. Model. Softw.* **2020**, *130*, 104740. [CrossRef]

40. Cappietti, L.; Simonetti, I.; Esposito, A.; Streicher, M.; Kortenhaus, A.; Scheres, B.; Schuettrumpf, H.; Hirt, M.; Hofland, B.; Chen, X. Large-scale experiments of wave-overtopping loads on walls: Layer thicknesses and velocities. *Ocean. Eng.* **2018**, *7*, 28. [CrossRef]
41. Kortenhaus, A.; Streicher, M.; Gruwez, V.; Altomare, C.; Hofland, B.; Chen, X.; Marinov, K.; Vanneste, D.; Willems, M.; Suzuki, T.; et al. WALOWA (WAve LOads on WAlls)-Large-scale Experiments in the Delta Flume on Overtopping Wave Loads on Vertical Walls. *Zenodo* **2019**. [CrossRef]
42. OpenFOAM Foundation. Available online: https://openfoam.org/ (accessed on 6 August 2019).
43. Berberović, E.; Van Hinsberg, N.P.; Jakirlić, S.; Roisman, I.V.; Tropea, C. Drop impact onto a liquid layer of finite thickness: Dynamics of the cavity evolution. *Phys. Rev. E* **2009**, *79*, 036306. [CrossRef]
44. Deshpande, S.S.; Anumolu, L.; Trujillo, M.F. Evaluating the performance of the two-phase flow solver interFoam. *Comput. Sci. Discov.* **2012**, *5*, 014016. [CrossRef]
45. Larsen, B.E.; Fuhrman, D.R.; Roenby, J. Performance of interFoam on the simulation of progressive waves. *Coast. Eng. J.* **2019**, *61*, 380–400. [CrossRef]
46. Roenby, J.; Larsen, B.E.; Bredmose, H.; Jasak, H. A new volume-of-fluid method in OpenFOAM. In Proceedings of the VII International Conference on Computational Methods in Marine Engineering, Gothenburg, Sweden, 13–15 May 2017.
47. Higuera, P. *phicau/olaFlow: CFD for Waves*, Zenodo: Geneva, Switzerland, 2018. [CrossRef]
48. Larsen, B.E.; Fuhrman, D.R. On the over-production of turbulence beneath surface waves in Reynolds-averaged Navier-Stokes models. *J. Fluid Mech.* **2018**, *853*, 419–460. [CrossRef]
49. Larsen, B.E. stabRAS_OF50. 2018. Available online: https://github.com/BjarkeEltardLarsen/StabRAS_OF50 (accessed on 18 September 2018).
50. DualSPHysics Team. DualSPHysics: GPU and OpenMP based Smoothed Particle Hydrodynamics. Available online: https://dual.sphysics.org/ (accessed on 14 August 2019).
51. Wendland, H. Piecewise polynomial, positive definite and compactly supported radial functions of minimal degree. *Adv. Comput. Math.* **1995**, *4*, 389–396. [CrossRef]
52. Monaghan, J.J. Smoothed particle hydrodynamics. *Annu. Rev. Astron. Astrophys.* **1992**, *30*, 543–574. [CrossRef]
53. Dalrymple, R.A.; Rogers, B. Numerical modeling of water waves with the SPH method. *Coast. Eng.* **2006**, *53*, 141–147. [CrossRef]
54. Molteni, D.; Colagrossi, A. A simple procedure to improve the pressure evaluation in hydrodynamic context using the SPH. *Comput. Phys. Commun.* **2009**, *180*, 861–872. [CrossRef]
55. Fourtakas, G.; Dominguez, J.M.; Vacondio, R.; Rogers, B. Local uniform stencil (LUST) boundary condition for arbitrary 3-D boundaries in parallel smoothed particle hydrodynamics (SPH) models. *Comput. Fluids* **2019**, *190*, 346–361. [CrossRef]
56. English, A.; Domínguez, J.M.; Vacondio, R.; Crespo, A.J.C.; Stansby, P.K.; Lind, S.J.; Gómez-Gesteira, M. Correction for dynamic boundary conditions. In Proceedings of the 14th International SPHERIC Workshop, Exeter, UK, 25–27 June 2019.
57. Altomare, C.; Domínguez, J.; Crespo, A.; González-Cao, J.; Suzuki, T.; Gómez-Gesteira, M.; Troch, P. Long-crested wave generation and absorption for SPH-based DualSPHysics model. *Coast. Eng.* **2017**, *127*, 37–54. [CrossRef]
58. Vacondio, R.; Altomare, C.; De Leffe, M.; Hu, X.Y.; Le Touzé, D.; Lind, S.; Marongiu, J.-C.; Marrone, S.; Rogers, B.; Souto-Iglesias, A. Grand challenges for Smoothed Particle Hydrodynamics numerical schemes. *Comput. Part. Mech.* **2020**, 1–14. [CrossRef]
59. Roselli, R.A.R.; Vernengo, G.; Altomare, C.; Brizzolara, S.; Bonfiglio, L.; Guercio, R. Ensuring numerical stability of wave propagation by tuning model parameters using genetic algorithms and response surface methods. *Environ. Model. Softw.* **2018**, *103*, 62–73. [CrossRef]
60. The SWASH team. *SWASH-User Manual (v4.01A)*; Delft University of Technology, Faculty of Civil Engineering and Geosciences, Environmental Fluid Mechanics Section: Delft, The Netherlands, 2017.
61. The SWASH team. *SWASH-User Manual (v5.01)*; Delft University of Technology, Faculty of Civil Engineering and Geosciences, Environmental Fluid Mechanics Section: Delft, The Netherlands, 2018.
62. Mansard, E.P.; Funke, E.R. The measurement of incident and reflected spectra using a least squares method. *Coast. Eng. Proc.* **1980**, *1*. [CrossRef]
63. Lykke Andersen, T. WaveLab. Available online: https://www.hydrosoft.civil.aau.dk/wavelab/ (accessed on 31 August 2019).

64. Jacobsen, N.G.; Van Gent, M.R.A.; Capel, A.; Borsboom, M. Numerical prediction of integrated wave loads on crest walls on top of rubble mound structures. *Coast. Eng.* **2018**, *142*, 110–124. [CrossRef]
65. Taylor, K.E. Summarizing multiple aspects of model performance in a single diagram. *J. Geophys. Res. Space Phys.* **2001**, *106*, 7183–7192. [CrossRef]
66. Willmott, C.J.; Matsuura, K.; Robeson, S.M. Ambiguities inherent in sums-of-squares-based error statistics. *Atmospheric Environ.* **2009**, *43*, 749–752. [CrossRef]
67. Jolliff, J.K.; Kindle, J.C.; Shulman, I.; Penta, B.; Friedrichs, M.A.; Helber, R.; Arnone, R.A. Summary diagrams for coupled hydrodynamic-ecosystem model skill assessment. *J. Mar. Syst.* **2009**, *76*, 64–82. [CrossRef]
68. Bullock, G.; Obhrai, C.; Peregrine, D.; Bredmose, H. Violent breaking wave impacts. Part 1: Results from large-scale regular wave tests on vertical and sloping walls. *Coast. Eng.* **2007**, *54*, 602–617. [CrossRef]
69. Brown, S.; Greaves, D.; Magar, V.; Conley, D. Evaluation of turbulence closure models under spilling and plunging breakers in the surf zone. *Coast. Eng.* **2016**, *114*, 177–193. [CrossRef]
70. Torres-Freyermuth, A.; Losada, I.J.; Lara, J.L. Modeling of surf zone processes on a natural beach using Reynolds-Averaged Navier-Stokes equations. *J. Geophys. Res. Space Phys.* **2007**, *112*. [CrossRef]
71. Torres-Freyermuth, A.; Lara, J.L.; Losada, I.J. Numerical modelling of short- and long-wave transformation on a barred beach. *Coast. Eng.* **2010**, *57*, 317–330. [CrossRef]
72. Van Gent, M.R.A. *Physical Model Investigations on Coastal Structures with Shallow Foreshores: 2D Model Tests on the Petten Sea-Defence*; Deltares: Delft, The Netherlands, 1999.
73. Hofland, B.; Chen, X.; Altomare, C.; Oosterlo, P. Prediction formula for the spectral wave period T m-1,0 on mildly sloping shallow foreshores. *Coast. Eng.* **2017**, *123*, 21–28. [CrossRef]
74. Vukčević, V.; Jasak, H.; Gatin, I. Implementation of the ghost fluid method for free surface flows in polyhedral finite volume framework. *Comput. Fluids* **2017**, *153*, 1–19. [CrossRef]
75. Whittaker, C.; Fitzgerald, C.; Raby, A.; Taylor, P.; Orszaghova, J.; Borthwick, A. Optimisation of focused wave group runup on a plane beach. *Coast. Eng.* **2017**, *121*, 44–55. [CrossRef]
76. Tromans, P.S. A new model for the kinematics of large ocean waves-Application as a design wave. In Proceedings of the 1st International Offshore and Polar Engineering Conference, Edinburgh, UK, 11–16 August 1991; Shell Research BV: Rijswijk, The Netherlands; Volume III, pp. 64–69.
77. Vacondio, R.; Rogers, B.; Stansby, P.; Mignosa, P. Variable resolution for SPH in three dimensions: Towards optimal splitting and coalescing for dynamic adaptivity. *Comput. Methods Appl. Mech. Eng.* **2016**, *300*, 442–460. [CrossRef]
78. Altomare, C.; Domínguez, J.M.; Crespo, A.J.C.; Suzuki, T.; Caceres, I.; Gómez-Gesteira, M. Hybridization of the wave propagation model SWASH and the meshfree particle method SPH for real coastal applications. *Coast. Eng. J.* **2015**, *57*, 1550024-1. [CrossRef]
79. Vandebeek, I.; Gruwez, V.; Altomare, C.; Suzuki, T. *Towards an Efficient and Highly Accurate Coupled Numerical Modelling Approach for Wave Interactions with a Dike on a Very Shallow Foreshore*; Coastlab: Santander, Spain, 2018; p. 10.
80. Vandebeek, I.; Toorman, E.; Troch, P. Numerical Simulation of Wave Propagation over a Sloping Beach Using a Coupled Rans-NLSWE Model. In Proceedings of the OpenFOAM Workshop, Shangai, China, 24–29 June 2018; pp. 1–4.
81. Willmott, C.J.; Robeson, S.M.; Matsuura, K. A refined index of model performance. *Int. J. Clim.* **2012**, *32*, 2088–2094. [CrossRef]

Publisher's Note: MDPI stays neutral with regard to jurisdictional claims in published maps and institutional affiliations.

Journal of
Marine Science and Engineering

Article

Coupled SPH–FEM Modeling of Tsunami-Borne Large Debris Flow and Impact on Coastal Structures

Anis Hasanpour, Denis Istrati * and Ian Buckle

Department of Civil and Environmental Engineering, University of Nevada, Reno, NV 89557, USA; hasanpour@nevada.unr.edu (A.H.); igbuckle@unr.edu (I.B.)
* Correspondence: distrati@unr.edu

Abstract: Field surveys in recent tsunami events document the catastrophic effects of large water-borne debris on coastal infrastructure. Despite the availability of experimental studies, numerical studies investigating these effects are very limited due to the need to simulate different domains (fluid, solid), complex turbulent flows and multi-physics interactions. This study presents a coupled SPH–FEM modeling approach that simulates the fluid with particles, and the flume, the debris and the structure with mesh-based finite elements. The interaction between the fluid and solid bodies is captured via node-to-solid contacts, while the interaction of the debris with the flume and the structure is defined via a two-way segment-based contact. The modeling approach is validated using available large-scale experiments in the literature, in which a restrained shipping container is transported by a tsunami bore inland until it impacts a vertical column. Comparison of the experimental data with the two-dimensional numerical simulations reveals that the SPH–FEM models can predict (i) the non-linear transformation of the tsunami wave as it propagates towards the coast, (ii) the debris–fluid interaction and (iii) the impact on a coastal structure, with reasonable accuracy. Following the validation of the models, a limited investigation was conducted, which demonstrated the generation of significant debris pitching that led to a non-normal impact on the column with a reduced contact area and impact force. While the exact level of debris pitching is highly dependent on the tsunami characteristics and the initial water depth, it could potentially result in a non-linear force–velocity trend that has not been considered to date, highlighting the need for further investigation preferably with three-dimensional models.

Keywords: tsunami; wave; bore; flooding; debris; SPH; numerical modeling; SPH–FEM coupling; fluid–structure interaction; coastal structures

Citation: Hasanpour, A.; Istrati, D.; Buckle, I. Coupled SPH–FEM Modeling of Tsunami-Borne Large Debris Flow and Impact on Coastal Structures. *J. Mar. Sci. Eng.* **2021**, *9*, 1068. https://doi.org/10.3390/jmse9101068

Academic Editors: Tomohiro Suzuki and Corrado Altomare

Received: 9 August 2021
Accepted: 21 September 2021
Published: 29 September 2021

1. Introduction

After the Indian Ocean Tsunami in 2004 and Great East Japan Tsunami in 2011, which resulted in significant destruction of coastline infrastructure and enormous financial losses [1], tsunamis have gained a great deal of attention. Due to the enormous amount of energy, tsunami waves can travel many kilometers and pose significant threats to coastal communities, taking away lives, and causing serious damage to coastal structures. In spite of the low frequency of the occurrence of such events, due to population growth, the urbanization trend and sea level rise, the exposure of coastal environments to extreme water hazards is increasing. Past tsunamis revealed that much of the nearshore infrastructure located in tsunami-prone areas is highly vulnerable to hydrodynamic loads. Extensive damage to coastal buildings, bridges, seawalls, and breakwaters was reported after the 2004 Indian tsunami and the 2011 Tohoku tsunami [2–5]. FEMA P-646 [3] reported that the Indian tsunami generated by a Mw9.1 subduction earthquake [3] caused extensive damage to critical infrastructure and total loss of life over 310,000 [6]. Flooding caused by the 2011 Tohoku tsunami affected areas that were 7 miles away from the shore and damaged over 162,000 buildings and more than 300 bridges [5–9].

Several numerical simulations and experimental studies have been carried out to date to advance the understanding of tsunami forces on coastal structures. Palermo et al. [10] described the tsunami-induced force components on nearshore structures and reported that the hydrodynamic and surge components were a function of the instantaneous velocity and water depth. Other studies focused on the experimental testing of tsunami bore loads on rectangular buildings [11], piloti-type buildings [12], walls with overhangs [13] and coastal bridges [14–17]. Some studies evaluated the effects of different wave forms on coastal decks, including unbroken, breaking and post-breaking waves and bores [14,18], and revealed fundamental differences in the effects caused by the two wave types. In fact, Istrati et al. [18] demonstrated that turbulent bores applied horizontal forces that were larger than the uplift forces by up to a factor of 2.2, while the unbroken waves exhibited an opposite trend with a ratio of horizontal/uplift force decreasing down to 0.54. This highlighted the need to be able to identify a priori the tsunami wave type that is expected to impact a specific coastal area in order to design properly a new structure or strengthen an existing one at that location. Fortunately, several simplified predictive load equations have been developed in the literature for a range of different types of structures, for both bores [19,20] and unbroken solitary waves [21,22].

Other critical aspects related to the hydrodynamic effects on coastal structures that have been identified by previous studies include the high aleatory variability of bore impact on structures [13,17], the importance of structural flexibility and dynamic fluid–structure interaction [17,23,24], the critical role of trapped air below elevated decks [21,25–28] and the demand on individual connections and columns [29]. For example, Robertson et al. [13] found a significant variability in the impulsive uplift pressures applied by a turbulent bore on the slab soffit of a vertical wall with an overhang, with the maximum pressures at selected locations ranging between 3 and 7.5 kPa among the different repetitions of the same bore. Similarly, Istrati [17] showed a large aleatory variability in the bore-induced horizontal force on a bridge deck with the standard deviation in the experimental tests being equal to approximately 20% of the average value. However, after the decomposition of the total force into a slamming and quasi-static component, it was revealed that the variability is generated by the slamming component. Moreover, several studies found that the air entrapment below elevated decks with girders can significantly increase the maximum total uplift forces generated by solitary waves [21,30]. However, the trapped air can also increase the overturning moment, while it has a complex and inconsistent effect on the total slamming and on the uplift demand in the offshore bearings and columns generated by bores [28].

In addition to the adverse tsunami-induced effects on coastal structures documented by the aforementioned studies focusing on 'clear-water' conditions, field surveys after recent tsunamis revealed the presence of waterborne debris (e.g., containers, cars, wooden poles), which can result in an increase in the uplift and drag forces and lead to significant lateral displacement and potential instability of the structure [1,31,32]. Consequently, the quantification of the impact forces caused by the waterborne debris attracted the attention of several research studies, both numerical [33–35] and experimental [36–39]. According to Haehnel and Daly [36], the peak debris impact force is a function of the impact velocity, the mass of the debris, and the effective stiffness. Como and Mahmoud [35] evaluated the debris impact on interior and exterior wood structural panels using fluid–structure interaction analyses and reported that the debris impact forces on an exterior panel increased with the wave height, while this was not the case for the interior panels. Ko [37] investigated experimentally the impact of a shipping container on a column and reported that the in-water applied load was 1.2-fold larger than the corresponding in-air condition, and had a longer duration. An equation to estimate the debris velocity based on the relative distance of debris pick-up location and the structure was developed in Shafiei et al. [38]. Derschum et al. [40] investigated the debris impact on structures and reported that the initial impoundment depth had a significant effect on the impact angle. Moreover, the same study observed a non-linear movement of the debris with a reduced impact velocity close to the structure, which was attributed to the formation of splitting streamlines and a stagnation zone in front of the coastal structure.

Yang [41] studied the tsunami-induced debris loading on bridge decks using the material point method. The results of the analyses showed that the presence of debris increases the applied loads on bridges, with the in-water analyses giving up to 35% larger debris impact forces than the in-air cases. Oudenbroek et al. [42] carried out an experimental and numerical study to evaluate the failure mechanism of selected bridges in Japan. The results confirmed that debris accumulation had significant effects on the hydraulic demand on bridges by increasing the uplift and drag forces, which could lead to the structural failure of the bridge. Istrati et al. [43] carried out a three-dimensional (3D) numerical investigation on the effects of tsunami-borne debris damming on coastal bridges. The results demonstrated that a shipping container trapped below the offshore overhang of a bridge has a negligible effect on the horizontal load but it can increase the overturning moment (pitching), which could consequently increase the probability of failure of the offshore bearings and connections that have to withstand this moment. If the container is trapped at locations offset from the mid-length of the span (e.g., close to the supports of the span), significant yaw and roll moments are generated due to 3D effects, which could lead to unequal distribution of the loads to the structural components of the two bent caps or abutments.

The majority of the numerical studies conducted to date on tsunami and extreme flooding effects on structures used mesh-based solvers that employed the finite volume [26,42] or the finite element method [23,44], while some of them used particle-based methods including smoothed particle hydrodynamics [45–47] or hybrid particle-finite methods such as PFEM [48]. SPH is a Lagrangian-based meshless technique, in which continuum properties of the fluid are discretized by a set of non-connected particles that carry individual material properties describing the medium, including, position, velocity, mass, density, pressure, and other physical quantities [49]. The Lagrangian nature of the SPH makes this method well suited to problems with large deformations and distorted free surface. The major advantage of using SPH is in dealing with free-surface problems where there is no need for special treatments for the free surface in order to simulate highly non-linear and potentially violent flows [50]. In recent years, SPH has been widely utilized to simulate a wide range of coastal and ocean engineering applications, such as solitary waves on beaches [51], breaking waves [52–55], wave–structure interaction and impact on coastal structures [45–47,56–60], and wave overtopping on offshore platforms [61]. Although the SPH method is one of the most matured forms of meshless techniques for cases of large deformations, such as during the breaking process of a wave, the computational accuracy and efficiency in simulating small deformation of solid bodies are lower than that of the FEM approach [62]. However, in the context of coastal structures and their dynamic response to large wave loads, the hybrid SPH–FEM approach could be utilized to take advantages of the ability of: (a) the SPH method to simulate complex free-surface flows and large fluid deformations, and (b) the FEM to simulate the dynamics of the structure [63,64].

The number of available numerical studies of tsunami-borne debris loading on coastal structures is quite limited, which can be attributed to the challenging multi-physics nature of the phenomenon. This phenomenon involves a complex fluid flow with turbulent wave breaking, a non-linear debris–fluid interaction with large deformations, a contact between the debris, the fluid and the coastal structure, and a dynamic structural response. Given (i) the catastrophic effects of large tsunami-borne debris, such as containers, in past tsunami events, and (ii) the good performance of SPH in other coastal engineering applications, the aim of this manuscript is to present a coupled SPH–FEM numerical modeling approach and evaluate its accuracy in simulating both the debris flow and the impact on a coastal structure. Therefore, this paper presents first a detailed explanation of the governing equations and modeling approach. Then, a comprehensive comparison of the coupled SPH–FEM models against experimental data available in the literature [37,65] is presented in terms of wave propagation and tsunami inland flow, debris motion and impact on a downstream coastal structure. Following the verification phase, a preliminary investigation focusing on the role of the debris restraints, the tsunami characteristics, and the initial water depth is conducted in order to shed light on the underlying physics of the debris–fluid interaction and impact on structures, and identify some critical parameters that govern this complex phenomenon.

2. Numerical Method

The SPH technique employed in this research is available in LS-DYNA [66] and is based on weakly compressible smoothed particle hydrodynamics (WCSPH). Several numerical investigations of fluid–structure interaction (FSI) were conducted using the SPH formulation in LS-DYNA to study large deformation problems [55,65,67]. For example, Pelfrene [55] used the SPH method for the simulation of free-surface water flow with focus on regular and breaking wave, and found that the solver is able to simulate the free-surface flow and capture the main features of the plunging breaking wave. Moreover, Grimaldi et al. [67] and Panciroli et al. [68] used the same method to investigate the impact of a solid body on water and both of them showed a good comparison with experimental results, justifying the selection of this solver for the numerical investigation conducted in the current study.

2.1. SPH Governing Equations

2.1.1. Kernel Approximation

In the SPH method, the particles are the computational framework on which the governing equations are resolved. This method is based on a quadrature formula on moving particles. Traditional SPH formulation exhibits very substantial pressure oscillation when modelling fluid flow. Instead of using the traditional computational grid, the conservation laws of continuum mechanics are defined by partial equations. For any SPH pseudo-particle, the function describing the field Ω is approximated in the form of a "kernel function", which is stated as the integral form of the product of any function and kernel function [66,69]:

$$< f(x) >= \int_{\Omega} f(x')W(x - x', h)dx', \tag{1}$$

where x and x' are the position vectors of the material points in a domain Ω; $f(x)$ is the continuous function of the field corresponding to the coordinate x; $f(x')$ is the value of quantity at the point x'; $W(x - x', h)$ is the bell-shaped smooth kernel function, where h is the smoothing length, defining the influence volume of the smooth function which varies in time and in space.

The smoothing function determines the range of computation with other particles and therefore has a significant effect on precision and accuracy of the analyses. The kernel function can be constructed taking in account a number of conditions and should satisfy the following properties [49]:

1. The smoothing function is normalized:

$$\int_{\Omega} W(x - x', h)dx' = 1, \tag{2}$$

2. There is a compact support for the smoothing function:

$$W(x - x', h) = 0, \; for \; |x - x'| > \kappa h, \tag{3}$$

where κ is the constant that determines the effective area of the smoothing function. This area is called the support domain.

3. $W(x - x', h)$ is non-negative for any x' within the support domain. This is necessary to achieve physically meaningful results in hydrodynamic computations.

4. The smoothing length increases as particles separate and reduces as the concentration increases.
5. With the smoothing length approaching zero, the kernel approaches the Dirac delta function:

$$\lim_{h \to 0} W(x - x', h) \ = \partial(x - x'), \tag{4}$$

6. The smoothing function should be an even function.

The forms of kernel functions usually include a quantic spline function [70], a cubic spline function [71] and a Gaussian kernel function [72]. By balancing the calculation accuracy and efficiency and considering that the commercial software LS-DYNA was used for the simulations throughout this study, the following B-spline smoothing function was adopted [66]:

$$W(x,h) = \frac{1}{h(x)^d}\theta(x), \tag{5}$$

where d is the number of space dimensions (2 or 3) and $\theta(x)$ is the cubic B-spline function defined by:

$$\theta(\mathrm{x}) = C * \begin{cases} 1 - \frac{3}{2}x_{xj}^2 + \frac{3}{4}x_{xj}^3 \ for \ x_{ij} \le 1 \\ \frac{1}{4}(2 - x_{ij})^3 \ for \ \ 1 < x_{ij} \le 2 \\ 0 \ for \ 2 < x_{ij} \end{cases} \tag{6}$$

where C is a constant for normalization, depending on the number of space dimensions and x_{ij} is the relative distance of particles i and j.

2.1.2. Particle Approximation

The second key aspect in SPH formulations is the particle approximation, which enables the system to be represented by a finite number of particles that carry an individual mass and occupy an individual space. For the SPH method, Equation (1) can be transformed into discretized forms by summing up the values of the field function within the support domain defined by the smoothing length h as follows [69]:

$$< f(x_i) >= \sum_{j=1}^{n} \frac{m_j}{\rho_j} f(x_i) \, W(x_i - x_j, h) = \sum_{j=1}^{n} \frac{m_j}{\rho_j} f(x_j) \, W_{ij}, \tag{7}$$

where $< f(x_i) >$ is the kernel approximation operator; $f(x_j)$ is the physical value at the jth position, i is the number of any particle in the domain; n is the total number of particles within the influence area of the particle i; and m_j and ρ_j are the mass and density associated with particle j. Thus, the value of particle i is approximated using the weighted average of the function values at all particles within the support domain of particle j. The partial approximation of the spatial derivation of a function can be expressed as in Das and Holm [73]:

$$< \nabla.f(x_i) > = \sum_{j=1}^{n} \frac{m_j}{\rho_j} f(x_i). \, \nabla \, W_{ij}, \tag{8}$$

where

$$\nabla \, W_{ij} = \frac{x_i \ - x_j}{r_{ij}} \frac{\partial W_{ij}}{\partial r_{ij}} \tag{9}$$

2.2. SPH for Viscous Fluid

The smoothing kernel and particle approximation can be used for discretization of partial differential equations (PDEs). The SPH formulation is derived by discretizing the Navier–Stokes equations spatially, thus leading to a set of ODEs which can be solved via time integration. Substituting the SPH approximations for a function and its derivative to the partial differential equations governing the physics of fluid flows, the discretization of these governing equations can be written as follows [74]:

$$
\begin{aligned}
\frac{d\rho_i}{dt} &= \sum_{j=1}^{n} m_j \left(x_i^{\beta} - x_j^{\beta}\right) \frac{\partial W_{ij}}{\partial x_i^{\beta}}, \\
\frac{dv_i^{\alpha}}{dt} &= \sum_{j=1}^{n} m_j \left(\frac{\sigma_i^{\alpha\beta}}{\rho_i^2} + \frac{\sigma_j^{\alpha\beta}}{\rho_j^2}\right) \frac{\partial W_{ij}}{\partial x_i^{\beta}}, \\
\frac{dv_i^{\alpha}}{dt} &= \sum_{j=1}^{n} m_j \left(\frac{\sigma_i^{\alpha\beta}}{\rho_i\rho_j} . \frac{\partial W_{ij}}{\partial x_i^{\beta}} - \frac{\sigma_j^{\alpha\beta}}{\rho_i\rho_j} . \frac{\partial W_{ij}}{\partial x_j^{\beta}}\right), \\
\frac{de_i}{dt} &= -\frac{p_i + \Pi_{ij}}{\rho_i^2} \sum_{j=1}^{n} m_j \left(v_i - v_j\right) \frac{\partial W_{ij}}{\partial x_i^{\beta}}, \\
\frac{dx_i^{\alpha}}{dt} &= v_i + \varepsilon \sum_{j=1}^{n} \frac{m_j}{\rho_j} \left(v_i - v_j\right) W_{ij},
\end{aligned}
\tag{10}
$$

where the superscripts α and β are the coordinate directions; g is the acceleration of gravity; σ is the particle stress; v is the particle velocity; e is the internal energy per unit mass; ε is the shear strain rate ($\varepsilon \cong 0.5$) and Π_{ij} is the Monaghan artificial viscosity [75]. In the analysis of the interaction between waves and structures, Π_{ij} can prevent the non-physical shock of the solution results in the impact area and effectively prevent the non-physical penetration of particles when they are close to each other. The role of artificial viscosity is to smoothen the shock over several particles and to allow the simulation of viscous dissipation, the transformation of kinetic energy to heat. Therefore, to consider the artificial viscosity, an artificial viscous pressure term Π_{ij} is added.

From Equations set (10), the following particle body forces can be derived [76]:

$$
F_i^{pressure} = -\sum_j m_j \frac{p_i + p_j}{2\rho_j} \nabla W(r_{ij}, h) F_i^{viscosity} = \mu \sum_j m_j \frac{v_i + v_j}{2\rho_j} \nabla^2 W(r_{ij}, h) \tag{11}
$$

where $r_{ij} = x_i - x_j$, and μ is the viscosity coefficient of the fluid. The pressure p_i is computed via the constitutive equation:

$$
p_i = K\left(\rho_i - \rho_0\right) \tag{12}
$$

where K is the stiffness of the fluid and ρ_0 is the initial density.

The acceleration of particle i is derived from:

$$
a_i = \frac{1}{\rho_i\left(F_i^{pressure} + F_i^{viscosity} + F_i^{external}\right)} \tag{13}
$$

where $F_i^{external}$ represents external forces such as body forces and forces due to contacts.

2.3. Sorting

In the SPH method, the location of neighboring particles is important. The sorting consists of finding which particles interact with which others at a given time. A bucket sort is used that consists of partitioning the domain into boxes where the sort is performed. With this partitioning, the closest neighbors will reside in the same box or in the nearest boxes. This method reduces the number of distance calculations and therefore the computational time [66].

2.4. Equation of State (EOS)

When the SPH method is applied in solving the FSI problems, the fluid is treated as weakly compressible, which means that an equation of state is utilized to determine the dynamic fluid pressure based on the variation in density and internal energy of particles.

The equation of state is originally applied to SPH by Monaghan [69] to model free-surface flows for water, which is stated in the following form:

$$P = k_0[(\frac{\rho}{\rho_0})^{\gamma} - 1] \tag{14}$$

where ρ_0 denotes the reference water density, ρ is the current density, γ is a constant parameter and often set to 7 for water, and k_0 is used to govern the maximum fluctuations of pressure, and is usually taken as follows [45]:

$$c_0 = \sqrt{\frac{\gamma k_0}{\rho_0}} \geq 10\ v_{\max} \tag{15}$$

where c_0 is the speed of sound in water at the reference density. In order to satisfy the Courant–Friedrichs–Lewy (CFL) condition, the real speed of sound should be at least 10-fold faster than the maximum fluid velocity. Satisfying this criterion will keep the density variations to within less than 1% and ensure low compressibility while allowing for a relatively large time step size.

2.5. Time Integration

The CFL condition requires the time step to be proportional to the smallest spatial particle resolution, which in SPH is represented by the smoothing length. LS-DYNA uses a simple and first-order scheme for integration. The time step is calculated by the following expression [66]:

$$\delta t = C_{CFL}\ min_i(\frac{h_i}{c_i + v_i}) \tag{16}$$

2.6. Contact Definitions

The interaction between the SPH and FE elements is defined using a penalty-based contact algorithm in which the SPH is always defined to be the slave part and the finite elements are defined to be the master. When a node is in contact with the surface, each slave node is checked for penetration. If the slave node penetrates, a restoring force is applied to prevent further penetration. This magnitude of this The restoring force is defined by [66]:

$$f = kdn \tag{17}$$

where d is the penetration distance, n is the surface normal vector and k is a penalty factor, comparable to a spring constant. The stiffness factor k for master segment s_i is given in terms of the bulk modulus K_i, the volume V_i, and the face area A_i of the element that contains s_i as:

$$k_i = \frac{A_i\ K_i\ f_{si}}{max\ (shell\ diagonal)} \tag{18}$$

where f_{si} is a scale factor for the interface stiffness and is normally defaulted to 10. The constant K should be set large enough to minimize penetration and instabilities, but it should not be too large that it generates artificially large forces. As the contact location and the direction may be difficult to predict, the automatic contacts are recommended, since they can detect the penetration at each time step, irrespective of whether it is coming from the slave or master part. The automatic contacts determine the contact surface by projecting normally from the shell mid-plane to a distance equal to half of the contact thickness. It must be noted that the solver can simulate the contact between flexible structures, rigid and flexible structures, or between rigid bodies only. Interestingly, even in the case of rigid bodies where the deformations are not calculated, it is possible to define a bulk modulus at the material level, which enables the user to adjust the contact parameters (e.g., contact stiffness) and avoid numerical spikes in the contact forces.

3. Experimental Work

The experimental study described in Ko and Cox [65], and Ko [37] was adopted for benchmarking the numerical SPH-FE modeling approach. These experiments had been conducted in the Large Wave Flume (LWF) at O.H. Hinsdale Wave Research Laboratory (HWRL) at Oregon State University. The flume is 104.24 m in length, 3.66 m in width, and 4.57 m in depth, and is shown in Figure 1. The LWF is equipped with a piston-type dry-back wavemaker that has a 4 m maximum hydraulic stroke actuator and a maximum speed of 4 m/s, which can generate both regular and random waves, as well as solitary waves, to simulate hurricane and tsunami waves. The LWF has adjustable bathymetry made of 20 square configurable concrete slabs. The flume includes a series of bolted holes with vertical patterns every 3.66 m along the flume for supporting test specimens, as well as, the concrete bathymetry slabs.

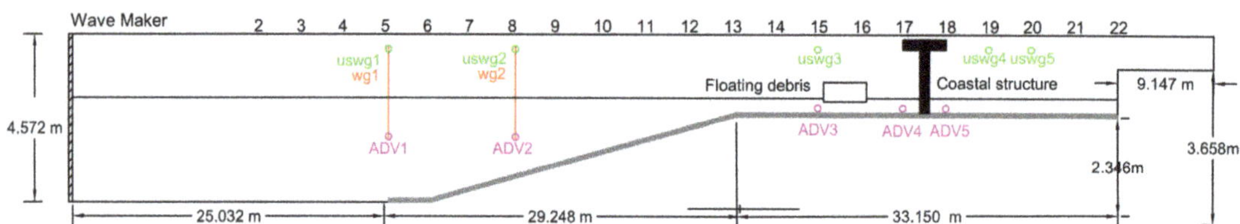

Figure 1. Cross-section of the Large Wave Flume (LWF) depicting the bathymetry, column location and flume instrumentation of the experiments of Ko and Cox [65] used for the validation study.

The coastal structure was represented by a column assembly located between bays 17 and 18, and was equipped with a load cell to measure the debris impact force. An aluminum 1:5 scale model of the standard intermodal container was used as the debris specimen. Given the fact that a standard container has dimensions of 6.1 m in length, 2.44 m in width, and 2.9 m in height, at 1:5 scale, the model had dimensions of 1.22 m × 0.49 m × 0.58 m, while the draft of the empty container was 9.1 cm. The longitudinal orientation is defined so that the major axis of the debris is parallel to the x axis and the direction of tsunami propagation. The debris specimen is located 3.5 m away from the column in the x direction. In order to control the movement of the debris and maintain the debris' orientation, guide wires had been installed in the LWF between bays 15 and 18. The debris specimen was allowed to move freely in the x and z directions, i.e., horizontal and vertical directions, but it was restrained in the y direction, which means that there was no translation across the flume width and no yaw.

To measure the free-surface time history and fluid velocity in the x direction, several resistance and ultrasonic wave gages, and acoustic doppler velocimeter (ADV) had been installed along the flume. The error function method proposed by Thomas and Cox [77] was used to generate the wave paddle displacement in the aforementioned experiments. To maximize the volume and duration of the tsunami inundation process, the full 4 m stroke of the wave maker had been used. The time for the wave paddle to travel the full 4 m stroke, i.e., error function period, is denoted as $T_{erf.}$

4. Coupled SPH–FEM Modeling

4.1. Numerical Settings

A numerical model of the experimental setup was developed using particles (SPH) for the fluid, and finite elements (FEM) for the flume walls, the debris specimen and the column. The bathymetry and experimental setup shown in Figure 1 was simulated via a two-dimensional model that represented a slice crossing through the mid-width of the flume, and cutting the debris and the column in two equal halves. This assumption was possible due to the debris restraints used in the experiments, which eliminated the movement of the debris across the flume width (normal to the tsunami flow). In this model, the wavemaker was represented with rigid shell elements with a prescribed horizontal motion equal to the input wavemaker displacements of the experiments found in Ko and Cox [65].

The influence of different SPH particle sizes on the computational time and numerical results was investigated by comparing sizes of 1, 2 and 3 cm, as shown in Appendix A. Decreasing the initial particle size improves the accuracy of the numerical results; however, too small sizes could cause numerical instability [78]. Considering the computational time, numerical accuracy, and calculation efficiency, the particle size of 1 cm was selected in the final numerical models presented herein. The debris and the column were generated using shell elements that had the same size with the fluid particles (1 cm), as shown in Figure 2. The final 2D numerical model consisted of 14,571 shell elements and 1,193,075 SPH particles. Given the fact that no significant deformations of the debris were observed in the selected experimental results [65] during the debris–flow interaction and impact on the structure, the debris was simulated as a rigid body. In the case of a 'rigid' assumption the properties of the rigid shell elements are not considered in the calculation of the time step of the explicit analyses, meaning that the time step is determined only by the remaining elements and particles. This in turn allows the explicit solver to use a significantly larger time step and reduce the required computational time per analysis.

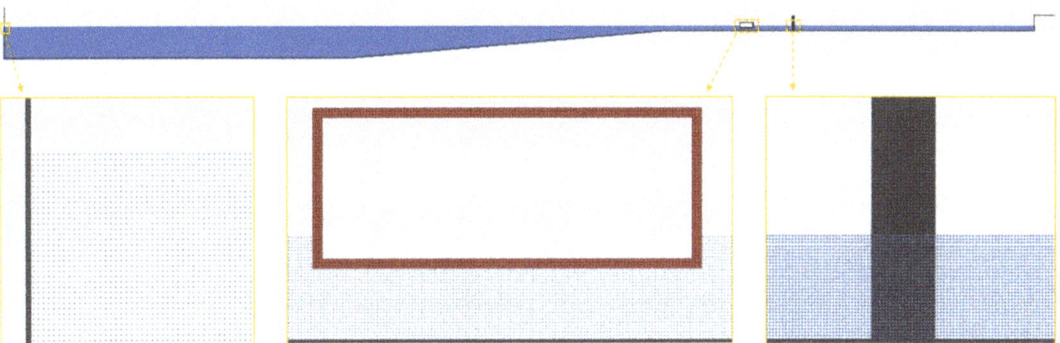

Figure 2. Numerical models with large debris: side-views of the debris, the column and the wave maker.

Given the 2D nature of the numerical model, several additional calculations and assumptions had to be made in order to ensure compatibility with the actual experiments. First, the 2D geometries of the container and the column were simulated (see Figure 2). However, since the sides of the container were not simulated in the 2D model, its thickness and density were adjusted in order to simultaneously satisfy the required draft of 9.1 cm and exhibit a rotational inertia that was equivalent to the one of the experimental specimen. Then, by assuming that the ratio of the debris' mass to the column's mass is identical in both the numerical and experimental models, the density of the column was determined accordingly. With the beneficial effect of the 'rigid' assumption in mind (in terms of computational time), and the expected small effect of the steel column deformation on the overall phenomenon (e.g., fluid flow around the column and the debris–fluid–column interaction), the column was assumed to be rigid as well. Despite this assumption, since the objective of this paper was to investigate the accuracy of predicting the impact forces, the contact stiffness was reasonably calculated by the solver via the bulk modulus (defined at the material level) and Equation (18) to avoid artificially high numerical spikes generated by a rigid-to-rigid body contact.

Key parts in the development of the numerical models included the definitions of the SPH–FEM and FEM–FEM contacts interfaces. In fact, different contact algorithms were employed to define the interaction (i) between the fluid (SPH) and the flume wall, the wavemaker and the debris specimen (FE), and (ii) between the debris and the column. The interaction between the FE and the SPH particles was defined in the numerical simulation via the *CONTACT_2D_NODES_TO_SOLID contact card, with the master-slave penalty algorithm, in which the shell elements were assigned the role of the master part and the SPH particles the slave part. Moreover, the contact between the floating debris and the column was determined via the *2D_AUTOMATIC_SURFACE_TO_ SURFACE contact type, which is a segment-based two-way contact. Such segment-based contacts are preferred over node-based contacts, since the penetration can be easily traced even if it happens at locations far from existing nodal locations, and consequently tend to be more accurate but computationally more expensive. It must be noted that the contact between the fluid and the column was not simulated in the 2D model because it would generate a reflection of the bore when it reached the structure, which is not realistic given the fact that in the actual experiments the bore can escape from the sides of the column.

Despite the speed-up of the numerical analyses due to the aforementioned assumptions, it was not feasible to run the 2D model on a regular desktop and therefore all the computational analyses were run on the high-performance computing (HPC) cluster at the University of Nevada, Reno, using up to 80 cores per analysis. The analysis time ranged between 15 and 21 h depending on the hydrodynamic characteristics (i.e., water depth), which affected the total number of SPH particles.

4.2. Accuracy of Numerical Modeling

4.2.1. Free Surface and Fluid Velocities

In order to quantify the accuracy of the numerical model, the results of the free-surface and fluid velocity histories at two different distances from the wavemaker, were compared with those measured in Ko and Cox [65]. Wave gage 1 (wg1) and adv1 are the closest to the wavemaker and are located at the end of the first horizontal part, with x = 24.930 m for both instruments and z = 1.240 m for the adv. In addition to this location where the wave has propagated only over a horizontal slab, the numerical results are also compared with the experimental results at x = 35.890 m (see wg2 and adv2), which is located along the sloped part. The z coordinate of adv2 is 1.236 m. Figure 3 shows the variation of the free-surface at (a) wg1 and (b) wg2, and the fluid velocity at (c) adv1 and (d) adv2 for all trials of the experimental tests with parameters of h_1 = 2.496 m and h_2 = 0.13 m, and T_{erf} = 30 s from [65], where h_1 and h_2 corresponds to the initial water depth offshore (at wavemaker location) and on the coast close to the debris, respectively. It can be observed that both the free-surface and fluid velocity histories computed by the numerical model are in good agreement with the experimental data, both in terms of the peak values and temporal evolution. There are some underpredictions of the maximum wave height and some overpredictions of the velocity but those do not exceed 6% for the selected wave. In addition to this good agreement, the encouraging thing is that the numerical model can predict the relative increase between wg2 and wg1 of the maximum free surface, indicating that the SPH–FEM coupled approach can capture the interaction of the fluid with the sloped part of the flume and result in a similar non-linear transformation of the wave during the shoaling process.

Figure 4 shows a comparison of the numerically predicted maximum values of the free surface at the locations (a) wg1 and (b) wg2 with the experimental data for all the tested tsunami waves (T_{erf} = 30 s, 40 s, 45 s) for h_1 = 2.496 m and h_2 = 0.13 m. Promising agreement with the experimental was achieved, with the maximum deviation from the average value of the measured data at wg1 being 6.3%, 11.7%, and 13.7% for T_{erf} = 30, 40 and 45 s, respectively. At wg2, the maximum differences are 4%, 15.9%, and 15.2%. As expected, both the numerical and experimental data show that the shorter $T_{erf,}$ i.e., T_{erf} = 30 s gives greater inundation depths at the two locations. This also seems to be true for the peak velocities, especially those measured at adv2, as shown in Figure 5. The latter figure presents the experimentally and numerically recorded maximum velocities at (a) adv 1 and (b) adv2 for the same hydrodynamic conditions as in Figure 4 (h_1 = 2.496 m). Similarly to the free surfaces, the SPH–FEM approach can predict reasonably the maximum fluid velocities with maximum deviations of 3%, 4.7%, and 2% at adv1 for the three different tsunami bores (T_{erf} = 30, 40, and 45 s), respectively. At adv2, the maximum differences are 6%, 14%, and 22%, respectively. It is noteworthy that (i) the experimentally recorded peak fluid velocities in Ko and Cox [65] had significantly larger variability than the measurements of the free surface, and (ii) the most sensitive experimental results corresponded to the measured velocities of the slower flows (e.g., for T_{erf} = 45 s), for which the numerical modeling yielded the largest deviations (underprediction of approximately 25%).

4.2.2. Debris Motion

Although the reasonable predictions of the free surface and fluid velocities achieved by the numerical simulations increase the confidence in the SPH solver and the coupled SPH–FEM approach, one of the most critical aspect for future risk assessment studies is the ability to estimate the debris transport and propagation overland. To this end, Figure 6 shows the velocity history of the debris specimen for one of the hydrodynamic conditions with h_2 = 0.13 m and T_{erf} = 30 s. The debris velocity was generated based on the publicly available videos on DesignSafe (i.e., Ko and Cox [65]), using a color-based tracking algorithm. However, it must be noted that the estimated velocities (from the videos of the experiments) could potentially entail some errors due to the fact that the ceiling cameras could not be perpendicular to the top surface of the debris for the whole propagation process (from its initial position up to the coastal structure). A more accurate estimate would require a perspective correction, as in Ko [37], which was not done herein due to the lack of adequate information.

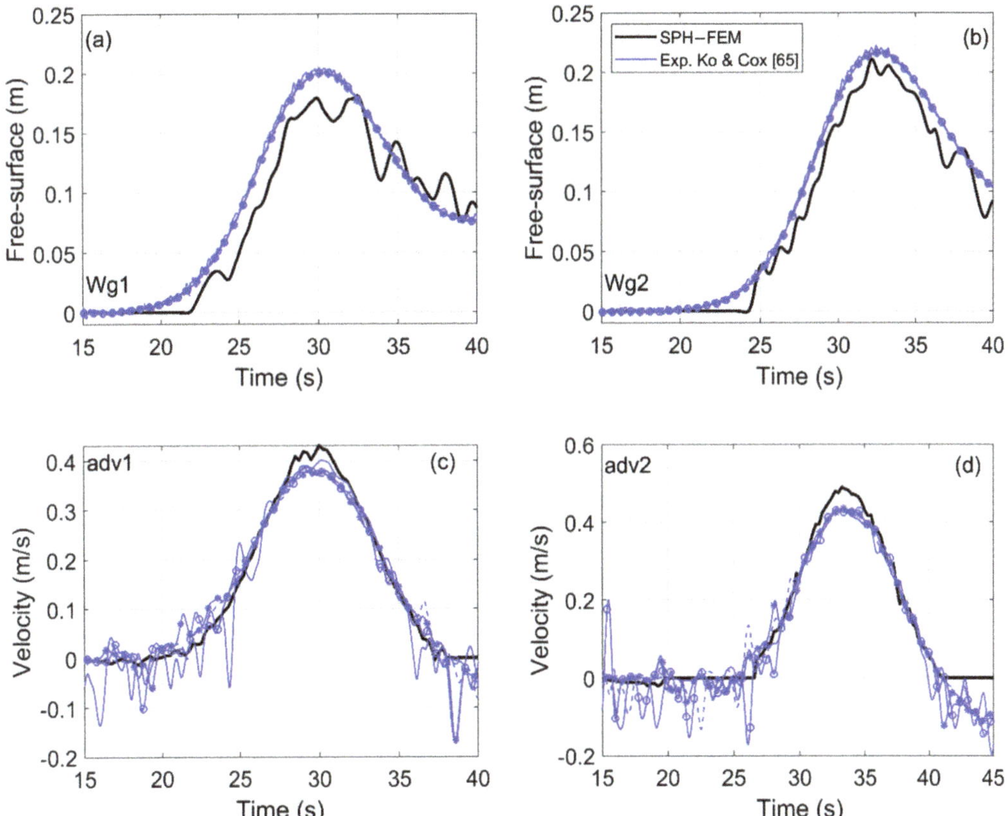

Figure 3. Variation of the free surface and fluid velocity at different locations along the flume. Experimental [65] and numerical results for h_1 = 2.496 m and T_{erf} = 30 s.

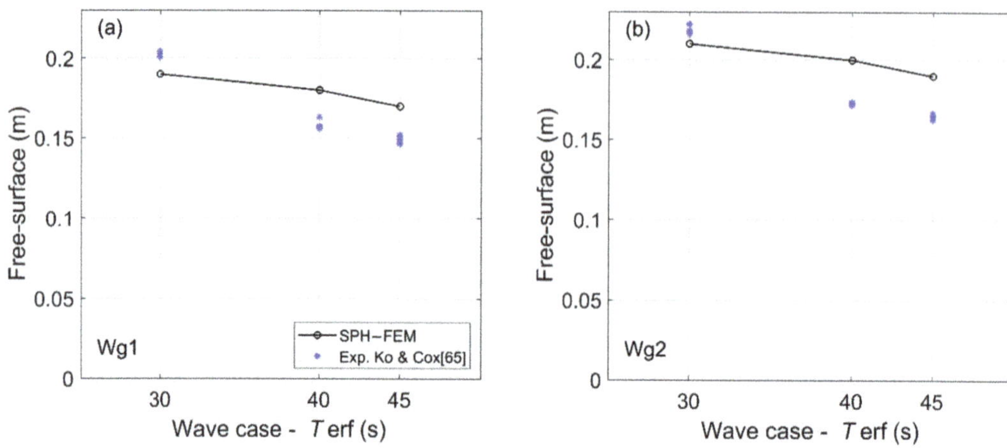

Figure 4. Maximum free-surface values of experiments [65] and numerical simulations for h_1 = 2.496 m, and three wave cases with T_{erf} = 30, 40 and 45 s.

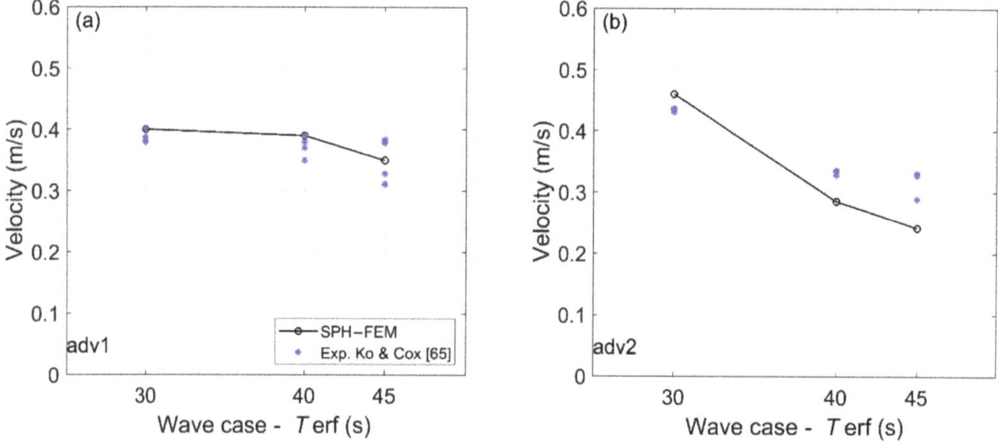

Figure 5. Maximum fluid velocities of experimental tests [65] and numerical simulations for h_1 = 2.496 m, and three wave cases with T_{erf} = 30, 40, and 45 s.

Since the debris specimen was modeled as a solid body, in order to evaluate the accuracy of the debris velocity from the SPH–FEM simulation, the velocity was output at four nodes of the debris, which corresponded to the external corners. Figure 6 presents the nodal velocities at the lower right and left corner together with the experimental velocity and a reasonable agreement is observed. The numerical and experimental simulations seem to give similar debris velocities when the bore reaches the container and starts transporting it. However, as the debris propagation continues, the numerical model seems to accelerate more and reach a larger velocity than the experimental specimen before the impact on the column. This in turn causes the numerical model of the container to cross the 3.5 m (initial distance between debris-column) faster and impact the column earlier than the experiment.

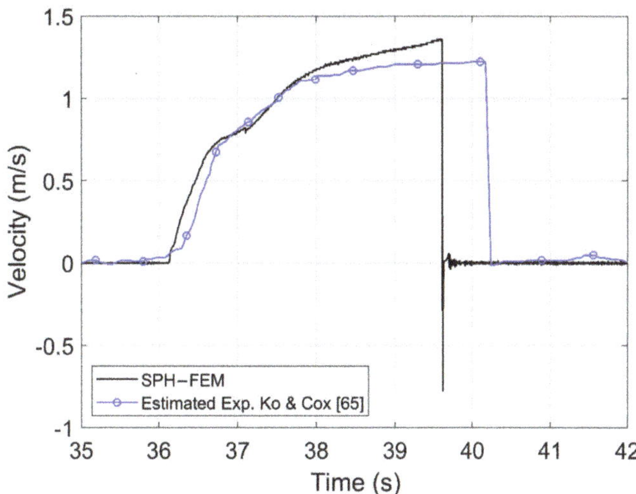

Figure 6. Debris velocity histories: Estimated based on the experimental tests of [65] and results from the numerical simulations for h_1 = 2.496 m, T_{erf} = 30 s.

One of the possible explanations for the observed differences lies in the 2D formulation of the current numerical model, which implies that the pressures applied from the bore on the offshore face of the debris (i.e., the mechanism causing the debris movement) are uniform across the debris width. However, in the actual experiments the pressures on the offshore face of the debris (and below its bottom side) are expected to be smaller at locations close to the sidewalls than the mid-width due to 3D effects. In other words, some of the bore pressures on the debris are relieved close to the vertical sides of the debris since the bore can propagate along these sides, given that there is a large area between the flume walls and the debris. Moreover, it must be clarified than in the presented numerical results the debris was restrained to move only in the horizontal direction, i.e., the direction of the bore propagation, eliminating any pitching that could potentially attenuate the debris transport.

The decision to restrain the debris was made based on the findings of Ko [37], who stated that there was significant variability in the experimentally recorded impact forces due to: (i) the "off-centered" impact of the debris on the load cell, and (ii) the effect of the pitch angle that was present in some trials but not in others. For example, it can be estimated from the figures of Ko and Cox [65] that for a debris velocity of approximately 1.4–1.5 m/s the "off-centered" impact can give maximum impact forces that are reduced by approximately 40% relative to the centered case. This explains why in the aforementioned study was decided to consider only the centered experimental results, and why in the present numerical investigation the debris was not allowed to pitch during the validation phase.

4.2.3. Debris Impact Force

Figure 7 shows the time histories of the debris impact forces on the column for two selected hydrodynamic conditions, with T_{erf} = 30 s and T_{erf} = 45 s for the same initial water depth h_2 = 0.13 m. Subfigures (a) and (b) correspond to T_{erf} = 30 s for the numerical and experimental data respectively, and subfigures (c) and (d) show the same results but for T_{erf} = 45 s.The agreement between the computations and the different trials of the physical tests is reasonable, both in terms of the peak impact force and in the overall trends. For instance, for the case of T_{erf} = 30 s, the SPH–FEM models predict a relatively higher impact force than the experiments by approximately 15%. However, this can be explained by the fact that as shown in Figure 6 the numerically predicted impact velocity was approximately

20% higher than the experimental one. Given the fact that the majority of the available simplified equations for debris impact loads, such as those presented in FEMA P646 [3] and ASCE [79], are a linear function of the impact velocity, it is reasonable to obtain larger impact forces from the numerical simulations since they predict larger velocities. Apart from the similarities in the results, there are also two noticeable differences:

- In the numerical simulations, the impact force on the column is applied earlier for T_{erf} = 30 s and later for T_{erf} = 45 s compared to the physical tests. However, these differences in the instants could be justified by the differences in the debris velocities, which were most likely overpredicted and underpredicted, respectively, as indicated by the trends in the maximum values. In other words, it is reasonable for the debris impact to occur earlier when the numerical models overpredict the magnitude of the impact, because the reason is the larger debris velocity.
- Immediately after the primary impact force, the column in the physical tests experienced a second short-duration impact force, which is relatively small compared to the main impact. However, this trend was not observed in the numerical results. In contrast, the simulations show a long duration load after the initial impact, which seems to have a nearly constant magnitude. This difference can be attributed again to the 2D simplification made in the numerical models, which are unable to allow the fluid to escape from the sides of the debris after the initial impact on the column and relieve the pressures applied on its offshore face. This leads consequently to the stagnation of the flow in front of the offshore face, resulting in a nearly steady-state horizontal damming load.

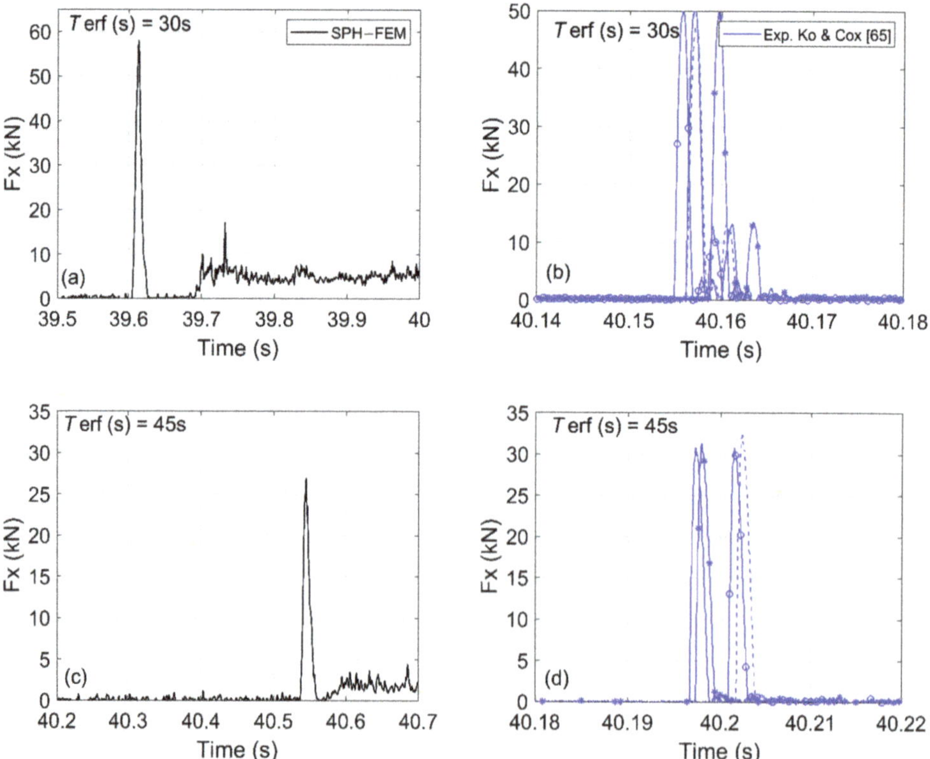

Figure 7. Comparison of the debris impact forces: experimental tests [65] and numerical simulations for h_1 = 2.496 m, and two wave cases with T_{erf} = 30 s and 45 s.

The trends of the peak values of the impact forces observed in the previous figure are also verified in Figure 8, which plots the corresponding maximum values from the experiments and the computational models for all three hydrodynamic flows (T_{erf} = 30, 40, and 45 s) with h_1 = 2.496 m and h_2 = 0.13 m. In fact, this figure reveals that the maximum deviation from the measured values is 14%, 25%, and 19% for the three flows, respectively, which is promising given the differences in the debris velocities. Moreover, another observation that reinforces the confidence in the SPH–FEM modeling approach is that it presents similar trends with the physical tests (as a function of $T_{erf,}$) since both of them give larger debris velocities and impact forces for the smallest T_{erf} = 30 s representing the more transient flow with the largest fluid velocities.

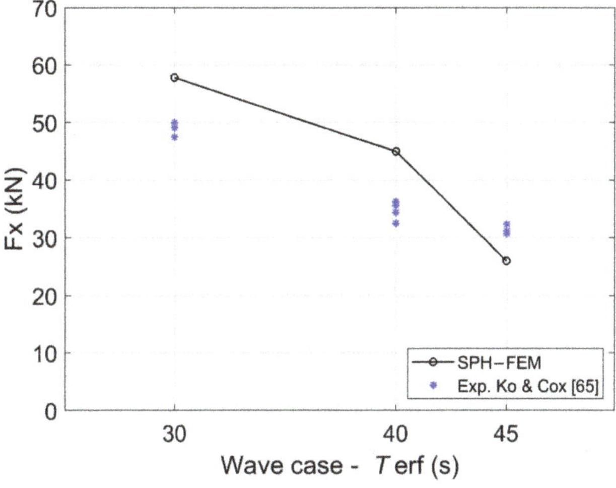

Figure 8. Maximum values of debris impact force on column: experimental tests [65] and numerical simulation for h_1 = 2.496 m, and three wave cases with T_{erf} = 30, 40, and 45 s.

5. Role of Debris Restraints

Given the fact that in the validation section the debris was restrained in order to make it more consistent with the experiments, it is essential to assess the role of the restraint for the debris–wave interaction, the debris transport overland and the impact on the structure. To this end, the restrained model of the previous section was compared with another model that allowed the debris to move freely in the 2D plane (horizontal and vertical displacement, and pitching). Figure 9 presents several selected snapshots of the tsunami flow with T_{erf} = 30 s for h_2 = 0.13 m as the free debris propagates towards the column location, from the instant that the bore starts moving the debris up to the instant of the second impact. Among these snapshots, 'e', 'f', 'g' correspond to following instants: (i) slightly before the first impact on the column, (ii) after the 1st impact, and (iii) at the instant of the 2nd impact on the column, respectively. As shown, the debris starts pitching in the clockwise direction up to the point that the onshore bottom corner tends to touch the bottom of the flume, after which it immediately changes the direction of pitching. It is also revealed that the large counter-clockwise pitching continues until the debris reaches the column location, which results in a non-normal impact angle and a consequently non-uniform contact of the onshore vertical face of the debris with the column, affecting the contact area and consequently the maximum impact forces. After the initial impact, the debris bounces back and re-impacts the column with a clockwise pitching angle, which generates a smaller magnitude impulse.

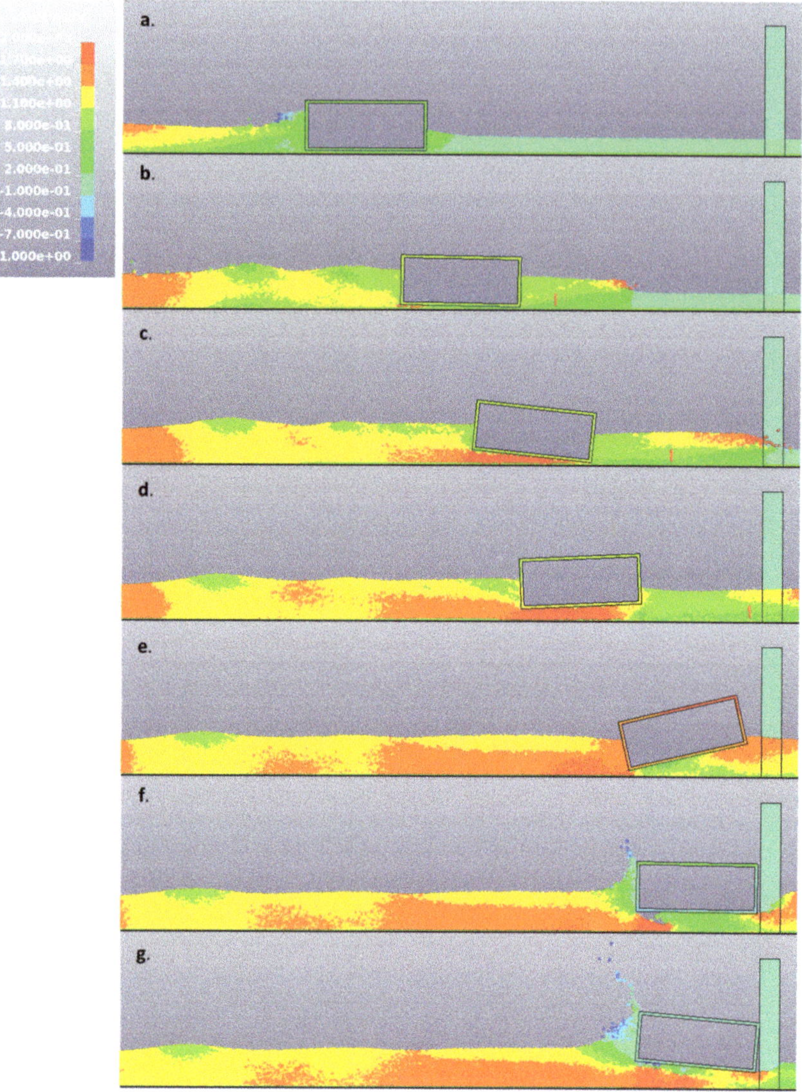

Figure 9. Selected instants of the debris–wave interaction and impact on the column of the free debris. Numerical results for the hydrodynamic case with h_1 = 2.496 m, h_2 = 0.13 m and T_{erf} = 30 s.

Figure 10 presents a quantitative comparison between the debris motion of the two cases. In the case of the free debris four locations are selected for presentation, including, the lower left corner (C1), the lower right corner (C2), the upper right corner (C3), and the upper left corner (C4). The x-displacement vs. y displacement curves for the free movement debris reveals that the two lower corners move in the opposite y direction as the debris moves along the x direction. A similar motion trend is observed for the two upper corners as well. The opposite movement in the y direction indicates the initiation of the debris rotation. To evaluate the level of the pitching, the rotation angle of the debris is calculated using the corner displacements and is presented on the right subplot of Figure 10. Based on this figure it becomes evident that the debris starts rotating clockwise after the bore reaches its location (negative rotation), then as it transports inland it changes the direction of rotation until it reaches the maximum rotation of approximately 37° slightly before the primary impact on the column, and then rotates in the opposite direction as it interacts with the column, until it stabilizes at a zero angle and the long duration damming process is initiated.

The debris velocity and impact forces histories for the free and restrained container models are presented in Figure 11. Interestingly, although it was expected the peak impact velocity to be affected by the pitching, the expectation was that the free debris would exhibit a more significant wave–structure interaction (due to the larger vertical displacements and rotation), which would dissipate more of the energy of the bore leading to a smaller horizontal debris velocity and a delayed arrival at the column location. While the delayed arrival did happen as expected and the debris velocity was indeed smaller than the respective velocity of the restrained model, this was true only at some instants during the debris transport inland and not at the instant of the impact on the column. The impact velocity of the free debris was surprisingly larger than the restrained one, which highlights the complexity and non-linearity of the debris interaction with the turbulent flow.

Another interesting finding can be reached from the subplot of the impact forces, according to which, the column experiences a significantly larger impact force from the restrained debris than the free case, although the latter one impacts the column with a larger velocity. The debris restraint seems to increase the maximum impact force by approximately a factor of 2.3 relative to the free case and reach the experimentally recorded values. The above trend could be attributed to the presence of debris pitching that affects the impact angle on the column, since as was observed in the swinging in-air tests of Ko [37] the cases with a zero pitch angle tended to give larger impact forces than the larger pitch angles, even when the debris impact velocities were the same.

The above finding indicates that future predictive equations might have to be a function of not only the impact velocity but also the impact angle, while design methodologies and risk assessment frameworks should be able to predict these two parameters. Ideally, the location of the impact on the coastal structure should also be estimated since the structural damage caused by the container could be very localized (close to the point of contact) if the debris impacts a coastal structure with a non-normal pitching angle.

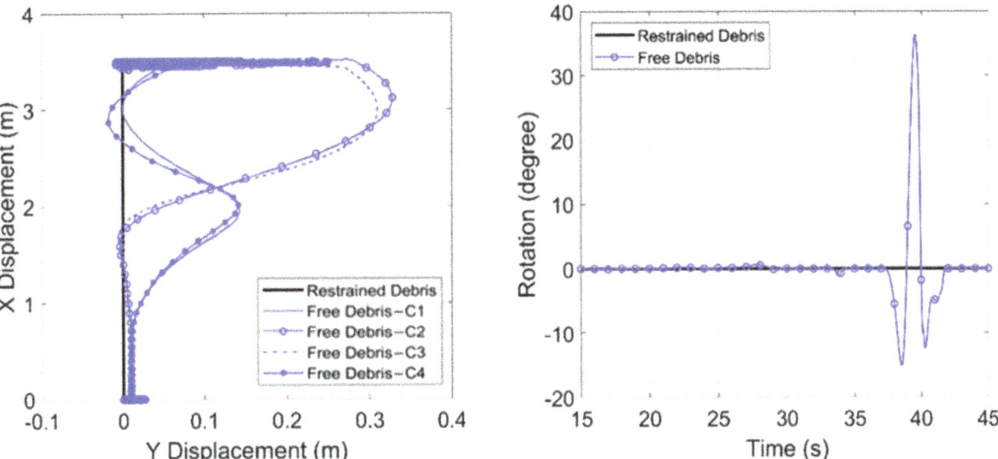

Figure 10. Debris trajectory (**left**) and rotation (**right**) for the restrained debris and free debris. Numerical results for the hydrodynamic case with h_1 = 2.496 m, and h_2 = 0.13 m and T_{erf} = 30 s.

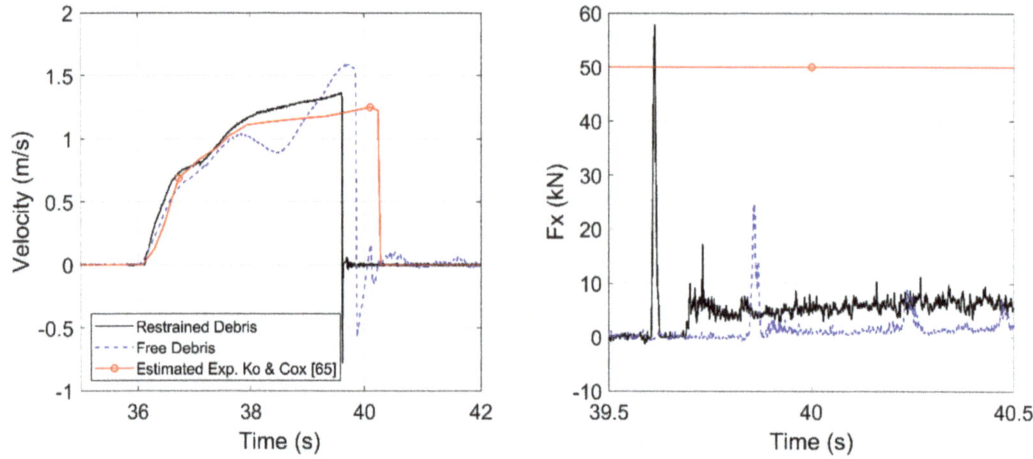

Figure 11. Debris velocity (**left**) and forces on the column (**right**). Experimental and numerical results (of restrained and free debris) for the hydrodynamic case h_1 = 2.496 m and T_{erf} = 30 s.

6. Effect of Hydraulic Conditions on Debris Motion and Impact Forces

6.1. Tsunami Flow Characteristics

To gain further insight into the debris–flow interaction and impact on coastal structures, this section will evaluate the effects of the flow characteristics by considering the three different cases with T_{erf} = 30 s, T_{erf} = 40 s, and T_{erf} = 45 s for h_1 = 2.496 m and h_2 = 0.13 m tested in [37]. In this section, the debris will be considered free in the 2D plane since it is considered more realistic, despite the fact that the restrained model captured better the experimental data. Comparison of the free surface and fluid velocity at a location close to the container, i.e., x = 62 m, is depicted in Figure 12. As discussed in [37], the error function (T_{erf}) affects the wave characteristics, and particularly the wave height and fluid velocity. The new figure shows that although there are similarities in the free-surface histories for all three waves, the smallest T_{erf} (which corresponds to the faster movement of the wavemaker) exhibits the largest peak values for the free surface and the fluid velocity, with the most noticeable differences observed in the fluid velocities. Interestingly, as the T_{erf} increases the flow height and fluid velocity is reducing, while the duration of the inundation is elongated, which indicates that the flow is switching from a highly transient bore to a more steady-state flow.

Figure 13 presents two selected snapshots of the debris–fluid interaction as the container propagates inland for two selected tsunami waves with T_{erf} = 30 s, and T_{erf} = 45 s, respectively. While both waves present similar trends in the debris motion, which comprises of a clockwise rotation followed by a counter-clockwise one, the faster moving bore results in larger particle velocities around the debris that in turn cause larger pitching of the container. This would indicate that the pitching of the debris is mainly caused by the larger velocities, and not that much by the differences in the free surface, which are much smaller. Figure 14 plots the vertical movement of the lower-right corner of the container and its rotation throughout the propagation inland. This figure illustrates that the debris flow is indeed highly dependent on the tsunami characteristics and that the fastest bore (T_{erf} = 30 s) can cause pitching angles that are approximately 85% larger than those of the slowest wave (T_{erf} = 45 s). This larger rotation leads to an increase in the upward vertical movement of the offshore face of the container, which enables it to impact structural locations at higher elevations.

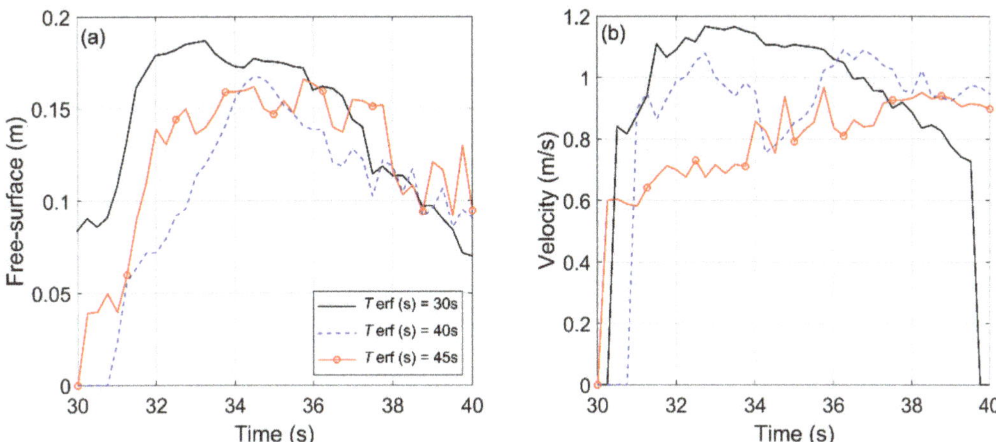

Figure 12. Variation of free surface (**left**) and fluid velocity (**right**) at x = 62 m: Numerical results for h_1 = 2.496 m and three wave cases with T_{erf} = 30, 40, and 45 s.

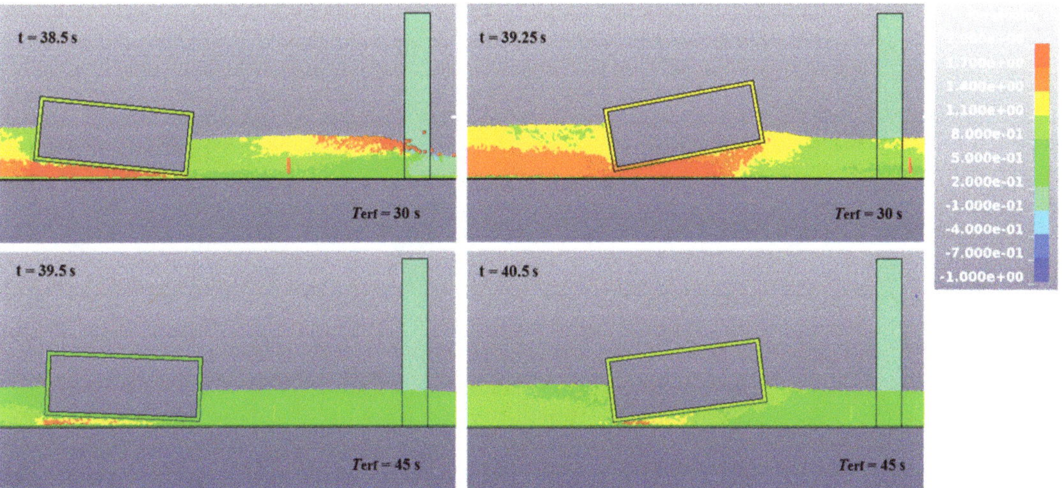

Figure 13. Snapshots of debris–tsunami interaction and impact on the column. Numerical results for h_1 = 2.496 m and two wave cases with T_{erf} = 30 s (**top**) and 45 s (**bottom**).

Combining the trends of the previous figures and snapshots, it is possible to identify three characteristic time instants in the motion of the debris for the case of T_{erf} = 30 s, as shown below:

- T1: This time instant represents the initiation of the debris rotation. It occurs slightly after the tsunami has started pushing the debris inland. After that initial contact with the bore, the debris starts accelerating and as the flow below the debris increases and the bore front surpasses the debris, the latter one starts rotating clockwise (see also snapshots b. and c. in Figure 10).
- T2: This instant corresponds to the largest clockwise rotation, at which point the lower right (onshore) corner has displaced downward so much that it impacts the floor of the flume. When this impact takes place, a restoring force is applied to the debris causing it to start rotating in the opposite direction (counter-clockwise).

- T3: After the primary debris impact on the flume floor and the initiation of the counter-clockwise rotation, the debris continues rotating in this direction until it reaches the maximum pitch angle, which tends to occur slightly before the primary debris impact on the column. At this instant, it is possible for the lower left (offshore) corner of the debris to touch the floor of the flume before it impacts the column. However, this will depend on the initial relative distance between the debris and the coastal structure, as well as, the hydrodynamic conditions. This means that instant T3 represents the maximum clockwise pitching angle, which might be close to the impact angle, but not necessarily the same. Future studies should investigate different debris–structure relative distances, and a larger range of hydrodynamic conditions in order to determine the dependence of the maximum pitching angle and the impact angle on these parameters. Ideally, such studies should employ three-dimensional models, which are expected to be more accurate than two-dimensional models.

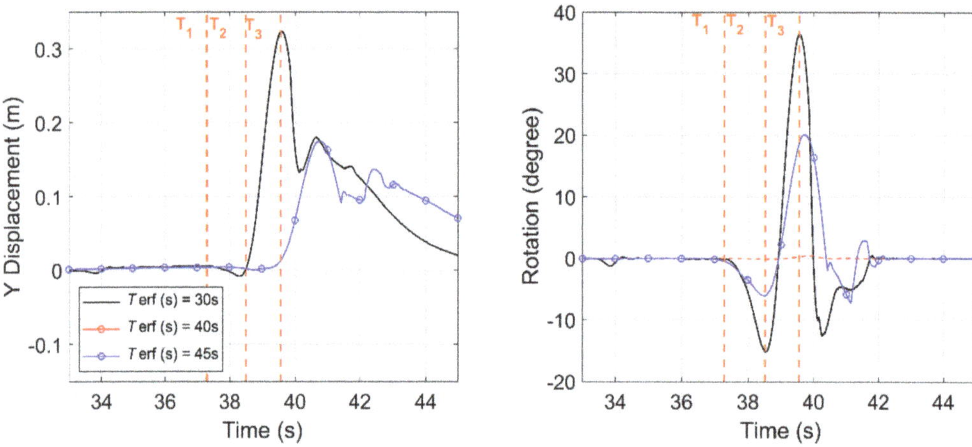

Figure 14. Numerical results of the free debris: vertical displacement (**left**) and rotation (**right**), for h_1 = 2.496 m and three wave cases with T_{erf} = 30, 40, and 45 s.

The debris impact forces on the column, including both time histories and the peak value versus the impact velocity, are presented in Figure 15 for all the tsunami waves of h_2 = 0.13 m. As expected, larger tsunami waves have more energy and higher particle velocities, which lead to higher values of the debris impact velocities and impulsive forces on the column. This is why the wave with T_{erf} = 30 s exerts the largest impact force on the column. Interestingly, the largest/fastest tsunami bores also result in higher damming loads, with T_{erf} = 30 s giving almost 3-fold larger values than T_{erf} = 45 s. However, this observation must be taken with caution, since, as explained earlier, the 2D nature of the numerical models can lead to over-prediction of the damming loads. Last but not least, the right subplot of Figure 15 reveals that contrary to the existing simplified equations of debris loads (e.g., FEMA P626 [3]), the impact forces might not necessarily be a linear function of the impact velocity, at least for the specific hydrodynamic conditions. In fact, when the debris velocity increases above a certain limit (e.g., 1.1 m/s for this water depth) the rate of the increase in the impact force with the velocity decreases, resulting in a non-linear increase in the force. This behavior can be explained by the trends observed in the previous figures, according to which, the largest impact velocity corresponds to the fastest bore (T_{erf} = 30 s) that cause significant pitching of the debris and non-normal impact on the column. Ultimately, the results presented herein indicate that the debris impact forces might be a function of both the debris velocity and pitching angle at the instant of the impact on the coastal structure. However, this indication must be further verified with validated three-dimensional models. Ideally, such future models should simulate the debris and the coastal structure as flexible bodies, since that will enable a more realistic prediction of the impact duration and determine its dependence on the debris velocity and pitch angle.

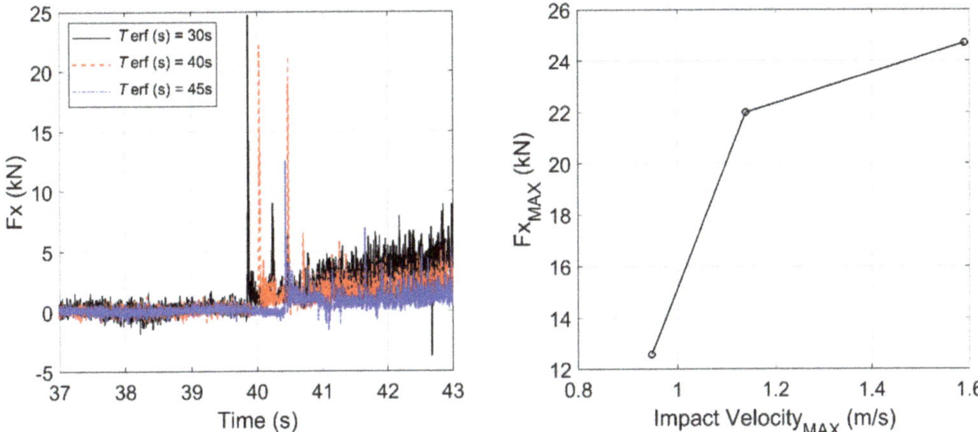

Figure 15. Free−debris histories (**left**) and maximum values of debris impact on the column vs. the impact velocity (**right**). Numerical results for h_1 = 2.496 m h_2 = 0.13 m, and T_{erf} = 30, 40, and 45 s.

The above findings seem to extend and complement the findings of previous studies, which have demonstrated the critical role of the relative tsunami–structure or debris–structure angle, in the estimation of hydrodynamic and debris loads, respectively. For example, Istrati and Buckle [44,80] demonstrated the dependence of the maximum tsunami loads on the skew angle of a bridge and the angle of attack (obliqueness of the wave), respectively, while Haehnel and Daly [36] and Derschum et al. [40] revealed experimentally the dependence of the debris loads on the yaw angle of the impact on the structure, for wooden logs and shipping containers, respectively. The former research study proposed a reduction coefficient for cases of oblique debris impact on a structure, and was later on the basis for the development of the orientation coefficient (equal to 0.65) in the Tsunami chapter of ASCE 7-16 [79]. Interestingly, the debris rotation and oblique collisions between adjacent containers were suggested as a critical factor in Stolle et al. [81], since these could result in energy losses and consequently reduce the impact forces on coastal structures. The latter authors simplified the estimation of the impact loads induced by multiple debris, by assuming them to be a function of an area coefficient that accounted for the impact debris geometry and compactness. However, they noted that accounting for the impact angle of the agglomeration in future studies, would be physically more realistic. Last but not least, the experimental observations of Shafiei et al. [38] are perhaps the closest in agreement with the findings of the current study, since they observed that the debris always impacted the vertical structure at non-zero pitching angle, which ranged between 3 and 10° and 15 and 30° for a rigid rectangular box and a disc, respectively. This pitching angle had a major effect because it governed the variability of the impact accelerations and caused a large vertical load (in the case of a horizontal tsunami flow) that was approximately 60% of the horizontal one.

6.2. Initial Water Depth

In addition to the effect of the tsunami characteristics, it is of interest to evaluate the effect of the initial water depth on the debris motion, and the debris interaction with the column and the bottom of the flume. For this purpose, two of the water depths tested in Ko [37], i.e., h_1 = 2.496 and 2.664 m and a new depth with h_1 = 2.8 m were considered for a range of tsunami flows. These offshore water depths translated into local initial water depths of 0.13, 0.30 and 0.43 m, respectively, at the debris location. Figure 16 illustrates (a) the variation of the free surface, (b) thefluid velocity, (c) the debris vertical displacement, (d) the debris rotation, (e) the debris velocity and (f) the impact force. The free-surface histories are plotted close to the offshore side of the debris, at x = 62 m, and are calculated relative to the initial water level, while the fluid velocities are plotted at the same x coordinate at the level of the initial free surface (i.e., 9.1 cm above the bottom of the debris). This means that the absolute elevation of the locations at which the fluid velocities are plotted are different for each water depth, but the relative distance from the bottom of the debris is the same. The figure reveals small differences in the maximum bore heights and nearly negligible differences in the fluid velocity histories close to the debris location, for three water depths. There are some differences in the free-surface history of the shallower water level relative to the two larger depths, with the most obvious difference in the bore front and the instant of the arrival at x = 62 m. In the case of the deeper water, the tsunami waves are arriving slightly faster, which is attributable to the increase in the wave celerity offshore caused by the increase in the water depth. Nonetheless, the tsunami bores at the debris location are similar enough for the three water depths, enabling the proper investigation of the effect of this parameter.

Interestingly, despite the smaller bore height in the case of the h_1 = 2.496 m relative to the other depths, this case exhibits the largest vertical displacement of the debris (at its onshore corner) and the largest pitching. In fact, the maximum pitching angle seems to consistently decrease with the increase in the water depth, with the shallowest case inducing an approximately 5-fold larger maximum pitching angle relative to the deepest case of 2.8 m. This demonstrates that the rotation of the debris is highly dependent on the initial water depth. Moreover, in addition to the differences in the magnitudes, it can be observed that the previously identified pattern in the debris motion, which involved an initial clockwise debris rotation followed by a counter-clockwise one before the impact on the column, is not consistent for all the water depths. For larger initial depths it is possible to notice an opposite sequence of debris rotations (i.e., for h_1 = 2.66 m) or just a clockwise rotation before the initial debris contact with the column (i.e., for h_1 = 2.8 m). Last but not least, in contrast to the differences in the vertical displacement and rotation of the debris, the debris horizontal velocities present more similarities. The major difference is observed in the fact that the deeper cases exhibit a gradual increase in the debris velocity, which becomes nearly constant as it approaches the coastal structure, while in the shallowest case more abrupt increases in the debris velocity are observed that result in a larger impact velocity on the column. Despite the larger debris impact velocity for h_1 = 2.496 m (for the specific tsunami bore), this case gives similar impact forces with the large water depths, which can be justified by the higher level of pitching in the former case.

Figure 17 presents selected snapshots for the three water depths and T_{erf} = 40 s, as the debris moves inland and impacts the coastal structure. This visual representation of the phenomenon verifies the previously observed trends, with the two larger water depths being associated with nearly a consistent debris orientation (small rotation) contrary to the shallow water that is dominated by debris pitching effects. Moreover, despite the similar bore velocities at x = 62 m (a few meters from the debris) observed in Figure 16, the fringe plots of the fluid particle velocities reveal that there are significant differences in the flow around the debris, since in the shallow case the flow seems to accelerate more below the debris. This is probably due to the fact that in the latter case the bore is more restricted and does not have as much space to propagate below the debris as in deeper waters, resulting in faster flows horizontally that tend to uplift on one side of the debris and consequently rotate it. Another major difference lies in the fact that when the initial water level is low, the pitching of the debris can move one of its corners downwards so much that it impacts the bottom of the flume. This contact between the debris and the flume complicates further the debris–fluid interaction and is a distinguishable feature of the small water depths only. Last, but not least the snapshots reveal major differences in the fluid flow below the debris after the initial impact on the column, which affects the number of secondary impacts and their magnitude, as well as, the damming loads. For example, smaller damming loads can be noticed in the case of larger water depths, because less of the bore gets reflected on the structure and more of it propagates onshore by moving below the debris. However, these differences might be exaggerated by the 2D assumption made in the development of the numerical models.

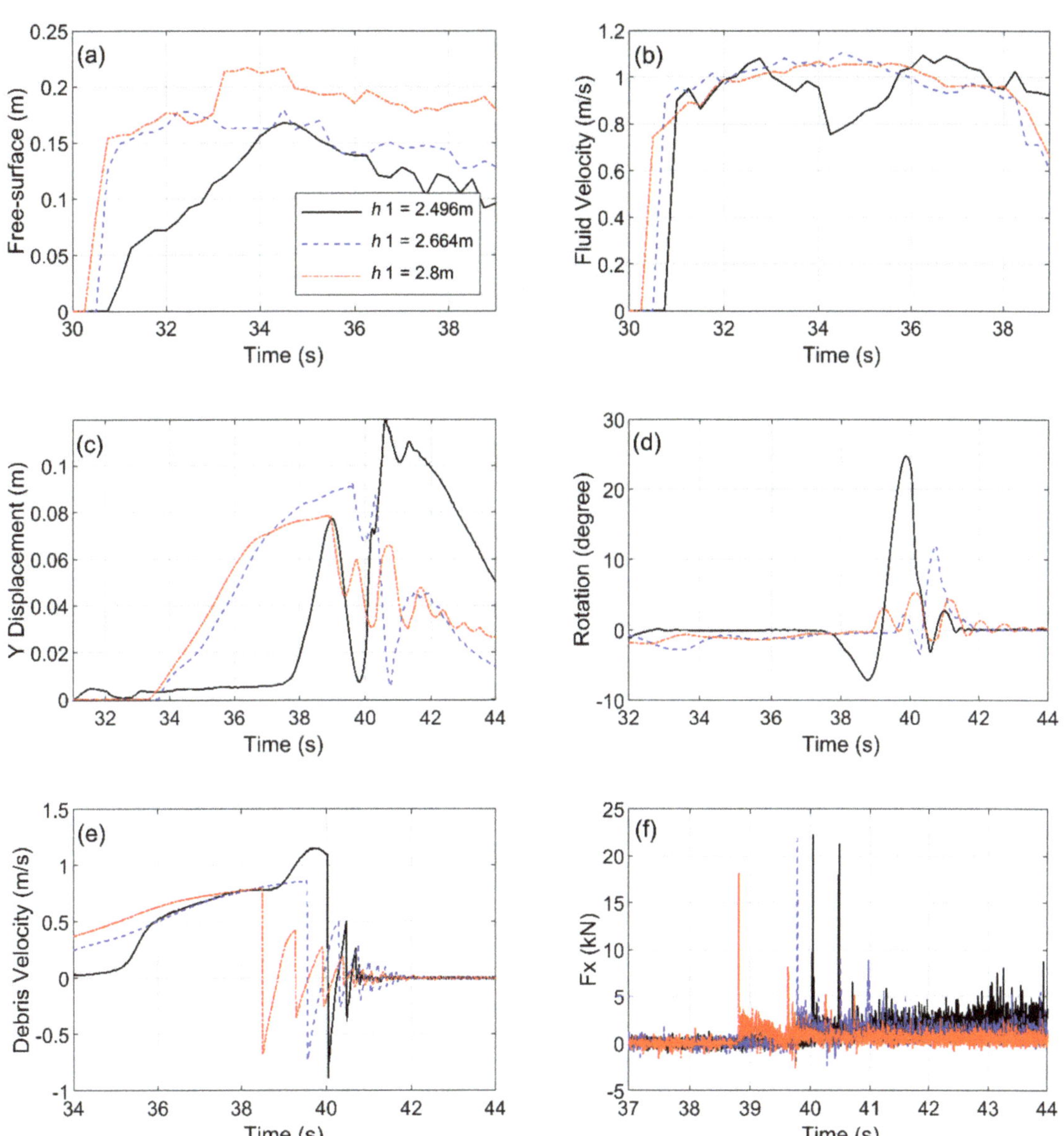

Figure 16. Time histories of the free surface, fluid velocity, motion of the debris (y displacement and rotation), debris velocity and debris impact force on the column. Numerical results of free debris for T_{erf} = 40 s and three water depths with h_1 = 2.496, 2.664 and 2.8 m.

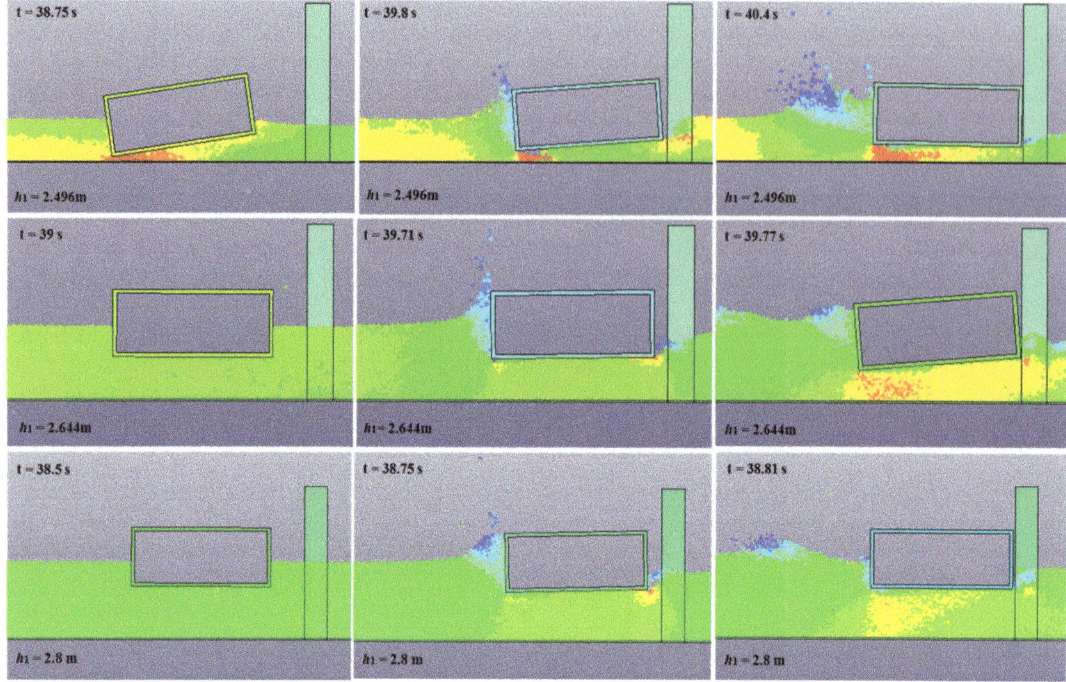

Figure 17. Snapshots of debris-tsunami interaction and impact on the column. Numerical results of free debris for T_{erf} = 40 s and three water depths with h_1 = 2.496 (**top**), 2.664 (**center**) and 2.8 m (**bottom**).

The maximum debris impact forces as a function of the impact velocities for all the water depths and bores, are presented in Figure 18. This figure reveals, that although for larger water depths (e.g., h_1 = 2.66 and 2.8 m) the relationship between the maximum debris impact force and impact velocity is nearly linear, which agrees with existing predictive equations (e.g., FEMA P646 [3] and ASCE [79]), for small water depth the trend seems to be non-linear. In fact, in the latter case, after the exceedance of the debris velocities above a certain limit, the impact force increases less than what a linear force–velocity assumption would suggest, indicating that predictive equations that utilize such an assumption might yield conservative results. To investigate whether such an indication is true, additional parametric investigations are required, followed by direct comparisons with available simplified equations, which is beyond to scope of this manuscript. However, the results presented herein raise the question on whether future predictive equations of debris loads should: (i) be a function of the water depth, so that for large initial water depths a linear impact force–velocity relationship is used and for shallow depths a non-linear one that will account for the possibility of debris pitching and non-normal impact angle on the structure, especially if the structure is located close to the location of the debris entrainment, and (ii) limit the applicability of the linear force–velocity equations to a certain velocity limit, above which, the rate of the force increase with the velocity will drop significantly.

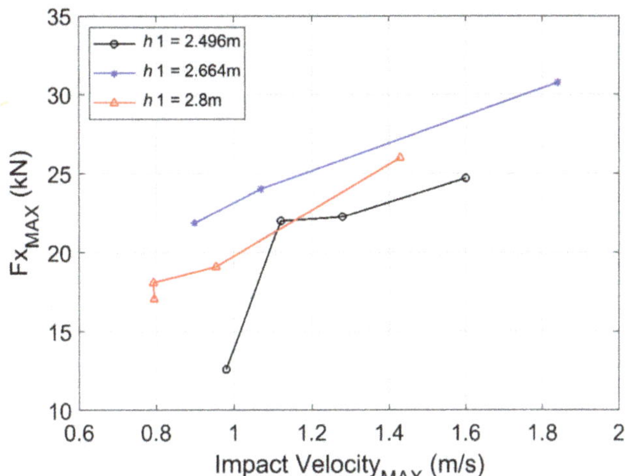

Figure 18. Maximum values of the debris impact force as a function of the impact velocity for three water depths and nine tsunami bores.

As explained earlier, the indications of this preliminary numerical investigation should be verified with follow-up expanded studies that will include a wider range of hydrodynamic conditions (e.g., faster flows and more water depths) and different relative distances between the debris and the coastal structure. The latter parameter is especially crucial, since as observed in Goseberg et al. [82] the debris tended to yaw towards an equilibrium position as it propagated inland, in which the long axis was perpendicular to the flow direction. It would be interesting to see if the pitching of the debris will be damped out in the case where it could propagate for a longer distance before impacting the structure. If such a behavior was true, then the significance of pitching effects documented herein might be applicable only to locations near the debris entrainment by the bore, and disappear after a certain propagation distance. Moreover, although the current 2D SPH–FEM models were validated against the restrained debris experiments of Ko [37] and Ko and Cox [65], future studies should employ 3D numerical models that will be able to simulate both the yaw and roll of the debris, in order to ensure that the influence of the pitching angle on the impact forces identified by the current study is not affected by the 2D assumption. Last but not least, given the observed complex interaction of the debris with a simplified vertical coastal structure, it would be interesting for future studies to investigate how to debris will interact with more complex structural geometries, such as elevated decks with overhangs, piloti-type buildings with columns or multiple structures in urban environments.

7. Summary and Conclusions

Given the documented catastrophic effects of large debris in past tsunamis, and the difficulty of simulating numerically such effects, this study presents a coupled SPH–FEM modeling approach and evaluates its capability to predict both the debris–fluid interaction and the impact on a coastal structure. In this modeling approach the fluid was modeled via particles, based on the weakly compressible smoothed particle hydrodynamics, while the wavemaker, flume, debris and structure were modeled with mesh-based finite elements. The interaction between the fluid and solid bodies was defined via node-to-solid contacts, while the interaction of solid bodies (e.g., debris–flume, debris–structure) was defined via a two-way segment-based contact that could trace the impact at any location of the body. For the validation of the coupled modeling, the large-scale experimental data from Ko and Cox [65] and Ko [37] were selected as a benchmark. In these experiments, a tsunami-like wave was generated offshore via a wavemaker, which then propagated over a complex bathymetry, forming a transient flow on the coast that entrained the debris and propagated it onshore until it impacted a vertical column. The fact that the debris was restrained experimentally to move only in the direction of the flow (i.e., no yaw and no movement across the flume width) enabled the development of a two-dimensional (2D) numerical model instead of a 3D one, which reduced consequently the required computational time.

The comparison between the experimental and numerical results revealed an overall acceptable accuracy of the SPH–FEM coupled modeling approach, with some parameters being estimated more accurately than other, as explained below:

- The free surface and fluid velocities had good agreement with the experimentally recorded results, both offshore and during the wave propagation along the slope. The deviation of the maximum wave height from the average value of the experimental tests ranged between 4% and 15.2%, while the deviation of the maximum fluid velocities was between 2% and 22%, depending on the location along the flume and the tsunami flow. These results showed that the numerical model can predict the relative increase in the free surface and fluid velocities as the wave propagates over the sloped part and undergoes a non-linear transformation, indicating that the fluid–flume contact worked properly.
- The SPH–FEM models estimate similar debris velocities with the experiments, especially when the bore reaches the container and starts transporting it. However, as the debris propagation inland continues, the numerical model tends to accelerate more and reach an impact velocity that is approximately 20% larger than in the experiments, leading consequently to some differences in the arrival time at the column location. One possible explanation for these differences lies in the 2D formulation of the current numerical model, which implies that the pressures applied from the bore on the offshore face of the debris is uniform across the debris width, which is not necessarily the case in real 3D environments.
- The deviation of the numerically predicted maximum debris impact forces on the column from the experimental data was in the range of 14–25% for the investigated hydrodynamic flows. However, these differences are consistent with the observed differences in the debris impact velocities. Overall, the numerical results presented similar trends with the physical tests since both gave larger impact forces for the more transient and faster tsunami flows.

Following the verification of the numerical modeling, a preliminary investigation was conducted with the aim to gain an insight into (i) the sensitivity of the results to the debris restraints, and (ii) the role of the hydrodynamic conditions. The analyses of the 2D free debris revealed that for small water depths, the debris starts rotating clockwise upon the arrival of the bore, then as it transports inland, it changes the direction of rotation until it reaches the maximum counter-clockwise pitching angle slightly before the primary impact on the column. The prominent pitching effect during the debris transport inland resulted in higher impact velocities relatively to the restrained debris, which was unexpected, highlighting the complexity and non-linearity of the debris interaction with the turbulent flow. Moreover, despite the larger impact velocity of the free debris, the applied impact forces were smaller by approximately 56% relative to the restrained one. This is attributed to the fact that the free debris continues pitching until it reaches the column location, which results in a non-normal impact angle and a consequently non-uniform contact. This non-normal pitching angle reduces the contact area between the debris and the column, and consequently the maximum impact forces.

The comparison of different tsunami flows revealed that the 2D motion of the debris is highly dependent on the tsunami characteristics, with the largest wave (i.e., faster flow inland) causing pitching angles that were approximately 85% larger than the smallest wave. In fact, there was some indication that the pitching of the debris was caused mainly by the high velocities of the turbulent flow passing below the debris, justifying why faster bores caused more debris pitching. Moreover, the comparison of similar tsunami flows for three different water depths demonstrated that the debris–fluid interaction is also dependent on the initial water level at the debris location, with the debris pitching decreasing consistently with the increase in the water depth. When the water level is low the bottom corners of the debris can impact the bottom of the flume if the pitching angle is large, complicating further the debris–fluid interaction and impact on the column. Interestingly, although for larger water depths the relationship between the maximum debris impact force and impact velocity is nearly linear, which agrees with existing predictive equations, for the small water depth a non-linear trend was observed. In fact, in the latter case, after the exceedance of the debris velocity above a certain limit, the impact force increased less than what a linear force–velocity relationship would suggest.

Despite the limited investigated hydrodynamic conditions, the observed trends raise the question on whether debris load equations should: (i) be a function of the water depth, so that for large initial water depths a linear impact force–velocity relationship is used and for shallow depths a more complex equation is developed that will account for the debris pitching angle at the instant of impact, and/or (ii) limit the applicability of the linear force–velocity equations to a certain velocity range, above which, the rate of the force as a function of the impact velocity will drop. Future studies should consider a wider range of hydrodynamic conditions (e.g., faster flows, more water depths) and different relative distances between the debris and the coastal structure in order to determine if the pitching effects observed herein are going to be damped out as the debris propagates over longer distances. Such studies should employ 3D numerical models that will be able to simulate both the yaw and roll of the debris, in order to ensure that the influence of the pitching angle is not exaggerated by the 2D SPH–FEM models used in this study.

Author Contributions: Conceptualization, D.I.; methodology, A.H. and D.I.; validation, A.H. and D.I.; formal analysis, A.H. and D.I.; investigation, A.H. and D.I.; resources, A.H., D.I. and I.B.; data curation, A.H.; writing—original draft preparation, A.H. and D.I.; writing—review and editing, D.I. and I.B.; visualization, A.H. and D.I.; supervision, D.I. and I.B.; project administration, D.I. and I.B.; funding acquisition, D.I. and I.B. All authors have read and agreed to the published version of the manuscript.

Funding: This research was funded by the Pacific Earthquake Engineering Research Center, headquartered at the University of California Berkeley, Under Research Agreement Number 1149-NCTRIB.

Data Availability Statement: The numerical data presented in this study are available on request from the corresponding author.

Acknowledgments: The authors acknowledge Michael Scott from Oregon State University and Christos Papachristos from the University of Nevada, Reno for their support of the research project, and the Engineering Computing Team at the University of Nevada, Reno for providing computational support and resources that have contributed to the research results reported within this paper.

Conflicts of Interest: The authors declare no conflict of interest.

Appendix A

The sensitivity of the numerical results to the SPH particle size was investigated for the fastest bore with T_{erf} = 30 s. This appendix (Figures A1–A3) shows the numerically predicted free surface, debris velocity and impact force for particle sizes equal to 1, 2 and 3 cm.

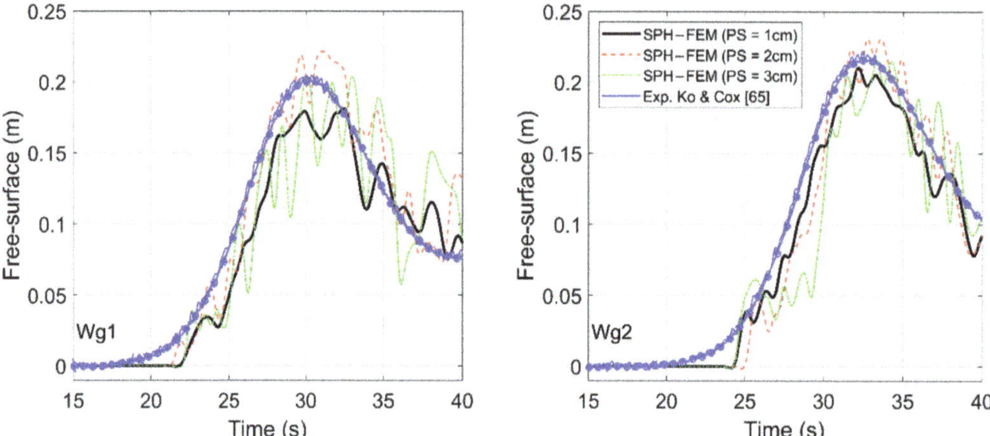

Figure A1. Variation of the free surface at different locations along the flume. Experimental [65] and numerical results for three particle sizes, for h_1 = 2.496 m and T_{erf} = 30 s.

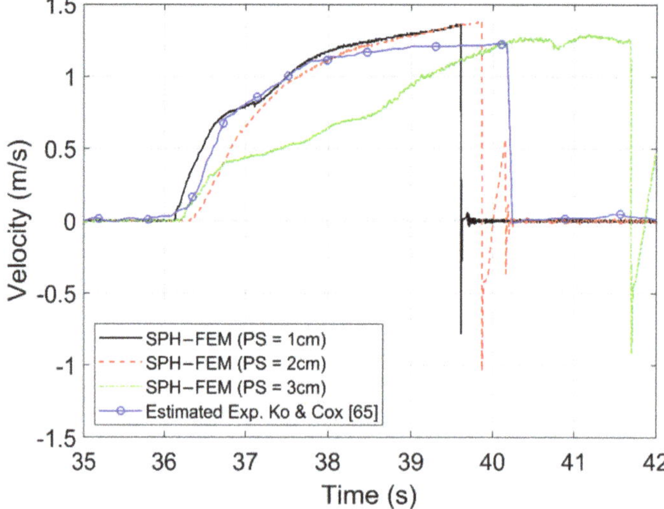

Figure A2. Debris velocity histories. Estimated based on the experimental tests of [65] and numerical results for three particle sizes, for h_1 = 2.496 m and T_{erf} = 30 s.

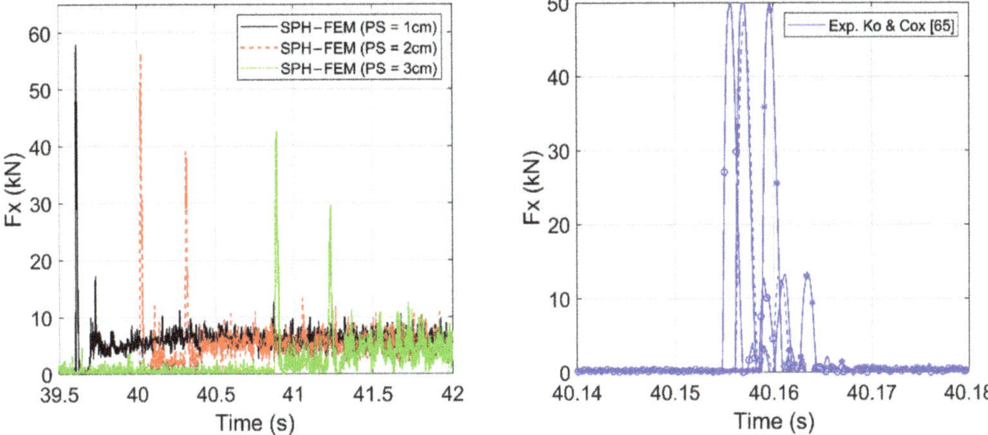

Figure A3. Debris impact forces. Experimental [65] and numerical results for three particle sizes, for h_1 = 2.496 m and T_{erf} = 30 s.

References

1. Yeh, H.; Sato, S.; Tajima, Y. The 11 March 2011 East Japan earthquake and tsunami: Tsunami effects on coastal infrastructure and buildings. *Pure Appl. Geophys.* **2013**, *170*, 1019–1031. [CrossRef]
2. Earthquake Engineering Research Center (EERI). *The Tohoku, Japan, Tsunami of March 11, 2011: Effects on Structures, Special Earthquake Report*; EERI: CA, USA, 2011. Available online: https://www.eeri.org/images/archived/wp-content/uploads/Tohoku_Japan_March_11_2011_EERI_LFE_ASCE_Tsunami_effects_on-Buildingssmallpdf.com-1.pdf (accessed on 20 September 2021).
3. FEMA P646. *Guidelines for Design of Structures for Vertical Evacuation from Tsunami*; Federal Emergency Management Agency: Washington, DC, USA, 2012.
4. Takahashi, S. Urgent survey for 2011 great east Japan earthquake and tsunami disaster in ports and coasts. *Tech. Note Port Airpt. Res. Inst.* **2011**, *1231*, 75–78.
5. Mikami, T.; Shibayama, T.; Esteban, M.; Matsumaru, R. Field survey of the 2011 Tohoku earthquake and tsunami in Miyagi and Fukushima prefectures. *Coast. Eng. J.* **2012**, *54*, 1250011-1–1250011-26. [CrossRef]
6. Canadian Association of Earthquake Engineering (CAEE). *Reconnaissance Report on the December 26, 2004 Sumatra Earthquake and Tsunami*; CAEE: British Columbia, CA, USA, 2005.
7. Akiyama, M.; Frangopol, D.M.; Strauss, A. Lessons from the 2011 Great East Japan Earthquake: Emphasis on life-cycle structural performance. In Proceedings of the Third International Symposium on Life-Cycle Civil Engineering, Vienna, Austria, 3–6 October 2012; pp. 13–20.
8. Kawashima, K.; Buckle, I. Structural performance of bridges in the Tohoku-oki earthquake. *Earthq. Spectra* **2013**, *29*, 315–338. [CrossRef]
9. Maruyama, K.; Tanaka, Y.; Kosa, K.; Hosoda, A.; Arikawa, T.; Mizutani, N.; Nakamura, T. Evaluation of tsunami force acting on bridge girders. In Proceedings of the Thirteenth East Asia-Pacific Conference on Structural Engineering and Construction, Sapporor, Japan, 11–13 September 2013. Available online: http://hdl.handle.net/2115/54508 (accessed on 15 July 2021).
10. Palermo, D.; Nistor, I.; Nouri, Y.; Cornett, A. Tsunami loading of near-shoreline structures: A primer. *Can. J. Civ. Eng.* **2009**, *36*, 1804–1815. [CrossRef]
11. Foster, A.S.J.; Rossetto, T.; Allsop, W. An experimentally validated approach for evaluating tsunami inundation forces on rectangular buildings. *Coast. Eng.* **2017**, *128*, 44–57. [CrossRef]
12. Honda, T.; Oda, Y.; Ito, K.; Watanabe, M.; Takabatake, T. An experimental study on the tsunami pressure acting on Piloti-type buildings. *Coast. Eng. Proc.* **2014**, *1*, 1–11. [CrossRef]
13. Robertson, I.N.; Riggs, H.R.; Mohamed, A. Experimental results of tsunami bore forces on structures. In Proceedings of the International Conference on Offshore Mechanics and Arctic Engineering, Shanghai, China, 6–11 January 2008; pp. 509–517. [CrossRef]
14. Araki, S.; Ishino, K.; Deguchi, I. Stability of girder bridge against tsunami fluid force. In Proceedings of the 32th International Conference on Coastal Engineering (ICCE), Shanghai, China, 30 June–5 July 2010.
15. Lau, T.L.; Ohmachi, T.; Inoue, S.; Lukkunaprasit, P. *Experimental and Numerical Modeling of Tsunami Force on Bridge Decks*; InTech: Rijeka, Croatia, 2011; pp. 105–130.

16. Hoshikuma, J.; Zhang, G.; Nakao, H.; Sumimura, T. Tsunami-induced effects on girder bridges. In Proceedings of the International Symposium for Bridge Earthquake Engineering in Honor of Retirement of Professor Kazuhiko Kawashima, Tokyo, Japan, March 2013. Available online: https://xueshu.baidu.com/usercenter/paper/show?paperid=a8a299a7f8922c5a6a312b0e8f9ad6b9&site=xueshu_se (accessed on 20 September 2021).
17. Istrati, D. Large-Scale Experiments of Tsunami Inundation of Bridges Including Fluid-Structure-Interaction. Ph.D. Thesis, University of Nevada, Reno, NV, USA, 2017. Available online: https://scholarworks.unr.edu//handle/11714/2030 (accessed on 15 July 2021).
18. Istrati, D.; Buckle, I.G.; Itani, A.; Lomonaco, P.; Yim, S. Large-scale FSI experiments on tsunami-induced forces in bridges. In Proceedings of the 16th World Conference on Earthquake, 16WCEE, Santiago, Chile, 9–13 January 2017; pp. 9–13. Available online: http://www.wcee.nicee.org/wcee/article/16WCEE/WCEE2017-2579.pdf (accessed on 15 July 2021).
19. Chock, G.Y.; Robertson, I.; Riggs, H.R. Tsunami structural design provisions for a new update of building codes and performance-based engineering. *Solut. Coast. Disasters* **2011**, 423–435. [CrossRef]
20. Azadbakht, M.; Yim, S.C. Simulation and estimation of tsunami loads on bridge superstructures. *J. Waterw. Port Coast. Ocean Eng.* **2015**, *141*, 04014031. [CrossRef]
21. McPherson, R.L. Hurricane Induced Wave and Surge Forces on Bridge Decks. Ph.D. Thesis, Texas A&M University, College Station, TX, USA, 2008.
22. Xiang, T.; Istrati, D.; Yim, S.C.; Buckle, I.G.; Lomonaco, P. Tsunami loads on a representative coastal bridge deck: Experimental study and validation of design equations. *J. Waterw. Port Coast. Ocean Eng.* **2020**, *146*, 04020022. [CrossRef]
23. Istrati, D.; Buckle, I.G. Effect of fluid-structure interaction on connection forces in bridges due to tsunami loads. In Proceedings of the 30th US-Japan Bridge Engineering Workshop, Washington, DC, USA, 21–23 October 2014. Available online: https://www.pwri.go.jp/eng/ujnr/tc/g/pdf/30/30-10-2_Buckle.pdf (accessed on 15 July 2021).
24. Choi, S.J.; Lee, K.H.; Gudmestad, O.T. The effect of dynamic amplification due to a structure's vibration on breaking wave impact. *Ocean Eng.* **2015**, *96*, 8–20. [CrossRef]
25. Bozorgnia, M.; Lee, J.J.; Raichlen, F. Wave structure interaction: Role of entrapped air on wave impact and uplift forces. In Proceedings of the International Conference on Coastal Engineering, 30 June–5 July 2010.
26. Seiffert, B.; Hayatdavoodi, M.; Ertekin, R.C. Experiments and computations of solitary-wave forces on a coastal-bridge deck. Part I: Flat plate. *Coast. Eng.* **2014**, *88*, 194–209. [CrossRef]
27. Xu, G.; Cai, C.S.; Chen, Q. Countermeasure of air venting holes in the bridge deck–wave interaction under solitary waves. *J. Perform. Constr. Facil.* **2017**, *31*, 04016071. [CrossRef]
28. Istrati, D.; Buckle, I. Role of trapped air on the tsunami-induced transient loads and response of coastal bridges. *Geosciences* **2019**, *9*, 191. [CrossRef]
29. Istrati, D.; Buckle, I.; Lomonaco, P.; Yim, S. Deciphering the tsunami wave impact and associated connection forces in open-girder coastal bridges. *J. Mar. Sci. Eng.* **2018**, *6*, 148. [CrossRef]
30. Seiffert, B.R.; Cengiz Ertekin, R.; Robertson, I.N. Effect of entrapped air on solitary wave forces on a coastal bridge deck with girders. *J. Bridge Eng.* **2016**, *21*, 04015036. [CrossRef]
31. Rossetto, T.; Peiris, N.; Pomonis, A.; Wilkinson, S.M.; Del Re, D.; Koo, R.; Gallocher, S. The Indian Ocean tsunami of december 26, 2004: Observations in Sri Lanka and Thailand. *Nat. Hazards* **2007**, *42*, 105–124. [CrossRef]
32. Robertson, I.N.; Carden, L.; Riggs, H.R.; Yim, S.; Young, Y.L.; Paczkowski, K.; Witt, D. Reconnaissance following the September 29, 2009 tsunami in Samoa. *Res. Rep.* **2010**. Available online: https://www.eeri.org/site/images/eeri_newsletter/2010_pdf/Samoa-Rpt.pdf (accessed on 15 July 2021).
33. Yeom, G.S.; Nakamura, T.; Mizutani, N. Collision analysis of container drifted by runup tsunami using drift collision coupled model. *J. Disaster Res.* **2009**, *4*, 441–449. [CrossRef]
34. Madurapperuma, M.A.K.M.; Wijeyewickrema, A.C. Inelastic dynamic analysis of an RC building impacted by a tsunami water-borne shipping container. *J. Earthq. Tsunami* **2012**, *6*, 1250001. [CrossRef]
35. Como, A.; Mahmoud, H. Numerical evaluation of tsunami debris impact loading on wooden structural walls. *Eng. Struct.* **2013**, *56*, 1249–1261. [CrossRef]
36. Haehnel, R.B.; Daly, S.F. Maximum impact force of woody debris on floodplain structures. *J. Hydraul. Eng.* **2004**, *130*, 112–120. [CrossRef]
37. Ko, H. Hydraulic Experiments on Impact Forces from Tsunami-Driven Debris. Master's Thesis, Oregon State University, Corvallis, OR, USA, 2013.
38. Shafiei, S.; Melville, B.W.; Shamseldin, A.Y.; Adams, K.N.; Beskhyroun, S. Experimental investigation of tsunami-borne debris impact force on structures: Factors affecting impulse-momentum formula. *Ocean Eng.* **2016**, *127*, 158–169. [CrossRef]
39. Stolle, J.; Goseberg, N.; Nistor, I.; Petriu, E. Debris impact forces on flexible structures in extreme hydrodynamic conditions. *J. Fluids Struct.* **2019**, *84*, 391–407. [CrossRef]
40. Derschum, C.; Nistor, I.; Stolle, J.; Goseberg, N. Debris impact under extreme hydrodynamic conditions part 1: Hydrodynamics and impact geometry. *Coast. Eng.* **2018**, *141*, 24–35. [CrossRef]
41. Yang, W.C. Study of Tsunami-Induced Fluid and Debris Load on Bridges using the Material Point Method. Ph.D. Thesis, University of Washington, Seattle, WA, USA, 2016. Available online: http://hdl.handle.net/1773/37064 (accessed on 15 July 2021).

42. Oudenbroek, K.; Naderi, N.; Bricker, J.D.; Yang, Y.; Van der Veen, C.; Uijttewaal, W.; Moriguchi, S.; Jonkman, S.N. Hydrodynamic and debris-damming failure of bridge decks and piers in steady flow. *Geosciences* **2018**, *8*, 409. [CrossRef]
43. Istrati, D.; Hasanpour, A.; Buckle, I. Numerical Investigation of Tsunami-Borne Debris Damming Loads on a Coastal Bridge. In Proceedings of the 17 World Conference on Earthquake Engineering, Sendai, Japan, 13–18 September 2020.
44. Istrati, D.; Buckle, I.G. Tsunami Loads on Straight and Skewed Bridges—Part 1: Experimental Investigation and Design Recommendations (No. FHWA-OR-RD-21-12). Oregon Department of Transportation. Research Section. 2021. Available online: https://rosap.ntl.bts.gov/view/dot/55988 (accessed on 15 July 2021).
45. St-Germain, P.; Nistor, I.; Townsend, R.; Shibayama, T. Smoothed-particle hydrodynamics numerical modeling of structures impacted by tsunami bores. *J. Waterw. Port Coast. Ocean Eng.* **2014**, *140*, 66–81. [CrossRef]
46. Pringgana, G.; Cunningham, L.S.; Rogers, B.D. Modelling of tsunami-induced bore and structure interaction. *Proc. Inst. Civ. Eng.-Eng. Comput. Mech.* **2016**, *169*, 109–125. [CrossRef]
47. Sarfaraz, M.; Pak, A. SPH numerical simulation of tsunami wave forces impinged on bridge superstructures. *Coast. Eng.* **2017**, *121*, 145–157. [CrossRef]
48. Zhu, M.; Elkhetali, I.; Scott, M.H. Validation of OpenSees for tsunami loading on bridge superstructures. *J. Bridge Eng.* **2018**, *23*, 04018015. [CrossRef]
49. Liu, G.R.; Liu, M.B. *Smoothed Particle Hydrodynamics: A Meshfree Particle Method*; World Scientific: Singapore, 2003.
50. Violeau, D. *Fluid Mechanics and the SPH Method: Theory and Applications*; Oxford University Press: Oxford, UK, 2012.
51. Monaghan, J.J.; Kos, A. Solitary waves on a Cretan beach. *J. Waterw. Port Coast. Ocean Eng.* **1999**, *125*, 145–155. [CrossRef]
52. Colagrossi, A.; Landrini, M. Numerical simulation of interfacial flows by smoothed particle hydrodynamics. *J. Comput. Phys.* **2003**, *191*, 448–475. [CrossRef]
53. Monaghan, J.J.; Kos, A.; Issa, N. Fluid motion generated by impact. *J. Waterw. Port Coast. Ocean Eng.* **2003**, *129*, 250–259. [CrossRef]
54. Crespo, A.J.C.; Gómez-Gesteira, M.; Dalrymple, R.A. 3D SPH simulation of large waves mitigation with a dike. *J. Hydraul. Res.* **2007**, *45*, 631–642. [CrossRef]
55. Pelfrene, J. Study of the SPH Method for Simulation of Regular and Breaking Waves. Master's Thesis, Gent University, Ghent, Belgium, 2011.
56. Dalrymple, R.A.; Knio, O.; Cox, D.T.; Gesteira, M.; Zou, S. Using a Lagrangian particle method for deck overtopping. *Ocean Wave Meas. Anal.* **2002**, 1082–1091. [CrossRef]
57. Gómez-Gesteira, M.; Dalrymple, R.A. Using a three-dimensional smoothed particle hydrodynamics method for wave impact on a tall structure. *J. Waterw. Port Coast. Ocean Eng.* **2004**, *130*, 63–69. [CrossRef]
58. Barreiro, A.; Crespo, A.J.C.; Domínguez, J.M.; Gómez-Gesteira, M. Smoothed particle hydrodynamics for coastal engineering problems. *Comput. Struct.* **2013**, *120*, 96–106. [CrossRef]
59. Altomare, C.; Crespo, A.J.; Domínguez, J.M.; Gómez-Gesteira, M.; Suzuki, T.; Verwaest, T. Applicability of Smoothed Particle Hydrodynamics for estimation of sea wave impact on coastal structures. *Coast. Eng.* **2015**, *96*, 1–12. [CrossRef]
60. Aristodemo, F.; Tripepi, G.; Meringolo, D.D.; Veltri, P. Solitary wave-induced forces on horizontal circular cylinders: Laboratory experiments and SPH simulations. *Coast. Eng.* **2017**, *129*, 17–35. [CrossRef]
61. Gómez-Gesteira, M.; Cerqueiro, D.; Crespo, C.; Dalrymple, R.A. Green water overtopping analyzed with a SPH model. *Ocean Eng.* **2005**, *32*, 223–238. [CrossRef]
62. Yang, Y.; Li, J. SPH-FE-Based Numerical Simulation on Dynamic Characteristics of Structure under Water Waves. *J. Mar. Sci. Eng.* **2020**, *8*, 630. [CrossRef]
63. Hu, D.; Long, T.; Xiao, Y.; Han, X.; Gu, Y. Fluid–structure interaction analysis by coupled FE–SPH model based on a novel searching algorithm. *Comput. Methods Appl. Mech. Eng.* **2014**, *276*, 266–286. [CrossRef]
64. Thiyahuddin, M.I.; Gu, Y.; Gover, R.B.; Thambiratnam, D.P. Fluid–structure interaction analysis of full scale vehicle-barrier impact using coupled SPH–FEA. *Eng. Anal. Bound. Elem.* **2014**, *42*, 26–36. [CrossRef]
65. Ko, H.; Cox, D. OSU: 1-D Hydraulic Experiment, Aluminum, Water depth = 249.6 cm. *DesignSafe-CI* **2012**. [CrossRef]
66. Hallquist, J.O. LS-DYNA theory manual. *Liverm. Softw. Technol. Corp.* **2006**, *3*, 25–31.
67. Grimaldi, A.; Benson, D.J.; Marulo, F.; Guida, M. Steel structure impacting onto water: Coupled finite element-smoothed-particle-hydrodynamics numerical modeling. *J. Aircr.* **2011**, *48*, 1299–1308. [CrossRef]
68. Panciroli, R.; Abrate, S.; Minak, G.; Zucchelli, A. Hydroelasticity in water-entry problems: Comparison between experimental and SPH results. *Compos. Struct.* **2012**, *94*, 532–539. [CrossRef]
69. Monaghan, J.J. Simulating free surface flows with SPH. *J. Comput. Phys.* **1994**, *110*, 399–406. [CrossRef]
70. Wu, Z. Compactly supported positive definite radial functions. *Adv. Comput. Math.* **1995**, *4*, 283–292. [CrossRef]
71. Monaghan, J.J.; Lattanzio, J.C. A refined particle method for astrophysical problems. *Astron. Astrophys.* **1985**, *149*, 135–143.
72. Gingold, R.A.; Monaghan, J.J. Smoothed particle hydrodynamics: Theory and application to non-spherical stars. *Mon. Notices R. Astron. Soc.* **1977**, *181*, 375–389. [CrossRef]
73. Das, J.; Holm, H. On the improvement of computational efficiency of smoothed particle hydrodynamics to simulate flexural failure of ice. *J. Ocean Eng. Mar. Energy* **2018**, *4*, 153–169. [CrossRef]
74. Petschek, A.G.; Libersky, L.D. Cylindrical smoothed particle hydrodynamics. *J. Comput. Phys.* **1993**, *109*, 76–83. [CrossRef]
75. Monaghan, J.J.; Gingold, R.A. Shock simulation by the particle method SPH. *J. Comput. Phys.* **1983**, *52*, 374–389. [CrossRef]

76. Xu, J.; Wang, J. Node to node contacts for SPH applied to multiple fluids with large density ratio. In Proceedings of the 9th European LS-DYNA Users' Conference, Manchester, UK, 3 June 2013; pp. 2–4.
77. Thomas, S.; Cox, D. Influence of finite-length seawalls for tsunami loading on coastal structures. *J. Waterw. Port Coast. Ocean Eng.* **2012**, *138*, 203–214. [CrossRef]
78. Swegle, J.W.; Attaway, S.W.; Heinstein, M.W.; Mello, F.J.; Hicks, D.L. An analysis of smoothed particle hydrodynamics (No. SAND-93-2513). *Sandia Natl. Labs* **1994**. [CrossRef]
79. American Society of Civil Engineers (ASCE). *Minimum Design Loads and Associated Criteria for Buildings and Other Structures;* ASCE/SEI 7-16; American Society of Civil Engineers (ASCE): Reston, VA, USA, 2016.
80. Istrati, D.; Buckle, I.G. Tsunami Loads on Straight and Skewed Bridges—Part 2: Numerical Investigation and Design Recommendations (No. FHWA-OR-RD-21-13). Oregon Department of Transportation. Research Section. 2021. Available online: https://rosap.ntl.bts.gov/view/dot/55947 (accessed on 15 July 2021).
81. Stolle, J.; Nistor, I.; Goseberg, N.; Petriu, E. Multiple Debris Impact Loads in Extreme Hydrodynamic Conditions. *J. Waterw. Port Coast. Ocean Eng.* **2020**, *146*, 04019038. [CrossRef]
82. Goseberg, N.; Stolle, J.; Nistor, I.; Shibayama, T. Experimental analysis of debris motion due the obstruction from fixed obstacles in tsunami-like flow conditions. *Coast. Eng.* **2016**, *118*, 35–49. [CrossRef]

Journal of
Marine Science and Engineering

Article

SPH Simulations of Real Sea Waves Impacting a Large-Scale Structure

Corrado Altomare [1,2], Angelantonio Tafuni [3,*], José M. Domínguez [4], Alejandro J. C. Crespo [4], Xavi Gironella [1] and Joaquim Sospedra [1]

1 Maritime Engineering Laboratory, Universitat Politècnica de Catalunya, BarcelonaTech, 08034 Barcelona, Spain; corrado.altomare@upc.edu (C.A.); xavi.gironella@upc.edu (X.G.); joaquim.sospedra@upc.edu (J.S.)
2 Department of Civil Engineering, Ghent University, B-9052 Ghent, Belgium
3 School of Applied Engineering and Technology, New Jersey Institute of Technology, Newark, NJ 07102, USA
4 Environmental Physics Laboratory, Campus Sur, Universidade de Vigo, 32004 Ourense, Spain; jmdominguez@uvigo.es (J.M.D.); alexbexe@uvigo.es (A.J.C.C.)
* Correspondence: atafuni@njit.edu; Tel.: +1-973-596-6187

Received: 18 September 2020; Accepted: 14 October 2020; Published: 21 October 2020

Abstract: The Pont del Petroli is a dismissed pier in the area of Badalona, Spain, with high historical and social value. This structure was heavily damaged in January 2020 during the storm Gloria that hit southeastern Spain with remarkable strength. The reconstruction of the pier requires the assessment and characterization of the wave loading that determined the structural failure. Therefore, a state-of-the-art Computational Fluid Dynamic (CFD) code was employed herein as an aid for a planned experimental campaign that will be carried out at the Maritime Engineering Laboratory of Universitat Politècnica de Catalunya-BarcelonaTech (LIM/UPC). The numerical model is based on Smoothed Particle Hydrodynamics (SPH) and has been employed to simulate conditions very similar to those that manifested during the storm Gloria. The high computational cost for a full 3-D simulation has been alleviated by means of inlet boundary conditions, allowing wave generation very close to the structure. Numerical results reveal forces higher than the design loads of the pier, including both self-weight and accidental loads. This demonstrates that the main failure mechanism that led to severe structural damage of the pier during the storm is related to the exceeded lateral soil resistance. To the best of the authors' knowledge, this research represents the first known application of SPH open boundary conditions to model a real-world engineering case.

Keywords: fluid–structure interaction; waves; smoothed particle hydrodynamics; SPH; Pont del Petroli; storm Gloria

---

1. Introduction

In January 2020, the sea storm Gloria [1] struck the Mediterranean coasts of Spain and France with remarkable strength. The combination of extreme wave conditions, wind velocities, and the long event duration caused severe damage to assets and infrastructures. Many coastal platforms suffered from extensive damage due to the impact of big waves and fatigue. One of these is the Pont del Petroli, a pier located in Badalona, in the northern area of Barcelona, Spain. It is a structure integrated into the coastline with high historical, scientific and social value, and it was heavily damaged by the storm Gloria. Changes in the bathymetry of the area where the pier is built, together with extreme wave conditions that were not accounted for in the design and building process of the pier, led to stresses that the structure was eventually unable to withstand. As a result, the main platform of the pier suffered from serious damage, most likely due to the induced shear stress in the concrete and exceeded soil bearing capacity. Moreover, one of the beams that form the footbridge was completely washed away.

Parts of the damaged structure were removed after the storm for safety reasons. The local authorities of Badalona intend to repair and rebuild the pier mainly for its social and historical value to the city of Barcelona. However, there is a lack of knowledge regarding the exact weather conditions that led to such high damage during storm Gloria.

With the aim to cover this gap, an experimental campaign is currently under development at the Maritime Engineering Laboratory at Universitat Politècnica de Catalunya—BarcelonaTech (LIM/UPC), in Barcelona, Spain. The main goal is to reproduce the loads exerted on the pier by wave conditions similar to those that manifested during the sea storm Gloria. The pier will be modeled in the large-scale wave flume of the Maritime Engineering Laboratory at UPC. In preparation for the experimental campaign, numerical simulations have been chosen as a means to obtain accurate predictions of water flow around the pier under certain extreme weather conditions.

The use of Computational Fluid Dynamics (CFD) for enabling the study of fluid–structure interaction (FSI) and wave–structure interaction (WSI) problems, as well as the structural analysis of onshore and offshore facilities, has a long history. The first methods were based on potential flow theory [2–6], usually resorting to a velocity potential that satisfies a set of simplified governing equations (often times represented by the Euler equations) throughout the fluid domain. The employment of these techniques would request assumptions such as irrotational, inviscid fluid flow, linear or nonlinear wave theory and small displacements. However, the solution of the Navier–Stokes (NS) equations, with the incorporation of an accurate treatment for the viscous terms as well as the air–water interface, is generally required for WSI problems involving violent breaking and extreme wave loads. These equations can generally be discretized on a 2- or 3-D grid by the use of well-established techniques such as the Finite Volume (FV) or Finite Element (FE) methods. With the advent of new and more powerful hardware, including Central Processing Unit (CPU) and Graphics Processing Unit (GPU), the efficiency and effectiveness of the aforementioned methodologies have improved significantly, allowing the study of real-life problems [7–11].

In the last few decades, numerous additional CFD algorithms have been devised and perfected to address the ever increasing complexity of computer-aided simulations for the solution of challenging engineering problems. Among these is Smoothed Particle Hydrodynamics (SPH) [12–14], a fully Lagrangian meshless method that has been adapted from an original astrophysics framework to subsequent use in problems of free-surface hydrodynamics [15–17]. SPH has numerous advantages when simulating problems in the context of offshore and other marine structure analyses. Due to their fully Lagrangian nature, SPH particles do not require explicit routines that model advection, therefore the treatment of free-surface boundary conditions becomes seamless also in the case of large deformations, such as during the breaking process of a wave. Moreover, SPH has excellent conservation properties, not only for energy and linear momentum, but also for angular momentum. These and other aspects make SPH particularly suited for studying WSI and FSI problems [18,19]. In [20], several test cases were investigated, including a dam break flow past a fixed rigid column and the dynamical response of a floating object to the impact of incoming waves. The results suggest excellent agreement of SPH simulations with experimental benchmark data. In the context of marine structures and their dynamical response to large wave loads, the work of [21] proposes a hybrid SPH-FE method where the strength of SPH in simulating complex free surface flow is mixed with the high accuracy of the FE method in simulating the dynamics of the structure. Therein, a dam break flow was also investigated but with the presence of a flexible structure downstream, showing very good agreement with other methods. The authors of [22] build on the idea of coupling with other methods by employing a state-of-the-art SPH code together with a lumped-mass mooring dynamics model for the simulation of floating moored devices in regular waves. The comparison with experiments suggests that SPH is an excellent candidate for simulating freely and moored floating objects undergoing WSI. However, the authors anticipate the need for more work to cover a larger range of sea states, with an emphasis on irregular waves. All the aforementioned cases employ a one-phase flow approach, where the liquid is generally (though not always) modeled as a weakly-compressible fluid while the gas phase is neglected. Nevertheless, the capability of SPH of handling two-phase flow is also an active area of research with significant contributions available in the literature. The interested reader is referred to the following studies for a representative (though non-exhaustive) list of computational analyses with multi-phase SPH models in the contest of FSI and WSI [19,23–25].

In the present work, the open-source SPH code DualSPHysics [26] was employed to simulate the local wave conditions that determined the failure of the Pont del Petroli, and to analyze the wave–structure interaction. In particular, DualSPHysics was employed to characterize the wave loads on each pier element, with particular attention to those that were severely damaged during the sea storm Gloria. DualSPHysics is currently one of the most advanced CFD models based on SPH, with the latest developments significantly improving the efficiency of the solver for use on GPU and multi-GPU, as well as for coupling with other schemes and techniques [27–29]. Results from this work will be instrumental in the subsequent experimental campaign that will be carried out for the final design of the upgraded Pont del Petroli.

A description of the damage brought by storm Gloria to the Pont del Petroli and the possible source of it is presented next.

2. Study Case

2.1. The Pont del Petroli Pier

The Pont del Petroli was built in the 1960s to enable the transfer of oil from tankers onto land. The pier extends for approximately 250 m at sea near the Badalona sandy beach, close to Barcelona, Spain (Figure 1). The pier went out of industrial usage in 1990. In 2001, there was a popular opposition to its demolition and a proposal to adapt the pier for public use. The Pont del Petroli was then handed over to the Badalona City Council in 2003, with the council starting construction work in 2009 to repurpose the jetty into a leisure facility for public use. Taking advantage of this remodeling, LIM/UPC designed, integrated and installed scientific equipment all along the pier, effectively creating a base for sea and atmospheric monitoring. Because of this, the pier is now the only of such equipped coastal stations on the Catalan coast, unique on the Eastern coast of Spain, and a very rare instance in the entire Mediterranean sea. The pier has a front platform located 6 m above the mean water level that can be reached via a 3 m-wide and 240 m-long footbridge. The platform consists of a 9.75 m $\times$ 6.75 m armored concrete slab supported by perpendicular beams, also built in concrete. The footbridge structure is made of π-shaped concrete beams, each being 15 m long, and is supported by one pile cap on each extreme. Between 2 and 4 metallic 14 inch diameter piles are connected to the pile caps to support the whole structure. In total, there are 16 groups of piles, 14 of which with only two piles connected to the pile cap, while the remaining two are connected with four piles. A total of 20 piles are placed below the platform, connected to the supporting beams.

Sketches of each element are depicted in Figure 2. The piles below the pile caps present an inclination of 7.5 degrees with respect to the vertical direction. The piles are embedded about 6 m deep into the sandy bottom, as detected from the original design report. The main exposed area, the self-weight and the total distributed load for each element that forms the pier are reported in Table 1. All values are extracted from the original report on the structural design of the pier and have been checked carefully. The water depth and beach profile at the Pont del Petroli location changed significantly since its date of construction. Initially conceived with a water depth of 12 m at the deepest point (i.e., at the toe of the platform), the pier experienced radical changes related to both the water depth and bottom slope. LIM/UPC carried out 19 bathymetric surveys in the period between 2011 and 2020, the last one being carried out right after storm Gloria. Results from these surveys are shown in Figure 3, where the original profile from the design report is highlighted in blue while the profiles right before and after the passage of storm Gloria are indicated in red. The x-distance is measured from the land-side—the beginning of the footbridge. The platform toe is located at $x = 240$ m. The most seaward footbridge beam and the platform are located between $x = 216$ and $x = 240$ m. The vertical axes express the distance from the mean water level, being positive if upwards. While for the rear part of the pier, from the fourth pile cap to the shoreline, the bottom slope did not show remarkable modifications and remained on average between 1:30 and 1:25, the front and most seaward part, from the fourth pile cap to the platform, showed the greatest change, especially in terms of the local water depth. The local sand accretion determined a reduction of the water depth from the initial 12 to almost 9 m. While an almost linear trend had been observed during the past years, a sudden change was surveyed after storm Gloria in early 2020. The measured water depth at the toe of the platform was observed to be 8 m, hence 4 m less than the design value.

Figure 1. 1980s (**left**) and 2017 (**right**) aerial photos of the Pont del Petroli. Originally constructed for industrial purposes, the pier is now open to the public. (Sources: Concepció Riba).

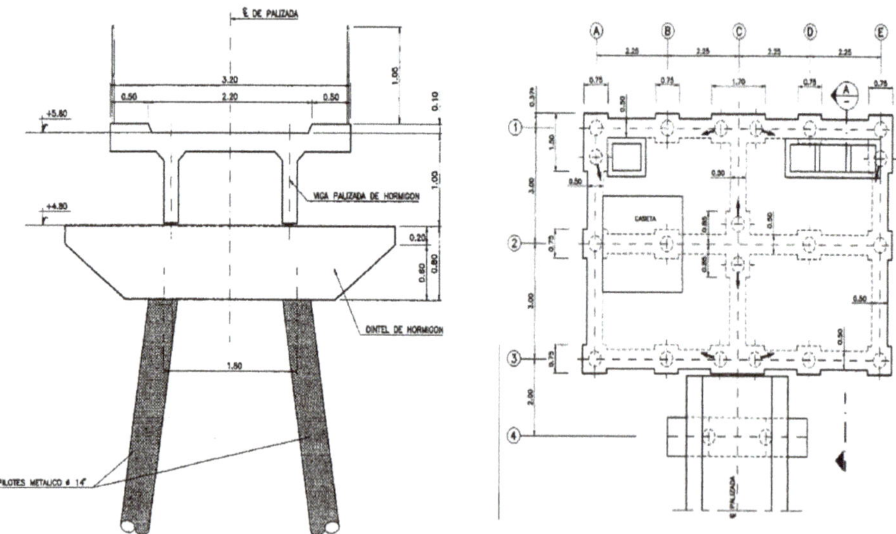

Figure 2. Geometrical details and top view of the platform, beam and pile cap forming the Pont del Petroli pier. The sketches are extracted from the first design report and provided by the Maritime Engineering Laboratory of Universitat Politècnica de Catalunya-BarcelonaTech (LIM/UPC) for the present manuscript. The sketches belong to the original project deposited in the 1965 archive "Colegio de Ingenieros de Caminos, Canales y Puertos de Madrid", PN 9280/65, under José Entrecanales Ibarra.

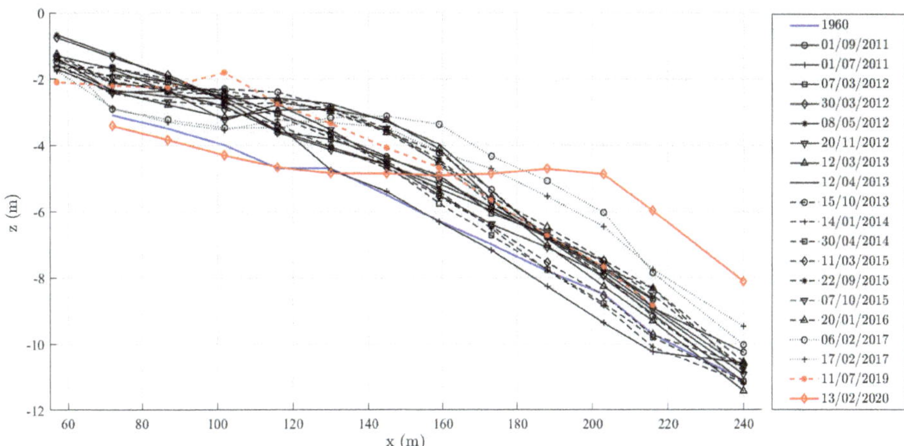

Figure 3. Beach profiles at Pont del Petroli. The original beach profile from the design report is indicated by a blue line. In red, the two profiles surveyed by LIM/UPC before and after storm Gloria.

Table 1. Main characteristics of the Pont del Petroli pier elements.

Element	Main Exposed Area (Volume)	Self-Weight	Distributed Vertical Load (Self-Weight + Accidental Load)	Final Design Vertical Load
Platform	9.75 m × 6.75 m	32.90 tons	2.00 tons/m^2	131.62 tons
π-shaped beam	15.00 m × 3.20 m	29.25 tons	3.60 tons/m^2	54.00 tons
Pile cap (2 piles)	4.40 m × 0.8 (×1.2) m	9.12 tons	12.30 tons/m^2	54.12 tons
Pile cap (4 piles)	4.40 m × 0.8 (×2.0) m	15.20 tons	15.80 tons/m^2	60.20 tons

2.2. Storm Gloria: Description and Damage to the Pont del Petroli

Since its refurbishment and opening to the public in 2009, the Pont del Petroli pier has been subjected to several storms, some of which caused serious damage and a subsequent need for major repair work (i.e., storms in 2107 and 2020). In particular, the sea storm Gloria left the pier badly damaged, forcing the local government to close public access to it. The pier platform and the first beam attached to it experienced the most severe damage. The platform was partly destroyed and the beam was washed away together with the first pile cap.

It is difficult to characterize the wave climate in the area surrounding the Pont del Petroli. The closest buoy to the pier is located at Puertos del Estado [30], located just outside the Barcelona harbor (Lat. 41.32° N, Long. 2.20° E), south of Badalona. However, the buoy stopped recording data during storm Gloria. The latest available data were recorded on 19 January 2020 at 21:54, Barcelona time: the maximum recorded wave height was 5.64 m. The most complete and available information to date on the characteristics of storm Gloria can be found in the readings by the buoy of Capo Begur (Lat. 41.90° N, Long. 3.66° E), several kilometers north-east of Badalona (Figure 4). At the same time that the Barcelona buoy stopped working, the maximum wave height recorded at by the Capo Begur buoy was 6.8 m, approximately half of the 13.8 storm peak occurred (and recorded) 1 day later. The significant wave height, $H_{m,0}$, recorded at Capo Begur was higher than 6.0 m for 44 h, with a peak value of 7.8 m. The main wave and wind direction was east-northeast. Of all the recordings from the Capo Begur buoy in the period between 2001 and 2017, only a handful of them show wave heights close to or larger than 7.0 m, as the values recorded during storm Gloria: $H_{m,0} = 7.4$ m from the north in 2003, $H_{m,0} = 7.2$ m from north-east in 2010 and $H_{m,0} = 6.9$ m from north-east in 2016. These values notwithstanding, there is no trace of extreme data similar to that of storm Gloria, especially in terms of the duration of the storm peak and the strong winds (greater than 70 km/h).

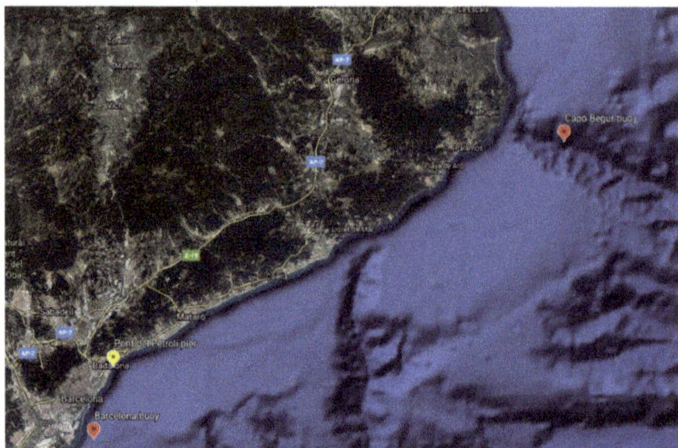

Figure 4. View of the Catalan coast between Barcelona and Capo Begur. (Map Data: Google Earth, SIO, NOAA, U.S. Navy, NGA, GEBCO).

The intensity of the sea storm Gloria caused damage to the measurement station installed by LIM/UPC at Pont del Petroli. Hence, there is almost no direct recorded information of the wave climate close to the pier in that occurrence. Due to the lack of local data, LIM/UPC carried out a study where the wave propagation was simulated by means of the numerical model SWAN [31] for several stretches of the Catalan coast between Barcelona and the town of Blanes, in Costa Brava. The results from the wave propagation model were used to characterize the wave climate at a water depth between 20 and 25 m; values that will be employed for the wave generation in the planned experimental campaign. The results showed values of $H_{m,0}$ ranging between 5.5 and 6.5 m. A visual representation of waves during the storm Gloria was made possible by pictures and videos from the media, press release and photo amateurs, suggesting that waves as large as 7 m and higher reached and hit the Pont del Petroli during the storm, see for example Figure 5.

Figure 5. A large wave hitting the platform of the Pont del Petroli during storm Gloria. (Source: Badalona City Council).

3. Numerical Model

3.1. The DualSPHysics Code

DualSPHysics [26] is an open-source code based on Smoothed Particle Hydrodynamics (SPH) that has been developed to study real engineering problems. Its high computational efficiency is mixed with its ability to be executed on both CPU and GPU with powerful parallel computing capabilities. The underlying rationale behind SPH is to discretize the fluid with a set of particles, whose physical quantities (position, velocity, density and pressure) are obtained via interpolations of the same quantities evaluated at the surrounding particles [17]. The weighted contribution of these neighbor particles is accounted for using a kernel function, W, with an area of influence that is defined using a characteristic smoothing length, h. In DualSPHysics, the quintic Wendland kernel [32] is used and defined to vanish beyond $2h$. Particles are initially separated by a uniform particle distance, dp, which is also used as a reference value to define the (constant) smoothing length. In this work, $h/dp = 2$, so that $2h = 4dp$, giving an idea of the number of neighbors per particle, at least during the initial time step.

The Navier–Stokes equations can be written in a discrete SPH formalism using $W_{a,b}$ as the kernel function, which depends on the normalized distance between particle a and the neighboring particles $b = 1, \ldots, N_a$:

$$\frac{d\mathbf{r}_a}{dt} = \mathbf{v}_a \tag{1}$$

$$\frac{d\mathbf{v}_a}{dt} = -\sum_{b=1}^{N_a} m_b \left[\left(\frac{p_b + p_a}{\rho_b\, \rho_a} \right) \nabla_a W_{a,b} - \Pi_{a,b} \right] + \mathbf{g} \tag{2}$$

$$\frac{d\rho_a}{dt} = \sum_{b=1}^{N_a} m_l \mathbf{v}_{ab} \cdot \nabla_a W_{a,b} + 2\delta h c_a \sum_{b=1}^{N_a} m_b \left(1 - \frac{\rho_a}{\rho_b} \right) \frac{\mathbf{r}_{a,b} \cdot \nabla_a W_{a,b}}{|\mathbf{r}_{k,l}|^2} \tag{3}$$

Here t is time, $\mathbf{r}$ is the position vector, $\mathbf{v}$ is the velocity vector, p is the pressure, ρ is the density, m is the mass, c_a is the speed of sound at particle a and g is the acceleration of gravity. For the term $\Pi_{a,b}$, the artificial viscosity proposed in [13] is used. Specifically, the constants suggested in [33,34] are chosen to guarantee a correct wave propagation. Moreover, the density diffusion term described in [35] is chosen, with a value $\delta = 0.1$ as recommended therein. The fluid is treated as weakly compressible, with an equation of state employed to calculate the pressure of the fluid as a function of its density. Hence the system in Equations (1) to (3) is closed by using:

$$p_a = \frac{{c_a}^2 \rho_0}{\gamma} \left[\left(\frac{\rho_a}{\rho_0} \right)^\gamma - 1 \right] \tag{4}$$

where $\gamma = 7$ is the polytropic exponent and ρ_0 is the fluid reference density. The speed of sound was set to be ten times the maximum fluid velocity, keeping density variations within 1% of ρ_0 and therefore preventing the introduction of major deviations from an incompressible approach. The calculated speed of sound at the start of the simulation was approximately 110 m/s, with an average 5 steps per second of simulation (about 242,000 steps total) and a runtime per physical second equal to 1600 s on average. The symplectic position Verlet time integrator scheme [36], which is second-order accurate in time, was used here to perform time integration of flow quantities. A variable time step was calculated according to the procedure in [37], involving the Courant–Friedrich–Lewy (CFL) condition, the force terms and the viscous diffusion term.

The solid boundary conditions, needed for idealizing the seabed and the coastal structures, are discretized by a set of boundary particles that differ from the fluid particles. The Dynamic Boundary Particles (DBPs) [38] are boundary particles that satisfy the same equations as fluid particles, however, they do not move according to the forces exerted on them. Instead, they remain either fixed in predefined position or move according to an imposed/assigned motion function (i.e., for moving objects like wave-makers). When a fluid particle approaches a boundary particle, and thus the distance between the two becomes smaller than the interaction distance, the density of the affected boundary particles increases, resulting in a pressure increase. This, in turn, results in a repulsive force being exerted on the fluid particle due to the pressure term in the momentum equation. DBPs have been successfully used for marine and coastal engineering problems [33,39–41] due their capability of discretizing complex 3-D geometries without the need of implementing cumbersome mirroring techniques [42] or complex semi-analytical wall boundary conditions [43].

One option to achieve wave generation in SPH would see the employment of a moving boundary that mimics the displacement of a wavemaker in experimental wave tanks. This approach has been used extensively in the literature [34,44–46]. Although the generation of waves with a wavemaker-like moving boundary is robust and often accurate, a computational domain size of at least 3 to 4 wavelengths is needed to accurately resolve the physics [47]. Combining this requirement with the need for an adequate level of resolution to properly capture the free surface deformation leads to limitations on the physical time that can be simulated, mainly due to the extremely large number of particles that occur. In light of these considerations, a technique based on open boundary conditions was herein adopted for the generation of waves, specifically by enforcement of user-defined inlet conditions. This novel approach has first been introduced in DualSPHysics in [29] and has proven successful in reducing the computational domain size down to only one wavelength. The interested reader can find detailed information on the SPH inlet/outlet algorithm in [42] and its application to wave generation in [29,46].

3.2. Validation

As described in previous sections, experimental testing on the Pont del Petroli pier is scheduled to follow the numerical modeling phase described in this work. No previous attempts at simulating the case proposed herein could be found in the literature, therefore it is imperative to assess the capability of DualSPHysics to tackle this problem by means of other similar research published in the literature, with a special focus on wave–structure interaction. In all the references listed in this section, the numerical results obtained with DualSPHysics are compared with reference solutions such as experimental data, analytical solutions and other model solutions, proving that the implemented SPH model is accurate when predicting wave loads, wave run-up, etc. The model setup for the numerical work carried out herein employed default or similar parameter settings to those reported in [33,39,48].

The work in [44] is one of the first examples where wave–structure interaction was studied with a forerunner of DualSPHysics, i.e., SPHysics [49,50] including proper validation with experimental data. Therein, numerical results were compared with field measurements of the movement of a caisson breakwater under the forcing of periodic waves. Promising agreement with experimental data was obtained for the displacement and the horizontal forces on the caisson. Modeling of wave loading on coastal structures was also presented in several other works [39,51–53]. The impact of tsunami waves was studied in [54,55] using DualSPHysics, showing a good agreement between numerical data and physical tests. Moreover, different theories for the generation of solitary waves have been tested in [51], with satisfactory agreement obtained between numerical surface elevations and wave loads with respect to experimental data. The authors of [56] present the first successful application not only of DualSPHysics but, more generally, of a SPH-based 3-D model for studying wave run-up on a real coastal defense and resolving fluid trajectories in between the breakwater armor blocks. Later, DualSPHysics was employed in the analysis of wave run-up for the design of coastal defenses. The authors of [57] first validate run-up time series with experiments using a porous breakwater for which an exact geometry was used. The authors of [58] employed the same model to deepen the knowledge of the influence of the curvature in a dike line on wave run-up, after proper validation against 3-D physical model tests from a wave basin facility. Finally, the interaction between waves and floating objects was studied in detail by [22], where the numerical results of nonlinear waves interacting with freely and moored floating objects are compared with experimental data. Good agreement was obtained for the motions of the floating body (heave, surge and pitch) and the mooring tensions.

3.3. Model Set-Up

The numerical model was conceived of to prepare the experimental testing that will be carried out at LIM/UPC. Therefore, the design of the different simulations was done in a way to accurately mimic the experimental facility where the physical testing will be carried out. The experimental campaign will be executed in the large-scale wave flume CIEM at LIM/UPC. The flume is 100 m long and 3 m wide, and equipped with a wedge-type wavemaker. Tests will be carried out with an initial water depth of 2.6 m, measured close to the wavemaker location. The water depth was chosen as a compromise between the technical capabilities of the wavemaker and the model scale. The model of the Pont del Petroli pier will be reproduced at a geometrical scale of 1:10, based on the Froude similitude. A beach with 1:15 slope extends between $x = 27.80$ m and $x = 60.13$ m, $x = 0$ being the position of the wavemaker at rest. The pier model will be placed starting from $x = 54.00$ m, being this the location of the front platform. At this location, the local water depth is 0.84 m, corresponding to 8.4 m in the prototype, similar to the post-Gloria survey. The rear part of the foreshore consists of a 1:25 slope up to the back of flume. A sketch of the model layout is depicted in Figure 6. The pier will be built with all the original details, from the platform to the third beam that forms the footbridge, being this the part of the structure that was exposed to the waves impact. The piles will not be modeled to follow a conservative approach where the loading on the platform, footbridge and pile cap will be maximized, skipping the partial sheltering effect of the piles. The main focus of the numerical analysis is to measure the water load on the platform, the first pile cap and the first footbridge beam. These are the most seaward elements that experienced heavy damage during storm Gloria. The pier model will be equipped with load cells and pressure sensors during the experimental campaign. The numerical model will thus provide important and necessary information for the measurement design and setup of the instrumentation.

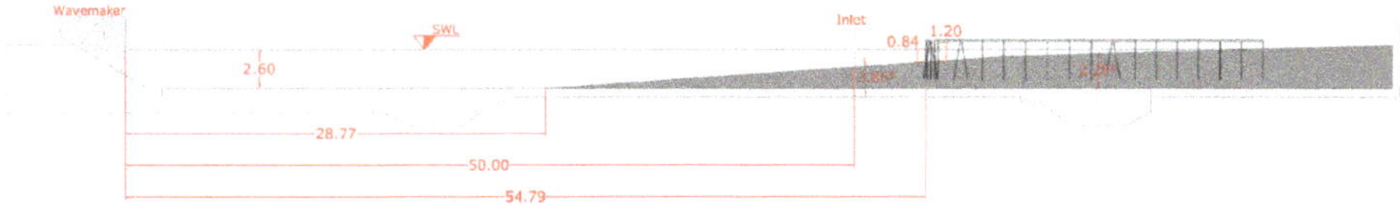

Figure 6. Sketch of the experimental layout. The 2-D numerical model mimics this layout to a great extent. Conversely, the 3-D model has an open boundary for wave generation at the inlet, i.e., at $x = 50$ m.

Prior to tackling the full 3-D problem, a 2-D model of the CIEM flume and Pont del Petroli pier was realized with DualSPHysics. The scope of this first analysis is twofold. On the one hand, it provides preliminary information on the expected loads on each pier element. On the other hand, it allows for studying the wave transformation and breaking process on the 1:15 beach slope for different wave conditions and initial water depths. The water surface elevation and velocity field that are needed as a forcing boundary condition in the full 3-D case are thus extracted from the 2-D cases. Waves are generated in the 2-D study mimicking the experimental wedge-type wavemaker. The whole flume length is therefore reproduced. An initial interparticle distance of $dp = 0.02$ m is employed. Such a resolution is chosen as a result of a sensitivity analysis where the dp parameter varies between 0.01 and 0.04 m. Simulations were run on a NVIDIA GeForce RTX 2080 (2944 CUDA cores, 1.80 GHz maximum clock rate). The total number of SPH particles was equal to 304,805. The runtime per physical second was approximately 71 s. To reduce computational time and thus avoid simulating very long irregular wave trains (usually equal to 1000 waves), only regular waves were modeled. Irregular waves will be simulated during the experimental campaign. Due to the lack of local data against which validating the wave propagation model, a range of wave conditions have been considered, i.e., wave height H varying between 6.1 and 9.0 m and wave period T varying between 9.6 and 12.7 s (these values are expressed in prototype scale). When modeling the entire flume length in 3-D, a particle resolution equal to 2 cm would lead to an excessive number of SPH fluid particles, approximately 150 times the number of particles in the 2-D case. This would imply a very expensive and time consuming computation despite the availability of powerful hardware. Therefore, an inlet boundary condition is employed to tackle the 3-D analysis. As seen in Figure 6, the inlet is located at $x = 50$ m, right before the breaking zone for the tested conditions, and only 4 m (40 m in real scale) from the pier head. At the inlet buffer, the water surface elevation and velocity have been enforced. The initial number of SPH fluid particles is equal to 1,729,260, with this number varying slightly during the simulation. About 831,000 new particles are generated during the execution. A physical time of 20 s was simulated, corresponding to 2–3 generated wave impacts on the structure. The runtime per physical second is about 26.7 min, varying slightly in each simulated test. For the 3-D model, a NVIDIA GeForce RTX 2080 Ti has been employed (4635 CUDA cores, 1.63 GHz maximum clock rate).

The water surface elevation to be imposed at the inlet has been measured from the 2-D model. Initially, the velocity has been calculated based on the linear shallow water theory:

$$u(t) = \eta(t)\sqrt{\frac{g}{d}} \tag{5}$$

where $\eta(t)$ is the water surface elevation, g is gravity and d is the initial water depth at the inlet position. Equation (5) returns a uniform velocity profile along the whole water depth. The value of the water depth at the inlet is equal to 1.16 m, expressed in model scale. The orbital velocity has been sampled in the 2-D model via 10 measurement points along the water depth, from the bottom to the free surface. These sampled velocities have been compared with the ones calculated via Equation (5), with lower peaks obtained for the SPH velocities. Therefore, Equation (5) has been corrected. The final equation for the velocity is as follows:

$$u(t) = \eta(t)\sqrt{\frac{g}{\eta(t) + d}} \tag{6}$$

Using findings from the 2-D simulations, two different wave conditions were selected as the worst case scenario for the case with a 8 m water depth at the platform toe. Values of wave height and period are reported in Table 2. Case A has a wave height of 0.65 m and period of 3.80 s in model scale, corresponding to real scale values of 6.5 m and 12.0 s respectively. For Case B, $H = 0.82$ m and $T = 4.00$ s, corresponding to $H = 8.2$ m and $T = 12.7$ s in real scale values, respectively. To investigate the variability of the forces exerted by the waves in a wider range of wave velocities, three different test cases were defined for each wave condition. Specifically, for each test the velocity calculated from Equation (6) was multiplied by a scaling factor, see Table 2.

Table 2. 3-D test cases chosen for the Smoothed Particle Hydrodynamics (SPH) model.

Test Case	Wave Conditions at Generation in 2-D (in Prototype Scale)	Scaling Factor for $u(t)$ from Equation (6)
A1	$H = 6.5$ m, $T = 12.0$ s	1.00
A2		1.05
A3		1.10
B1	$H = 8.2$ m, $T = 12.7$ s	1.00
B2		0.95
B3		1.05

4. Results and Discussion

Four snapshots of the results from the simulation of test cases A1 and B1 are depicted in Figures 7 and 8, respectively, corresponding to the impact of the first simulated wave on the pier. The colors represent the magnitude of horizontal velocity. The two test cases differ in the wave breaking process. In A1, the wave is still shoaling on the 1:15 (Figure 7a) when it bumps into the pier platform (Figure 7b). In B1, the wave starts to break before the structure. A characteristic plunging breaker profile can be identified in Figure 8a: the wave crest, characterized by very high velocities, is curling over and dropping onto the wave trough. During this process, the plunger encounters the pier platform where it is transferring most of its momentum (Figure 8b). Large splashes are produced as seen in Figure 8c, and part of the wave energy is finally transferred above the platform deck for wave overtopping (Figure 8d). The waves overtopping the platform are then falling onto the platform, exerting a downwards force on the platform deck. Wave overtopping can also be noted for case A1 (Figure 7c), though volumes are smaller than those in case B1.

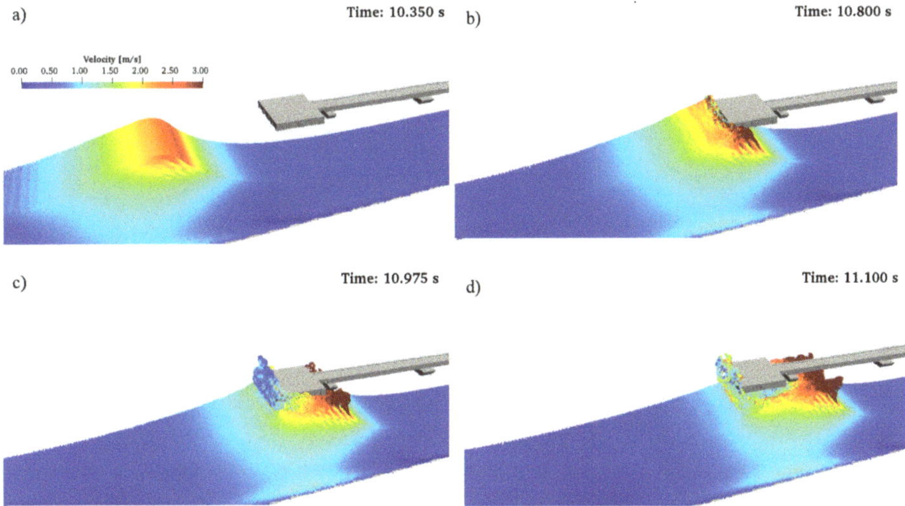

Figure 7. Snapshots of the SPH horizontal velocity contours for the test case A1. (**a**) Time: 10.350 s; (**b**) Time: 10.800 s; (**c**) Time: 10.975 s; (**d**) Time: 11.100 s.

Horizontal and vertical forces were measured on the platform: the first pile cap and the first footbridge beam. The beam was washed away during storm Gloria. The SPH model results will help understanding the mechanism that led to the failure. The Badalona City Council has commissioned a survey right after the storm to check the status of the pier, also including those elements that fell into the sea, namely the first pile cap (including the two piles) and the first footbridge beam. From the damage report, it appears that the whole system was composed of two piles and the pile cap failed most probably because the bearing capacity of the soil was exceeded by the action exerted by the waves on the piles and the pile cap. The survey revealed that the pile cap is still connected to the piles and that the whole system hit the base of the second group of piles after them, and is currently lying on the sea bottom. The bending of the system composed by the two piles and the pile cap towards the beach freed the footbridge beam that moved "sliding" towards the base of the platform, bumping onto the two piles supporting the platform and damaging them. The footbridge beam does not show any damage and is lying on the sea bottom. This fact confirms its failure due to rigid body displacement. It is very likely that the sliding of the beam was also eased by the vertical uplifting force exerted by the waves under the beam, which reduced its effective weight, and hence the friction between the beam and the pile cap. Despite the aforementioned considerations, the damage report does not present any calculation supporting the hypothesis of a bearing failure and consequent sliding of the beam into the sea.

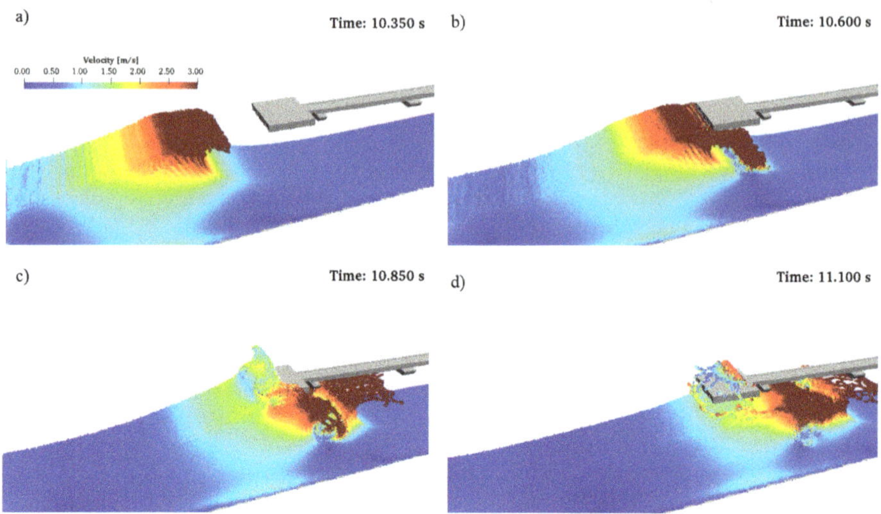

Figure 8. Snapshots of the SPH horizontal velocity contours for the test case B1. (**a**) Time: 10.350 s; (**b**) Time: 10.600 s; (**c**) Time: 10.850 s; (**d**) Time: 11.100 s.

Forces measured by DualSPHysics are initially normalized by the total vertical load used for the design of each pier element, including self-weight and accidental load, as reported in Table 1. Horizontal and vertical forces on the platform are shown in Figure 9 for all test cases. The time is normalized by the wave period T on the x-axis. For the vertical forces, the normalized value is lower or very close to 1, the worst case corresponding to case B3. The forces exerted by the waves are therefore lower than the ones used for the platform design. The horizontal forces are very high, up to 2.5 times the vertical force. The horizontal force is mainly exerted on the front concrete beam located below the platform deck and will be partly transferred to the pile cap right after the platform. The platform front beam did not show heavy damage, therefore it can be argued that the platform was stiff enough to withstand a similar load. The vertical force on the first footbridge beam is depicted in Figure 10, normalized by the design total load. The measured vertical force is lower or equal to the design load, while the horizontal force is negligible. Nevertheless, the first beam was actually washed away during storm Gloria. The load exerted on the pile cap are also plotted in Figure 10. Contrarily to the observations made for the pier platform, it can be seen that test cases A1–A3 led to the maximum loads on the pile caps. As argued before, it is possible that the failure of the beam is likely a consequence of the failure of the first pile cap. Based on the results of the damage report from the Badalona City Council, it is highly probable that the failure of the pile cap was a consequence of the bearing failure at the foundation of the two piles beneath it. Therefore, instead of using the deign vertical loads to normalize the forces exerted on the pile cap, a different procedure was followed.

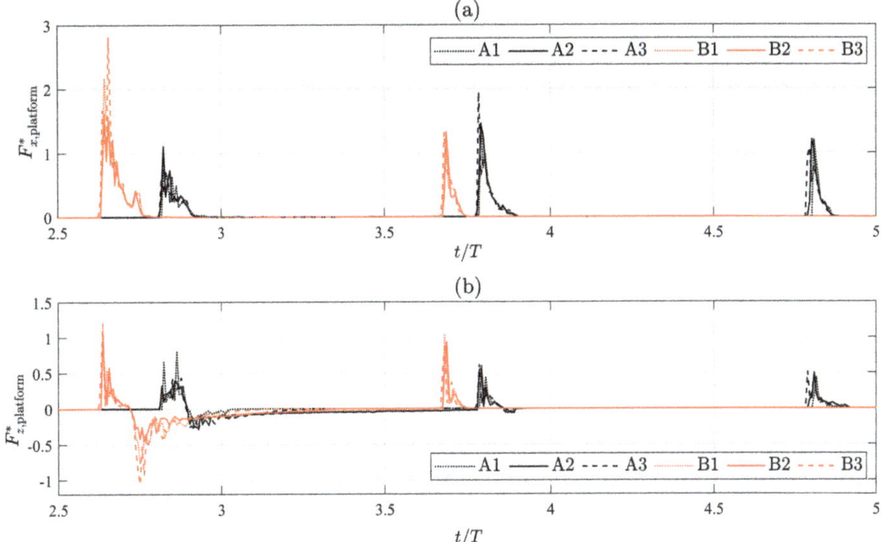

Figure 9. Time history of the normalized (**a**) horizontal and (**b**) vertical forces on the pier platform for different tested wave conditions.

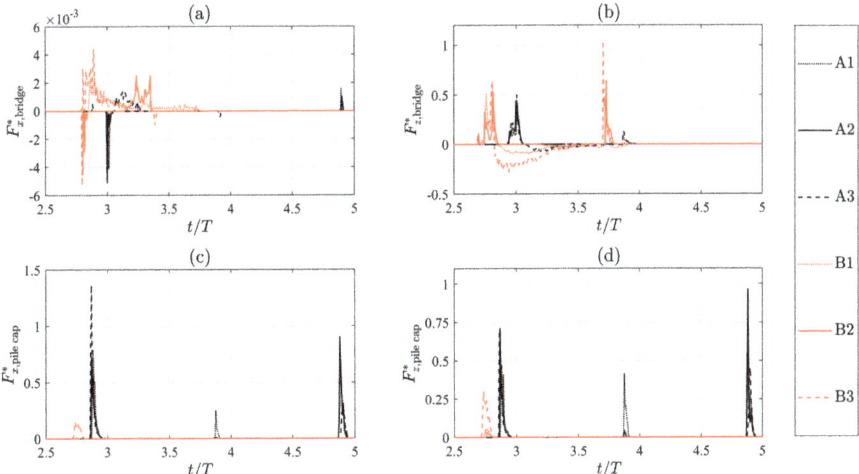

Figure 10. Time history of the normalized forces on the beam and pile cap for different tested wave conditions: (**a**) horizontal force on the pier beam, (**b**) vertical force on the pier beam, (**c**) horizontal force on the pile cap and (**d**) vertical force on the pile cap.

4.1. Derived Pile Axial Loads and Soil Bearing Capacity

The vertical and horizontal forces exerted by the waves on the pile cap will eventually generate new stresses on the piles and pile foundation. Even though the piles are not modeled numerically, it is always possible to calculate the forces and moments induced by the waves on the pile base and foundation. The value of the pile tolerable load is reported in the pier design report: specifically, the tolerable normal load calculated at the Ultimate Limit State (N_U) is equal to 81.80 metric tons. The bearing capacity of the soil foundation is calculated in terms of the maximum axial load. This load is evaluated considering the total vertical load coming from the pile cap and the footbridge beams on each pile, and the wind action. The latter is the only horizontal action considered during the design and acts perpendicularly to the footbridge, hence perpendicularly to the considered wave action. The wind generates a moment that is transferred from the pile cap to the piles (and to their foundation) as a pair of axial forces. The pile driving into the soil is equal to 6 m, as reported from the design. The method proposed by [59] is employed to calculate the bearing capacity at each pile, Q_f. This amounts to 107 metric tons.

In the design report, the wave action was neglected, most likely due to considerations that the piles are very slender and, consequently, the inertia or drag forces were assumed to be negligible. Furthermore, the action of the waves on the pile caps was not foreseen. When the design and remodel had been performed, the water depth below the platform and footbridge was actually larger than the one surveyed after storm Gloria. The deeper water and the unforeseen conditions generated by storms like Gloria might justify the choice to neglect any action on the pile caps, which are located 6 m above the mean water level. Having neglected the action of waves on the pile cap, the lateral resistance to the piles is not assessed. Waves acting on the pile cap generate a moment that is transferred from the piles to the soil. This bending actions should be balanced by the lateral soil resistance, calculated for cohesionless soils like sand. Here, a simple method proposed by [60] is employed for a characterization of the lateral resistance. This is evaluated in terms of ultimate horizontal load that the piles driven in the sand can bear. The expression for the ultimate horizontal load, H_U, is:

$$H_U = \frac{DL^3 K_p \gamma}{2(e+L)} \tag{7}$$

where L is the embedded pile length into the sand, e is the eccentricity of the horizontal load, D is the pile diameter, γ is the effective soil weight and K_p is the passive earth pressure coefficient. The values of γ and K_p are assessed based on the standard penetration test (SPT) N value, as included in the design report, resulting in $\gamma = 1750$ kg/m^3 and $K_p = 4.40$. The value of L is 6 m and the eccentricity happens to correspond to the level arm, i.e., the distance between the pile head and base, equal to 12 m in the present case. The calculated value of H_U is 32.82 metric tons (for a 2 pile cap system).

The normal load at the pile base can be derived from the forces on the pile cap as follows:

$$N_{\text{waves}} = \frac{1}{2} \frac{\left(V_{\text{pile cap}} - W_{\text{pile cap}} - \frac{1}{2} W_{\text{beam}}\right)}{\cos 7.5^\circ} \tag{8}$$

where $V_{\text{pile cap}}$ is the vertical force measured on the pile cap and exerted by the wave action, $W_{\text{pile cap}}$ is the weight of the pile cap and W_{beam} is the footbridge beam weight. The vertical load exerted by the waves is actually directed upwards, similarly to an uplifting force, whereas the weight of the structure is acting downwards, against the wave action. Finally, the action on the pile cap is expressed in terms of the normalized variables $N^* = N_{\text{waves}}/N_U$ and $H^* = 0.5\, H_{\text{pile cap}}/H_U$, where $H_{\text{pile cap}}$ is the horizontal force measured on the pile cap. Results are reported in Figure 11: the maximum value of N^* is smaller than 0.3 for test case A2, whereas larger values with a maximum of $H^* = 2.2$ are seen for test case A3. These results show that the ultimate lateral resistance is exceeded before the axial pile resistance. For test cases B1–B3, both N^* and H^* results lower than 1, in stead. Although the piles are not modeled directly in DualSPHysics, the action exerted by the waves on the piles was assessed. To this end, the modified Morison's equation was employed [61], to take into account the slamming force due to the breaking waves:

$$F_{w,\text{pile}} = F_D + F_I + F_S \tag{9}$$

where F_D and F_I are the drag and inertia forces based on Morison's equation, respectively. The inertia component can be neglected, being the Keulegan–Carpenter number for the present cases larger than

150. From [62,63], the expression for the maximum drag force for breaking wave conditions can be derived:

$$F_D = C_D \rho g D {H_b}^2 (1-\lambda)^2 K_D \tag{10}$$

being C_D the drag coefficient (equal to 1 for the present case), H_b the breaking wave height, λ the curling factor, K_D a non-dimensional factor for the maximum drag force, ρ the water density, and D the pile diameter. The maximum slamming force can be expressed as:

$$F_S = 0.5 C_S \rho D {C_b}^2 \lambda \eta_b \tag{11}$$

where $C_S = \pi$ is the slamming coefficient, C_b is the breaking wave celerity equal to $[g(d_b + \eta_b)]^{0.5}$, d_b and η_b are the water depth and wave crest height at breaking [64], respectively. Equations (9) to (11) have been applied to the six text cases reported in the present work. The total force calculated by Morison's equation has been normalized by the ultimate horizontal load, leading to values of $F_{w,\mathrm{pile}}/H_u = 0.6 - 0.8$. Hence, for the final pile stability assessment, the drag and slamming forces must be taken into account. In fact, for test cases A1–A3 drag and slamming forces result in the same order of magnitude of those transmitted from the pile caps. For test cases B1–B3, if drag and slamming forces are added to the calculated values of the horizontal force measured on the pile cap, then the total horizontal force on each pile will exceed the lateral soil resistance, leading once again to pile instability. The aforementioned results suggest to carry out future analyses on the on the wave forces exerted on the piles, which is outside the scope of the present work.

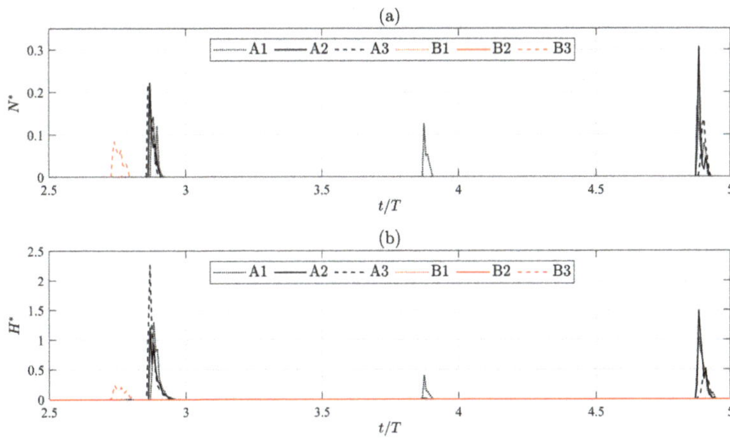

Figure 11. Time history of the normalized pile axial load (**a**) and lateral soil resistance (**b**).

From all aforementioned results and discussions, it can be argued that the first failure mechanism that led to the failure of the whole system (pile cap, piles and beam) is the foundation failure, corroborating the hypotheses in the damage report from the Badalona City Council.

5. Conclusions

The SPH-based open source code DualSPHysics is applied to model the interaction between sea waves and the Pont del Petroli pier, in Badalona (Spain). The pier experienced severe damage after the sea storm named Gloria struck the coast of Spain in January 2020. One beam forming the pier footbridge was washed away together with one pile cap. Moreover, several piles show large damaged areas. The numerical model employed herein allows us to: (a) characterize the wave loads exerted on the pier and help with the design and setup of the upcoming experimental campaign in the large-scale wave flume at LIM/UPC; (b) provide preliminary information regarding the main failure mechanisms that led to the observed damage. The numerical model was employed to simulate conditions similar to the ones caused by storm Gloria. Due to the lack of information on local wave conditions during the storm in the Badalona area, a preliminary wave propagation study was carried out with the SWAN model. Visual observations at Pont del Petroli showed waves as high as 7–8 m, overtopping the pier platform and footbridge. The bathymetric survey carried out after the storm revealed a drastic modification of the sea bottom below the pier with sand accretion that led to a reduction of the water depth by 1–2 m on average.

Initially, a 2-D analysis was performed for different wave conditions and water depths, with the wave height varying between 6.1 and 9.0 m and the wave period ranging between 9.6 and 12.7 s. Initial water depths at the pier toe between 8 and 10 m were considered. The 2-D model results were reported in terms of the exerted loads and wave breaking patterns. The latter ones were compared with visual observations during the storm Gloria. Then, two wave conditions were selected for further 3-D simulations, i.e., wave height values of 6.5 and 8.0 m, corresponding to wave periods of 12.0 and 12.7 s, respectively. The water depth at the pier toe for the 3-D simulations corresponds to the post-Gloria survey, i.e., about 8 m in prototype. While the 2-D discretization is made to closely mimic the layout of the wave flume at LIM/UPC and employed the same wave generation system (i.e., a wedge-type wavemaker), an inlet boundary condition is chosen for the 3-D simulations. Free surface elevation and velocity for the inlet area have been extrapolated from the 2-D results. This way, the 3-D model is more efficient, since the inlet is placed in the vicinity of the pier and the wave breaking point, allowing to optimize the computational effort while retaining a high accuracy.

Horizontal and vertical forces have been measured on three elements: the pier head platform, the first pile cap and the first seawards π-shaped beam forming the footbridge. Existing formulas for wave loads on exposed jetties [65,66] could not be directly applied since they were derived for jetties on the horizontal bottom. In the present work, the piles are not modeled numerically, however the forces exerted on the pile cap were used to calculate the expected loads at the pile base and foundation. The direct action exerted by the waves on the piles was computed by means of Morison's formula for slamming loads and breaking waves [62]. The forces measured on the platform and the footbridge beam were compared with the design loads, including self weight and accidental loads. The measured vertical forces are comparable or lower than the design ones. Larger forces are measured horizontally, especially for the most extreme wave conditions ($H = 8.0$ m, $T = 12.7$ s). The obtained values were employed to specify the required features of load cells and pressure sensors to be used during the experimental campaign at LIM/UPC: nominal force/pressure, breaking load, sensitivity, accuracy and measuring ranges. Snapshots of the SPH results show that for such large waves, the breaking process starts before the platform. The wave, already a plunger, is thus impacting the platform on its front, transferring a big portion of its momentum to the structure. Part of the energy is transmitted via overtopping on the platform deck and the footbridge, leading to vertical forces directed downwards. For the case with $H = 6$ m and $T = 12$ s, the waves reach the platform with no apparent breaking. The loads on the platform are smaller than the ones of the larger wave conditions, however the pile cap shows larger impacts. The forces exerted on the pile cap were used to calculate the axial load that the pile cap would transfer to the two piles below it. The horizontal force on the pile cap was compared with the ultimate lateral resistance of the soil for the embedded pile length and eccentricity. For the comparison the contribution of drag and slamming forces on the piles due to the wave action was considered, too. The results prove that the exerted horizontal force is far larger than the later soil resistance, while the axial pile loads are smaller than their tolerable loads.

The numerical results from this SPH simulation campaign suggest that the exceeded lateral soil resistance might be the cause of the heavy damage observed after storm Gloria, supporting the hypothesis in the damage report by the Badalona City Council. The lateral soil resistance was exceeded by the wave force exerted on the piles and the pile cap, leading to the overturn of the system piles + pile cap. As a consequence, the footbridge beam lost its support and slid into the sea.

To the best of the authors' knowledge, this work represents the first known application where a SPH inlet boundary condition is employed to model a real-world engineering test case. Findings in this manuscript will provide the basis for the proper design of an experimental campaign that will be carried out at the Maritime Engineering Laboratory of Universitat Politècnica de Catalunya-BarcelonaTech (LIM/UPC) with the aim of upgrading the design of the Pont del Petroli.

Author Contributions: Conceptualization, X.G. and C.A.; methodology, C.A. and J.M.D.; software, J.M.D., A.T. and A.J.C.C.; validation, C.A. and A.J.C.C.; formal analysis, C.A.; investigation, C.A., X.G. and J.S.; resources, X.G.; data curation, C.A.; writing—original draft preparation, A.T., C.A. and J.S.; writing—review and editing, A.T.; visualization, A.J.C.C. and C.A.; supervision, X.G.; project administration, X.G.; funding acquisition, C.A. All authors have read and agreed to the published version of the manuscript.

Funding: This research was funded by European Union's Horizon 2020 research and innovation programme under the Marie Sklodowska-Curie grant agreement No.: 792370.

Conflicts of Interest: The authors declare no conflict of interest. The funders had no role in the design of the study; in the collection, analyses, or interpretation of data; in the writing of the manuscript, or in the decision to publish the results.

References

1. Amores, A.; Marcos, M.; Carrió, D.S.; Gómez-Pujol, L. Coastal impacts of storm Gloria (January 2020) over the north-western Mediterranean. *Nat. Hazards Earth Syst. Sci.* **2020**, *20*, 1955–1968. [CrossRef]
2. Reed, H.L. Wave interactions in swept-wing flows. *Phys. Fluids* **1987**, *30*, 3419–3426. [CrossRef]
3. Rainey, R.C.T. A new equation for calculating wave loads on offshore structures. *J. Fluid Mech.* **1989**, *204*, 295–324. [CrossRef]
4. Ferrant, P. Runup On a Cylinder Due to Waves and Current: Potential Flow Solution With Fully Nonlinear Boundary Conditions. *Int. Soc. Offshore Polar Eng.* **2001**, *11*, 33–41.
5. Chalikov, D.; Sheinin, D. Modeling extreme waves based on equations of potential flow with a free surface. *J. Comput. Phys.* **2005**, *210*, 247–273. [CrossRef]
6. Garriga, O.S.; Falzarano, J.M. Water Wave Interaction on a Truncated Vertical Cylinder. *J. Offshore Mech. Arct. Eng.* **2008**, *130*, 031002. [CrossRef]
7. Casadei, F.; Halleux, J.; Sala, A.; Chillè, F. Transient fluid—Structure interaction algorithms for large industrial applications. *Comput. Methods Appl. Mech. Eng.* **2001**, *190*, 3081–3110. [CrossRef]
8. Wang, Q.; Goosen, J.; van Keulen, F. An efficient fluid—Structure interaction model for optimizing twistable flapping wings. *J. Fluids Struct.* **2017**, *73*, 82–99. [CrossRef]
9. Martínez-Ferrer, P.J.; Qian, L.; Ma, Z.; Causon, D.M.; Mingham, C.G. An efficient finite-volume method to study the interaction of two-phase fluid flows with elastic structures. *J. Fluids Struct.* **2018**, *83*, 54–71. [CrossRef]
10. Zhan, L.; Peng, C.; Zhang, B.; Wu, W. A stabilized TL—WC SPH approach with GPU acceleration for three-dimensional fluid—Structure interaction. *J. Fluids Struct.* **2019**, *86*, 329–353. [CrossRef]
11. Liang, H.; Ouled Housseine, C.; Chen, X.; Shao, Y. Efficient methods free of irregular frequencies in wave and solid/porous structure interactions. *J. Fluids Struct.* **2020**, *98*, 103130. [CrossRef]
12. Gingold, R.A.; Monaghan, J.J. Smoothed particle hydrodynamics: Theory and application to non-spherical stars. *Mon. Not. R. Astron. Soc.* **1977**, *181*, 375–389. [CrossRef]
13. Monaghan, J.J. Smoothed particle hydrodynamics. *Annu. Rev. Astron. Astr.* **1992**, *30*, 543–574. [CrossRef]
14. Liu, G.R.; Liu, M.B. *Smoothed Particle Hydrodynamics: A Meshfree Particle Method*; World Scientific: Singapore, 2003.
15. Violeau, D.; Rogers, B.D. Smoothed particle hydrodynamics (SPH) for free-surface flows: Past, present and future. *J. Hydraul. Res.* **2016**, *54*, 1–26. [CrossRef]
16. Gotoh, H.; Khayyer, A. Current achievements and future perspectives for projection-based particle methods with applications in ocean engineering. *J. Ocean Eng. Mar. Energy* **2016**, *2*, 251–278. [CrossRef]
17. Violeau, D. *Fluid Mechanics and the SPH Method: Theory and Applications*; Oxford University Press: Oxford, UK, 2012.
18. Tafuni, A.; Sahin, I. Non-linear hydrodynamics of thin laminae undergoing large harmonic oscillations in a viscous fluid. *J. Fluids Struct.* **2015**, *52*, 101–117. [CrossRef]

19. Shi, Y.; Li, S.; Chen, H.; He, M.; Shao, S. Improved SPH simulation of spilled oil contained by flexible floating boom under wave—Current coupling condition. *J. Fluids Struct.* **2018**, *76*, 272–300. [CrossRef]
20. Pan, K.; IJzermans, R.H.A.; Jones, B.D.; Thyagarajan, A.; van Beest, B.W.H.; Williams, J.R. Application of the SPH method to solitary wave impact on an offshore platform. *Comp. Part. Mech.* **2016**, *3*, 155–166. [CrossRef]
21. Yang, Y.; Li, J. SPH-FE-Based Numerical Simulation on Dynamic Characteristics of Structure under Water Waves. *J. Mar. Sci. Eng.* **2020**, *8*, 630. [CrossRef]
22. Domínguez, J.M.; Crespo, A.J.; Hall, M.; Altomare, C.; Wu, M.; Stratigaki, V.; Troch, P.; Cappietti, L.; Gómez-Gesteira, M. SPH simulation of floating structures with moorings. *Coast. Eng.* **2019**, *153*, 103560. [CrossRef]
23. Gong, K.; Shao, S.; Liu, H.; Wang, B.; Tan, S.K. Two-phase SPH simulation of fluid—Structure interactions. *J. Fluids Struct.* **2016**, *65*, 155–179. [CrossRef]
24. Peng, C.; Xu, G.; Wu, W.; sui Yu, H.; Wang, C. Multiphase SPH modeling of free surface flow in porous media with variable porosity. *Comput. Geotech.* **2017**, *81*, 239–248. [CrossRef]
25. Mokos, A.; Rogers, B.D.; Stansby, P.K. A multi-phase particle shifting algorithm for SPH simulations of violent hydrodynamics with a large number of particles. *J. Hydraul. Res.* **2017**, *55*, 143–162. [CrossRef]
26. Crespo, A.; Domínguez, J.; Rogers, B.; Gómez-Gesteira, M.; Longshaw, S.; Canelas, R.; Vacondio, R.; Barreiro, A.; García-Feal, O. DualSPHysics: Open-source parallel CFD solver based on Smoothed Particle Hydrodynamics (SPH). *Comput. Phys. Commun.* **2015**, *187*, 204–216. [CrossRef]
27. Altomare, C.; Tagliafierro, B.; Suzuki, T.; Domínguez, J.M.; Crespo, A.J.C.; Briganti, R. Relaxation zone method in SPH-based model applied to wave-structure interaction. In Proceedings of the International Ocean and Polar Engineering Conference, Sapporo, Japan, 10–15 June 2018; pp. 204–216.
28. Altomare, C.; Viccione, G.; Tagliafierro, B.; Bovolin, V.; Domínguez, J.M.; Crespo, A.J.C. *Computational Fluid Dynamics—Basic Instruments and Applications in Science*; Chapter Free-Surface Flow Simulations with Smoothed Particle Hydrodynamics Method Using High-Performance Computing; IntechOpen: London, UK, 2017; pp. 73–100.
29. Verbrugghe, T.; Domínguez, J.; Altomare, C.; Tafuni, A.; Vacondio, R.; Troch, P.; Kortenhaus, A. Non-linear wave generation and absorption using open boundaries within DualSPHysics. *Comput. Phys. Commun.* **2019**, *240*, 46–59. [CrossRef]
30. Gómez Lahoz, M.; Carretero Albiach, J.C. Wave forecasting at the Spanish coasts. *J. Atmos. Ocean Sci.* **2005**, *10*, 389–405. [CrossRef]
31. Booij, N.; Holthuijsen, L.; Ris, R. The "Swan" Wave Model for Shallow Water. In Proceedings of the 25th International Conference on Coastal Engineering, Orlando, FL, USA, 2–6 September 1996; pp. 668–676.
32. Wendland, H. Piecewise polynomial, positive definite and compactly supported radial functions of minimal degree. *Adv. Comput. Math.* **1995**, *4*, 389–396. [CrossRef]
33. Rota Roselli, R.A.; Vernengo, G.; Altomare, C.; Brizzolara, S.; Bonfiglio, L.; Guercio, R. Ensuring numerical stability of wave propagation by tuning model parameters using genetic algorithms and response surface methods. *Environ. Model. Softw.* **2018**, *103*, 62–73. [CrossRef]
34. Altomare, C.; Domínguez, J.; Crespo, A.; González-Cao, J.; Suzuki, T.; Gómez-Gesteira, M.; Troch, P. Long-crested wave generation and absorption for SPH-based DualSPHysics model. *Coast. Eng.* **2017**, *127*, 37–54. [CrossRef]
35. Fourtakas, G.; Dominguez, J.M.; Vacondio, R.; Rogers, B.D. Local uniform stencil (LUST) boundary condition for arbitrary 3-D boundaries in parallel smoothed particle hydrodynamics (SPH) models. *Comput. Fluids* **2019**, *190*, 346–361. [CrossRef]
36. Leimkuhler, B.; Matthews, C. *Molecular Dynamics*; Springer International Publishing: Cham, Switzerland, 2016.
37. Monaghan, J.J.; Cas, R.; Kos, A.; Hallworth, M. Gravity currents descending a ramp in a stratified tank. *J. Fluid Mech.* **1999**, *379*, 39–70. [CrossRef]
38. Crespo, A.J.C.; Gomez-Gesteira, M.; Dalrymple, R.A. Boundary conditions generated by dynamic particles in SPH methods. *Comput. Mater. Contin.* **2007**, *5*, 173–184.
39. Altomare, C.; Crespo, A.; Domínguez, J.M.; Gómez-Gesteira, M.; Suzuki, T.; Verwaest, T. Applicability of Smoothed Particle Hydrodynamics for estimation of sea wave impact on coastal structures. *Coast. Eng.* **2015**, *96*, 1–12. [CrossRef]

40. Tafuni, A.; Sahin, I. Hydrodynamic Loads on Vibrating Cantilevers Under a Free Surface in Viscous Fluids With SPH. In *ASME International Mechanical Engineering Congress and Exposition*; Volume 7B: Fluids Engineering Systems and Technologies; American Society of Mechanical Engineers: New York, NY, USA, 2013.
41. Mogan, S.C.; Chen, D.; Hartwig, J.; Sahin, I.; Tafuni, A. Hydrodynamic analysis and optimization of the Titan submarine via the SPH and Finite—Volume methods. *Comput. Fluids* **2018**, *174*, 271–282. [CrossRef]
42. Tafuni, A.; Domínguez, J.M.; Vacondio, R.; Crespo, A.J.C. A versatile algorithm for the treatment of open boundary conditions in Smoothed particle hydrodynamics GPU models. *Comput. Methods Appl. Mech. Eng.* **2018**, *342*, 604–624. [CrossRef]
43. Ferrand, M.; Laurence, D.R.; Rogers, B.D.; Violeau, D.; Kassiotis, C. Unified semi-analytical wall boundary conditions for inviscid, laminar or turbulent flows in the meshless SPH method. *Int. J. Numer. Methods Fluids* **2013**, *71*, 446–472. [CrossRef]
44. Rogers, B.D.; Dalrymple, R.A.; Stansby, P.K. Simulation of caisson breakwater movement using 2-D SPH. *J. Hydraul. Res.* **2010**, *48*, 135–141. [CrossRef]
45. Antuono, M.; Colagrossi, A.; Marrone, S.; Lugni, C. Propagation of gravity waves through an SPH scheme with numerical diffusive terms. *Comput. Phys. Commun.* **2011**, *182*, 866–877. [CrossRef]
46. Verbrugghe, T.; Domínguez, J.M.; Altomare, C.; Tafuni, A.; Troch, P.; Kortenhaus, A. Application of open boundaries within a two-way coupled SPH model to simulate non-linear wave-structure interactions. *Coast. Eng. Proc.* **2018**, *1*, 14. [CrossRef]
47. Altomare, C.; Domínguez, J.M.; Crespo, A.J.C.; Suzuki, T.; Caceres, I.; Gómez-Gesteira, M. Hybridization of the Wave Propagation Model SWASH and the Meshfree Particle Method SPH for Real Coastal Applications. *Coast. Eng. J.* **2015**, *57*, 1550024-1–1550024-34. [CrossRef]
48. Lowe, R.; Buckley, M.; Altomare, C.; Rijnsdorp, D.; Yao, Y.; Suzuki, T.; Bricker, J. Numerical simulations of surf zone wave dynamics using Smoothed Particle Hydrodynamics. *Ocean Model.* **2019**, *144*, 101481. [CrossRef]
49. Gomez-Gesteira, M.; Rogers, B.; Crespo, A.; Dalrymple, R.; Narayanaswamy, M.; Dominguez, J. SPHysics—Development of a free-surface fluid solver—Part 1: Theory and formulations. *Comput. Geosci.* **2012**, *48*, 289–299. [CrossRef]
50. Gomez-Gesteira, M.; Crespo, A.; Rogers, B.; Dalrymple, R.; Dominguez, J.; Barreiro, A. SPHysics—Development of a free-surface fluid solver—Part 2: Efficiency and test cases. *Comput. Geosci.* **2012**, *48*, 300–307. [CrossRef]
51. Domínguez, J.M.; Altomare, C.; Gonzalez-Cao, J.; Lomonaco, P. Towards a more complete tool for coastal engineering: Solitary wave generation, propagation and breaking in an SPH-based model. *Coast. Eng. J.* **2019**, *61*, 15–40. [CrossRef]
52. González-Cao, J.; Altomare, C.; Crespo, A.; Domínguez, J.; Gómez-Gesteira, M.; Kisacik, D. On the accuracy of DualSPHysics to assess violent collisions with coastal structures. *Comput. Fluids* **2019**, *179*, 604–612. [CrossRef]
53. St-Germain, P.; Nistor, I.; Townsend, R.; Shibayama, T. Smoothed-Particle Hydrodynamics Numerical Modeling of Structures Impacted by Tsunami Bores. *J. Waterw. Port Coast. Ocean Eng.* **2014**, *140*, 66–81. [CrossRef]
54. Cunningham, L.S.; Rogers, B.D.; Pringgana, G. Tsunami wave and structure interaction: An investigation with Smoothed-particle hydrodynamics. *Proc. Inst. Civ. Eng.—Eng. Comput. Mech.* **2014**, *167*, 126–138. [CrossRef]
55. Pringgana, G.; Cunningham, L.S.; Rogers, B.D. Modelling of tsunami-induced bore and structure interaction. *Proc. Inst. Civ. Eng.—Eng. Comput. Mech.* **2016**, *169*, 109–125. [CrossRef]
56. Altomare, C.; Crespo, A.; Rogers, B.; Dominguez, J.; Gironella, X.; Gómez-Gesteira, M. Numerical modelling of armour block sea breakwater with Smoothed particle hydrodynamics. *Comput. Struct.* **2014**, *130*, 34–45. [CrossRef]
57. Zhang, F.; Crespo, A.; Altomare, C.; Domínguez, J.; Marzeddu, A.; ping Shang, S.; Gómez-Gesteira, M. A numerical tool to simulate real breakwaters. *J. Hydrodyn.* **2018**, *30*, 95–105. [CrossRef]
58. Subramaniam, S.; Scheres, B.; Schilling, M.; Liebisch, S.; Kerpen, N.; Schlurmann, T.; Altomare, C.; Schüttrumpf, H. Influence of Convex and Concave Curvatures in a Coastal Dike Line on Wave Run-up. *Water* **2019**, *11*, 1333. [CrossRef]

59. Meyerhof, G.G. Closure of "Compaction of Sands and Bearing Capacity of Piles". *J. S. Mech. Fdtn. Div. ASCE* **1959**, *85*, 1–29.
60. Broms, B. Lateral Resistance of Piles in Cohesionless Soils. *J. Soil Mech. Found. Div.* **1964**, *90*, 123–156.
61. Goda, Y.; Haranaka, S.; Kitahata, M. Study of impulsive breaking wave forces on piles. *Rep. Port Harb. Res. Inst. Jpn.* **1966**, *5*, 1–30.
62. Goda, Y. Wave Forces on a Vertical Circular Cylinder: Experiments and a Proposed Method of Wave Force Computation. *Rep. Port Harb. Res. Inst. Jpn.* **1964**, *8*, 1–74.
63. Departmet of the Army, US Army Corps of Engineers. *Shore Protection Manual*; CERC: Washington, DC, USA, 1984; Volume 1.
64. Tanimoto, K.; Takahashi, S.; Kaneko, T.; Shiota, K. Impulsive breaking wave forces on an inclined pile exerted by random waves. In Proceedings of the 20th International Conference on Coastal Engineering, Taipei, Taiwan, 9–14 November 1986; pp. 2288–2302.
65. Cuomo, G.; Tirindelli, M.; Allsop, W. Wave-in-deck loads on exposed jetties. *Coast. Eng.* **2007**, *54*, 657–679. [CrossRef]
66. Liu, Q.; Sun, T.; Wang, D.; Wei, Z. Wave uplift force on horizontal panels: A laboratory study. *J. Oceanol. Limnol.* **2019**, *37*, 1899–1911. [CrossRef]

MDPI
St. Alban-Anlage 66
4052 Basel
Switzerland
Tel. +41 61 683 77 34
Fax +41 61 302 89 18
www.mdpi.com

Journal of Marine Science and Engineering Editorial Office
E-mail: jmse@mdpi.com
www.mdpi.com/journal/jmse

www.ingramcontent.com/pod-product-compliance
Lightning Source LLC
LaVergne TN
LVHW070141170726
843515LV00007B/1845
* 9 7 8 3 0 3 6 5 3 0 4 0 6 *